Archives of Virology

Supplementum 1

C. H. Calisher (ed.)

Hemorrhagic Fever with Renal Syndrome, Tick- and Mosquito-Borne Viruses

Springer-Verlag Wien New York

C. H. Calisher, Ph.D.
Division of Vector-Borne Viral Diseases, Center for Infectious Diseases
Centers for Disease Control, Fort Collins, CO, U.S.A.

Notice: The U.S. Department of the Army generously provided funding to partially offset the costs of the symposium held in Dubrovnik and supported the attendance of certain of its employees at that symposium and at the symposium held in Moscow. For the latter, it must be said that "The views, opinions, and/or findings contained herein are those of the authors and should not be construed as an official Department of the Army position, policy, or decision, unless so designated by other documentation."

Printed on acid-free paper

With 75 Figures

ISSN 0939-1983
ISBN-13:978-3-211-82217-3   e-ISBN-13:978-3-7091-9091-3
DOI: 10.1007/978-3-7091-9091-3

Contents

Abstracts

Arch Virol (1990) [Suppl 1]: 1–3

Editor's comments

C. H. Calisher

Center for Infectious Diseases, Centers for Disease Control, Fort Collins, Colorado, U.S.A.

The papers in this Special Issue of *Archives of Virology* were among many presented at two symposia: "The Second Symposium on Arboviruses in Mediterranean Countries" (Dubrovnik, Yugoslavia, September 24–29, 1989) and "An International Symposium on Arboviruses and Arboviral Infections" (Moscow, U.S.S.R., October 3–5, 1989).

This issue could not have been completed without the cooperation of the organizers of those symposia, particularly Drs. David H. L. Bishop and Patricia A. Nuttall, NERC Institute of Virology, Oxford, England; Academician Dmitri K. Lvov and Dr. Alexander M. Butenko, D.I. Ivanovsky Institute of Virology, Moscow, U.S.S.R.; and Drs. Thomas P. Monath, Virology Division, U.S. Army Medical Research Institute of Infectious Diseases, Ft. Detrick, Frederick, Maryland, U.S.A. and Frederick A. Murphy, Centers for Disease Control, Atlanta, Georgia, U.S.A. For their counsel, helpful suggestions, and gracious support in a miscellany of areas, I offer my thanks. To the authors, who tolerated my editorial style, met unrealistic deadlines, and generally were cooperative and cheerful through it all, I owe everything.

As newly appointed Special Issues Editor, along with Dr. H. D. Klenk, it is my responsibility to see that manuscripts are submitted according to the "Instructions to Authors" issued by the journal. That is quite a different situation from the usual, in which submitted manuscripts are refereed anonymously and the Editor acts in accord with the referee's recommendations. In this situation, there is much more work involved in trying to make all manuscripts match the style and standards of the journal—rejection of a manuscript is a last resort, rarely done in fact.

As Editor, and as an arbovirologist interested in the systematics of my field, I must comment on the viral nomenclature and taxonomic usage contained in certain of the papers in this volume. At the symposium held in Dubrovnik, I presented a paper "Classification and taxonomy of arboviruses: a useful occupation or the sign of obsessive-compulsive behavior?". I noted that the purpose of precision in nomenclature and universality in taxonomy is unambiguous communication. I explained my position on definition of the arbovirological terms *serogroup*, *complex*, *virus*, *subtype*, and

Table 1. A suggested system of arboviral antigenic classification

---

Serogroup   2 or more viruses, distinct from each other by quantitative serologic criteria (4-fold or greater differences between homologous and heterologous titers of both serum samples) in 1 or more tests, but related to each other or to other viruses by some (any) serologic method

Complex   Composed of viruses closely related within a serogroup but distinct from each other

Virus (type)   Belong to the same antigenic complex, are antigenically related but easily separable (4-fold or greater differences between homologous and heterologous titers of both serum samples) by 1 or more standard serologic tests (hemagglutination-inhibition, complement-fixation, fluorescent antibody, neutralization, enzyme-linked immunosorbent assay, etc.)

Subtype   Separable from other closely related isolates by at least a 4-fold difference between the homologous and heterologous titers of 1 but not both of the 2 serum samples tested.

Variety   Subtypes differentiable only by the application of special tests or reagents, including studies with monoclonal antibodies selected for epitope specificity, with special reagents (such as 1-dose antiserum or antibody to a specific glycoprotein), or with specialized tests (such as kinetic hemagglutination inhibition or absorption-hemagglutination inhibition).

---

variety (Table 1), which are consensus definitions used by the Subcommittee on InterRelationships Among Catalogued Arboviruses (Dr. Robert E. Shope, Chairman) of the American Committee on Arthropod-borne Viruses. Further, I urged that my colleagues avoid the crude patois of imprecise vernacular usage, e.g., there is no such entity as dengue virus (there are four different viruses that cause dengue); there is no such entity as hemorrhagic fever with renal syndrome virus (there is a disease syndrome by this name and there are Hantaan, Seoul, and Puumala viruses). Consistent usage of proper virologic names, whether formally or vernacularly, would improve communication between laboratory virologists, clinicians, public health workers, et al., and would eliminate the small enclaves of virologists who become isolated by their unique jargons. In the future, this journal and all virology journals will become much more demanding that authors use only the universal nomenclature and taxonomic terms established by the International Committee for Taxonomy of Viruses and by W.H.O. specialty groups. Examples of appropriate and inappropriate terms pertinent to this volume are included in Table 2.

Table 2. Examples of appropriate nomenclature and taxonomic usage

Disease	Inaccurate name for etiologic agent	Appropriate formal usage (family, genus) and universal vernacular usage (virus)[a]
Tick-borne encephalitis	TBE virus	*Flaviviridae*, *Flavivirus*, Russian spring-summer encephalitis virus
Tick-borne encephalitis	TBE virus	*Flaviviridae*, *Flavivirus*, Powassan virus
Dengue	dengue virus dengue virus	*Flaviviridae*, *Flavivirus*, dengue 1 virus *Flaviviridae*, *Flavivirus*, dengue 2 virus
Western equine encephalitis	WEE virus (western type) WEE virus (eastern type)	*Togaviridae*, *Alphavirus*, western equine encephalitis virus *Togaviridae*, *Alphavirus*, Highlands J virus
Eastern equine encephalitis	EEE virus	*Togaviridae*, *Alphavirus*, eastern equine encephalitis virus
Hemorrhagic fever with renal syndrome	HFRS virus (eastern type)	*Bunyaviridae*, *Hantavirus*, Hantaan virus
Nephropathia Epidemica	HFRS virus (western type)	*Bunyaviridae*, *Hantavirus*, Puumala virus
Bluetongue	bluetongue virus bluetongue virus	*Reoviridae*, *Orbivirus*, bluetongue 1 virus *Reoviridae*, *Orbivirus*, bluetongue 2 virus
Vesicular stomatitis	vesicular stomatitis virus vesicular stomatitis virus	*Rhabdoviridae*, *Vesiculovirus*, vesicular stomatitis Indiana virus *Rhabdoviridae*, *Vesiculovirus*, vesicular stomatitis New Jersey virus

[a] If the virus type is not known, it is acceptable to use the generic expression "a flavivirus belonging to the TBE antigenic complex"; "a flavivirus belonging to the dengue antigenic complex"; "an alphavirus belonging to the WEE antigenic complex"; "a hantavirus"; "an orbivirus most closely related to viruses of the bluetongue antigenic complex"; "a vesiculovirus"; or other, similarly accurate and inclusive expressions

Arch Virol (1990) [Suppl 1]: 5–18

Geographical distribution of hemorrhagic fever with renal syndrome and hantaviruses

H. W. Lee, P. W. Lee, L. J. Baek, and **Y. K. Chu**

WHO Collaborating Center for Virus Reference and Research (hemorrhagic fever with renal syndrome), Institute for Viral Diseases, Korea University, Seoul, Korea

Accepted March 8, 1990

Summary. World-wide, about 150,000 people are hospitalized with hemorrhagic fever with renal syndrome (HFRS) (3–10% fatality) each year. The etiologic agents of HFRS are Hantaan, Seoul, and Puumala viruses of the genus *Hantavirus*, family *Bunyaviridae*. A severe form of HFRS, caused by Hantaan virus, occurs in Asia and eastern parts of Europe, a moderate form, caused by Seoul virus, occurs in Asia, and a mild form, caused by Puumala virus, occurs in Europe. Hantaan virus occurs in Asia and in eastern parts of Europe, Seoul-like viruses occur world-wide, Puumala virus occurs in Europe, and Prospect Hill and Leaky viruses (other hantaviruses) have been isolated in the U.S.A. The reservoirs of hantaviruses are rodents and other small mammals. Serologic studies of 42 hantaviruses isolated from HFRS patients and from animals indicated that there are 6 or 7 serotypes. In the 1990s, it is highly possible to identify HFRS and HFRS-like illnesses caused by hantaviruses in parts of the world where HFRS is not known because of the availability of serodiagnostic tests.

Introduction

Hemorrhagic fever with renal syndrome (HFRS) is an acute infectious viral disease characterized by abrupt high fever, various hemorrhagic manifestations, a flushed face, and transient renal and hepatic dysfunctions. Various clinical forms of HFRS occur not only in Eurasia but also in Southeast Asia and in Africa. About 150,000 people are hospitalized each year with HFRS, 3–10% of them die [9].

Since the isolation of the etiologic agent of Korean hemorrhagic fever (KHF) [7, 8], now called Hantaan virus [5], and improvements in serodiagnostic tests for infection with Hantaan virus, numerous Hantaan-related viruses have been isolated from rodents. In addition, it has been confirmed serologically that KHF-like illnesses in various parts of the world are caused by Hantaan or Hantaan-related viruses [9].

In 1982, the WHO recommended naming KHF-like diseases caused by Hantaan and Hantaan-related viruses "Hemorrhagic fever with renal syndrome" [4]. More recently, Hantaan group viruses have been classified as members of a newly established genus *Hantavirus* in the family *Bunyaviridae* [12]. Of these, Hantaan, Seoul, and Puumala viruses are known to cause illnesses in humans; the pathogenicities of Prospect Hill [5], Leaky [1], and isolates of Seoul-like viruses transmitted from rodents to humans is unknown.

It is increasingly likely that HFRS or HFRS-like illnesses caused by hantaviruses will be observed in Africa, Americas, Australia, and other areas that thus far have not been shown to be endemic for HFRS. This is because many hantaviruses have been isolated from rodents in these areas and HFRS has been recently confirmed serologically in Malaysia [15], Sri Lanka [9], and Central African Republic [2].

This paper summarizes present knowledge of the geographic distribution of HFRS patients, virus isolates, human and rodent antibodies against the hantaviruses, and serologic relationships of hantaviruses isolated by laboratories in many parts of the world.

Materials and methods

Collaborations

Since 1980, the WHO Centre for Hemorrhagic Fever with Renal Syndrome has been collaborating with laboratories throughout the world in order to do a seroepidemiologic survey of hantavirus infections in humans and animals. Investigators at the Center collected and received hantaviruses from patients and from rodents. In addition, we received monoclonal antibodies prepared against Hantaan virus.

Viruses

The 42 hantaviruses studied were from HFRS patients and rodents collected in different parts of the world. These were used after propagation in Vero E6 cell cultures. ID_{50} of the viruses in Vero E6 cells was $10^{5.0}$–$10^{6.0}$/ml.

Serum samples

Sera from normal healthy persons including laboratory workers, HFRS patients, rodents and other mammals in different parts of the world were tested for antibodies against hantaviruses.

Antisera against hantaviruses

Two 4- to 5-week-old S.D. rats were immunized with each virus by giving them a single intramuscular inoculation of 0.5 ml supernatant fluid from infected Vero E6 cell cultures.

Whole blood was collected by cardiac puncture 28 days after inoculation of the virus and serum was separated from the clot, stored at −60 °C, and then tested for antibody against the homologous virus and other hantavirus isolates. Serum samples from convalescent-phase HFRS patients in Korea and Finland, and monoclonal antibodies to Hantaan virus were used for comparative serologic studies of hantaviruses.

Antibodies

Neutralizing (N) antibodies to hantaviruses were measured by plaque reduction neutralization test (PRNT) [10] and monoclonal antibodies to hantaviruses were titrated by indirect immunofluorescent antibody technique (IFAT), as described previously [11].

Results

Global distribution of humans and rodents infected with hantaviruses

Seroepidemiological surveys show that hantavirus infections are distributed throughout much of the world, as demonstrated by the presence of antibodies against hantaviruses in sera from humans and rodents [9], as shown in Fig. 1.

HFRS patients have been documented clinically and serologically throughout Eurasia and, recently, in Africa. Nine countries in Asia are focally enzootic for hantaviruses: Japan, South Korea, North Korea, China, Mongolia, U.S.S.R., Hong Kong, Malaysia and Sri Lanka; 15 countries in Europe: U.S.S.R., Finland, Sweden, Norway, Denmark, Bulgaria, Hungary, Albania, Federal Republic of Germany, France, Belgium, Netherlands, England, Yugoslavia and Greece; and 1 country in Africa: Central African Republic [2].

Severe and moderate clinical forms of HFRS occur in Asia and in Balkan countries. Annually, between 70,000 and 130,000 people are hospitalized with HFRS in China, about 500–800 in South Korea, several hundred in North Korea, and several hundred in the U.S.S.R. Recently, several cases of HFRS were documented in Malaysia and Sri Lanka. Most HFRS patients in Asia live in rural areas but there have been many infections acquired in urban areas of Japan, Korea, China and Hong Kong. In Europe, HFRS is usually a more mild illness, known as Nephropathia Epidemica, in which renal involvement dominates and hemorrhagic features are less prominent; the fatality rate is about 0.2%. However, the severe type occurs in parts of Yugoslavia and Greece.

Infections of laboratory workers with Hantaan and Seoul viruses have been reported from Korea, Japan, China, Belgium, and England [9]. Thus, seropositive humans and wild rodents have been determined almost world-wide.

Field rodents infected with hantaviruses were demonstrated in Asia, Europe, Africa and the Americas, specifically: Korea, China, U.S.S.R.,

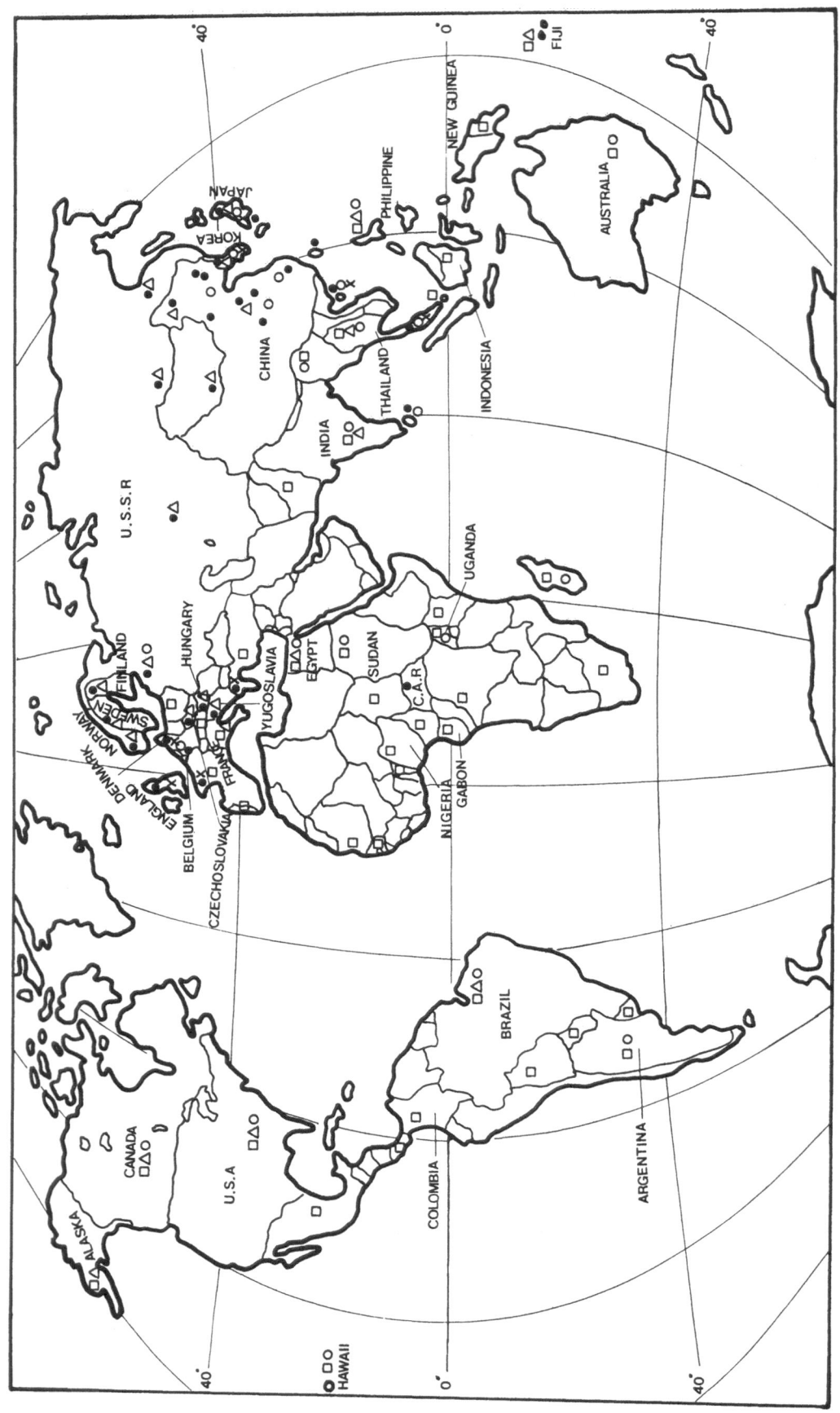

Fig. 1. Global distribution of humans with HFRS and humans and rodent infected with hantaviruses, based on demonstration of antibodies against hantaviruses. ── Seoul virus, ── Hantaan virus, ─·─ Puumala virus, ···· Maagi virus, ● Prospect Hill virus, ○ Prospect Hill virus, △ Leaky virus

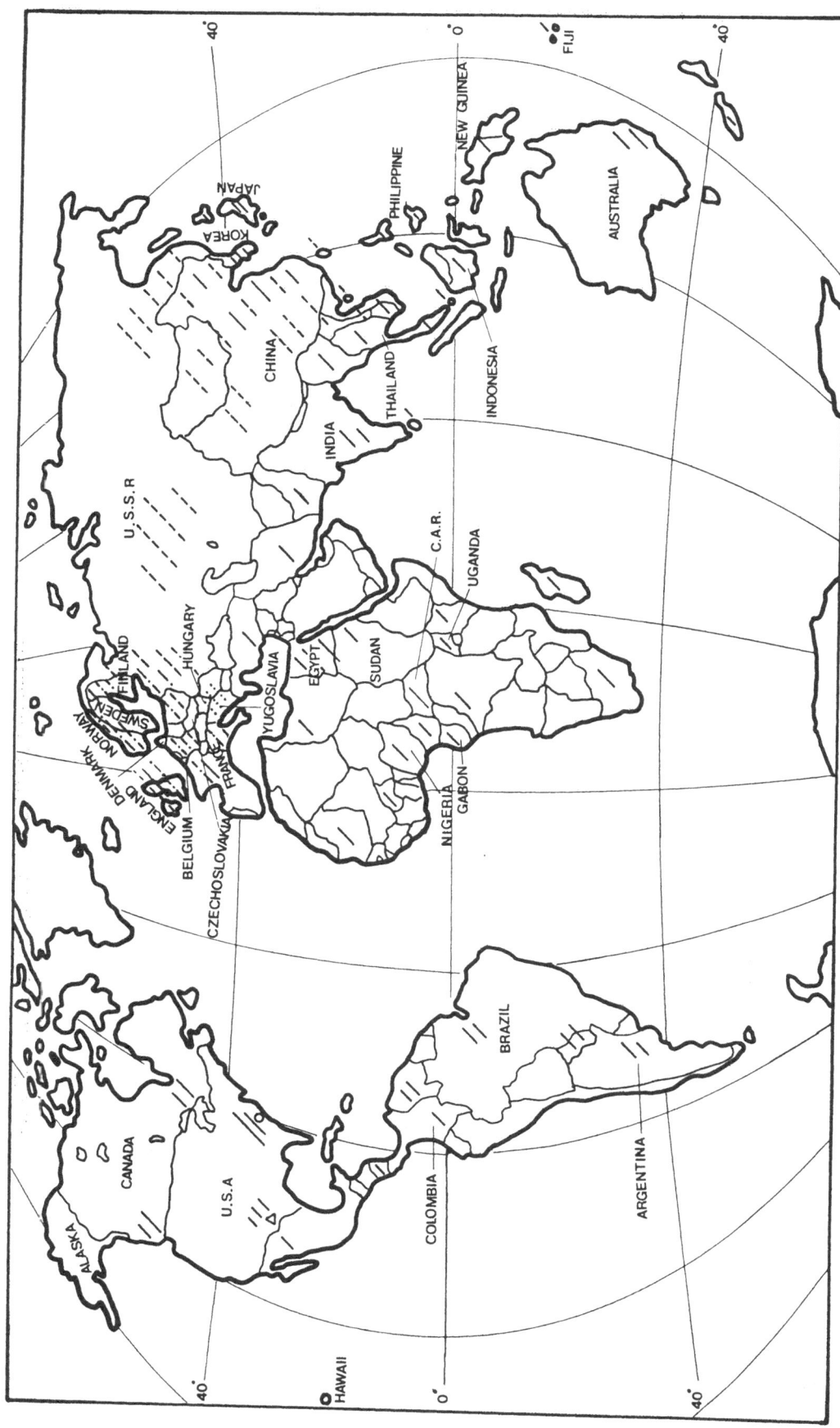

Fig. 2. Global distribution of hantaviruses based on demonstration of antibodies against the viruses in humans and rodents. ● Patient, ☐ seropositive person, △ infected field rodents, ○ infected house rats, × infected laboratory rats

Sweden, Finland, Norway, Yugoslavia, Greece, Egypt, U.S.A., Brazil, and Argentina. Urban rats infected with hantaviruses are in 10 Asian countries: Japan, Korea, China, Hong Kong, Malaysia, Sri Lanka, India, Singapore, Fiji, Philippines; in 3 European countries: Belgium, Federal Republic of Germany, Italy; and in the U.S.A. Laboratory rats infected with hantaviruses have been determined in 11 countries of the world: Japan, Korea, China, U.S.S.R., Belgium, England, Malaysia, Hong Kong, Singapore, Hawaii, Argentina.

Global distribution of hantaviruses

The demonstration of N antibodies against different hantaviruses in sera from humans and rodents from different parts of the world indicate the geographical distribution of hantaviruses (Fig. 2).

Hantaan virus is found in Korea, China, Mongolia, and the Far-East of the U.S.S.R., Seoul and related viruses exist essentially world-wide: 13 Asian countries (Japan, Korea, China, Hong Kong, Philippines, Malaysia, Singapore, India, Sri Lanka, Fiji, Thailand, Vietnam, Taiwan), 4 North and Central American countries (Canada, U.S.A., Mexico, Panama), 6 South American countries (Brazil, Bolivia, Colombia, Argentina, Uruguay, Paraguay), and 13 African countries (Egypt, Sudan, Uganda, Kenya, Benin, Cameroun, Mauritania, Senegal, Tchad, Central African Republic, Gabon, Madagascar, Nigeria), 4 countries in Europe (Belgium, Netherlands, Federal Republic of Germany, Italy).

Puumala virus is in Europe (U.S.S.R., Scandinavian countries, Finland, Belgium, Federal Republic of Germany, France and England). Prospect Hill and Leaky viruses have been found only in the U.S.A.

Maagi virus, first isolated from *Apodemus agrarius* collected in Maagi, Kyunggido, Korea, 1980 (H.W. Lee, unpubl. data) is found in Korea and probably in Yugoslavia and Greece, based on demonstration of specific antibodies against the virus in sera from HFRS patients.

Serologic relationships between hantaviruses by neutralization tests and monoclonal antibody assays

Neutralization tests were used to serologically classify 42 strains of hantaviruses isolated from HFRS patients and from rodents in different parts of the world. N antibody titers of rat antiserum to each strain of hantaviruses were measured against 6 ostensibly different serotypes of hantaviruses (Table 1). There were 6 strains of Hantaan virus and 1 of Seoul virus from 7 HFRS patients. Ten strains of Hantaan virus and one strain of Maagi virus were identified among 11 isolates from *Apodemus* mice.

Eight strains of Seoul virus and 6 strains of Seoul-like virus (equal N antibody titers against Hantaan and Seoul viruses) were identified from 14

Table 1. Serologic relation of hantaviruses by PRNT isolated from HFRS patients and rodents in the world

Rat antiserum to virus	PRN antibody titer against hantavirus					
	Hantaan virus	Seoul virus	Puumala virus	Prospect Hill virus	Maagi virus	Leaky virus
Human isolates						
ROK 79/89	*1,280*	20	<20	20	1,280	<20
ROK 84/105	*1,280*	20	<20	<20	320	n.d.
LEE# 188604	*320*	10	<20	<20	320	n.d.
US 84/2	*640*	20	<20	<20	320	n.d.
Chen	*320*	20	<20	20	640	n.d.
ROK 79/90	*320*	20	<20	20	640	n.d.
Hubei/1	80	*5,120*	<20	<20	320	n.d.
Apodemus isolates						
76/118	*5,120*	80	<20	20	5,120	<20
78/197	*1,280*	20	<20	<20	1,280	n.d.
83/18	*1,280*	20	<20	<20	1,280	n.d.
83/138	*5,120*	80	<20	<20	1,280	n.d.
Yugo/2508/84	*5,120*	80	<20	<20	5,120	n.d.
A9	*1,280*	20	<20	<20	640	n.d.
Jinhae 87/494	*1,280*	80	<20	<20	1,280	n.d.
Jinhae 87/502	*5,120*	80	<20	<20	5,120	n.d.
83/14	*320*	20	<20	<20	320	n.d.
Jinhae 87/526	*640*	80	<20	<20	1,280	n.d.
Maagi	320	20	<20	<20	*5,120*	n.d.
House rat isolates						
80/39 (#211808)	20	*1,280*	<20	<20	320	<20
I/RN/82/216	320	*20,480*	<20	<20	320	n.d.
JTRN/82/11	320	*20,480*	<20	<20	320	n.d.
Hong Kong R/14	320	*1,280*	20	20	n.d.	n.d.
Thailand #605	20	*1,280*	<20	20	80	n.d.
Brazil 2/4	80	*5,120*	<20	<20	80	n.d.
Singapore R/36	80	*1,280*	<20	<20	n.d.	n.d.
R22	80	*5,120*	20	<20	n.d.	n.d.
Girard Point	320	640	<20	<20	320	n.d.
Egypt R/12915	320	320	<20	20	n.d.	n.d.
Tchoupitoulas	320	320	<20	20	80	n.d.
JTRN/82/17	320	320	<20	<20	n.d.	n.d.
TR/352	80	80	<20	<20	n.d.	n.d.
I/RN/82/3	320	320	<20	<20	n.d.	n.d.
Lab. rat isolates						
KSNUSD 84/30	*1,280*	160	<20	<20	640	n.d.
KSNUSD 84/34	80	*1,280*	<20	<20	n.d.	<20
B/1	80	*5,120*	20	<20	<20	n.d.
SR/11 #191811	80	*320*	20	<20	n.d.	n.d.
Hamster isolate						
SNUS/Hamster 85/4	80	*5,120*	<20	<20	20	n.d.

Table 1 (continued)

Rat antiserum to virus	PRN antibody titer against hantavirus					
	Hantaan virus	Seoul virus	Puumala virus	Prospect Hill virus	Maagi virus	Leaky virus
Bandicota isolate						
Thailand #749	80	*1,280*	<20	<20	320	n.d.
Clethrionomys isolates						
Hällnäs B	<20	80	*320*	80	80	<20
USSR/CLS1/452	*320*	20	20	20	n.d.	n.d.
Microtus isolate						
Prospect Hill	20	20	320	*5,120*	n.d.	<20
M. musculus isolate						
Leaky	<20	<20	<20	<20	n.d.	*320*

n.d. Not done

isolates from house rats. Of three isolates from laboratory animals, 1 from a hamster and 1 from a bandicoot (*Bandicota indica*) were Seoul virus, the third isolate was Hantaan virus. One Puumala and 1 Hantaan virus were from *Clethrionomys* mice. One isolate from *Microtus pennsylvanicus* was Prospect Hill virus and one isolate from *Mus musculus* was Leaky virus.

By neutralization, there are 6 distinct serotypes (Hantaan, Seoul, Puumala, Prospect Hill, Maagi, and Leaky) and 1 undistinguishable type isolate (related closely to both Hantaan and Seoul viruses) among 42 hantavirus isolates from humans and animals. However, antisera made with 5 isolates from rats (Egypt R.12915, Tchoupitoulas, JTRN/82/17, TR-352, and I/RN/82-3) showed equal antibody titers to Hantaan and Seoul viruses; these isolates require further study.

Results of antigenic comparison of 38 hantaviruses isolates from HFRS patients and from animals in different parts of the world by IFA, using 8 monoclonal antibodies and 3 sera from HFRS patients are shown in Table 2. Ten Hantaan virus strains and Maagi virus were clearly distinguishable from each other and from other hantaviruses using monoclonal antibody BB01-BB08. Twenty-three strains of Seoul virus could be separated into two closely related clusters using monoclonal antibodies HC02-BE08 and HC02-BD05. Three (JTRN82/17, TR-352, I/RN/82/3) of 4 strains of Seoul virus that did not react with monoclonal antibodies HC02-BE08 and HC02-BD05 had shown equal N antibody titers to Hantaan and Seoul viruses by N (Table 1). However, Hubei/1 strain, 1 of 4 that did not react with monoclonal antibodies HC02-BZ08 and HC02-BD05, was essentially identical to Seoul virus by N (Table 1). Maagi virus can be distinguished from Hantaan virus using monoclonal antibodies HC02-BE08, HC02-BD05, and HC02-BD05, as shown in Table 2. Puumala, Prospect Hill, and Leaky viruses were not reactive with the monoclonal antibodies but could be differentiated from other hantaviruses by using convalescent sera from KHF and NE patients (Table 2).

Discussion

It has only been within the past decade that concrete progress has been made in the knowledge of the etiology and epidemiology of HFRS. Isolation of Hantaan virus, the prototype virus of HFRS, and development of an IFAT in 1976 for serologic diagnosis of the disease has led to the recognition that hantaviruses are wide-spread throughout the world, being isolated from 16 rodent species and 4 species of insectivores [3, 13, 14]. Recently, Chinese scientists claimed that they have isolated Hantaan virus from domestic cats and wild rabbits in China (Song Gan, pers. comm., 1989). Hantaviruses cause chronic asymptomatic infections in their reservoir hosts and in experimental animals, but Hantaan, Seoul and Puumala viruses cause diverse clinical symptoms in humans. It is quite likely that in the near future HFRS or HFRS-like diseases will be identified in many parts of the world where hantaviruses exist because of new knowledge and the availability of serologic diagnostic tests [11].

Seoul virus in Korea, Japan, China, and Sri Lanka causes HFRS but Seoul or Seoul-like viruses apparently do not cause disease in the Americas or in Africa. Further studies are needed to determine the human pathogenicity of Seoul or Seoul-like viruses from rodents in parts of the world where HFRS is not recognized in humans.

Presently, neutralization is the most sensitive and specific test for determining the infecting serotype. Our tests indicate that there are at least 6 hantavirus serotypes but we could not distinguish some strains of Seoul virus isolated from house rats (Table 1). It may be that cross reactions between rat antisera against the viruses cause such confusion, therefore further studies with antisera produced in rabbits or guinea pigs are needed. Recently, it was shown that Thottapalayam virus [5] from an Indian shrew has a serologic relation with hantaviruses (C.H. Calisher, pers. comm., 1990) but it is too preliminary to determine the status of this virus.

The monoclonal antibodies produced with Hantaan virus were useful for differentiation of Hantaan, Maagi and Seoul viruses. The results clearly showed that there are 2 serologic subtypes or varieties of Seoul virus; and 4 strains did not react with monoclonal antibodies HC02-BZ08 and HC02-BD05 (Table 2). These results might be significant because, of these 4 Seoul virus strains, 3 (JTRN82/17, TR352, I/RN/82/3) were strains isolated from house rats and antibody to these 3 isolates had equal N antibody titers against Seoul and Hantaan viruses. However, by N tests Seoul virus Hubei/1, isolated from an HFRS patient in China, is different from these 3 Seoul-like viruses (JTRN/82/17, TR352, I/RN/82/3) isolated from house rats in Japan

Table 2. Comparative IFA titers of monoclonal antibodies against different strains of Hantavirus isolated from HFRS patients and rodents in the world

Serotype by PRNT	No.	Virus strain	Monoclonal antibodies from Hantaan virus								HFRS patient's serum		
			BB01-BB08	HC02-BE08	HC02-BD05	FD03-AA11	FD03-AF03	Mc-Ab 33-B	Mc-Ab 40-A	Mc-Ab 80-A	KHF-85-26	NE/Fin 85-797	NE/Fin 85-802
Hantaan virus	1	KHF 83/61	8,192	256	2,048	256	16	64	32	8,192	4,096	1,024	32
	2	Lee # 188604	8,192	1,024	1,024	128	128	—	256	8,192	2,048	2,048	64
	3	ROK 84/105	65,536	256	1,024	64	—	256	256	1,024	8,192	2,048	64
	4	US 84/2	16,384	1,024	1,024	64	—	256	—	1,024	2,048	2,048	64
	5	76/118 # 050323	8,192	256	1,024	128	128	64	32	4,096	2,048	2,048	64
	6	USSR/CI S1/452	65,536	2,048	512	64	32	256	256	1,024	8,192	2,048	128
	7	KSNUSD 84/30	65,536	4,096	4,096	1,024	64	256	256	2,048	8,192	2,048	64
	8	79/89	8,192	64	256	128	—	64	—	8,192	8,192	1,024	32
	9	83/14	4,096	64	64	128	—	32	—	8,192	4,096	4,096	32
	10	83/183	4,096	64	—	128	—	64	32	8,192	8,192	1,024	64
Maagi virus	1	Maagi	16,384	—	—	—	—	256	—	1,024	8,192	2,048	256
Seoul virus	1	Thailand # 605	—	4,096	4,096	—	—	—	128	4,096	4,096	2,048	128
	2	Thailand # 749	—	4,096	4,096	1,024	1,024	—	128	4,096	8,192	2,048	128
	3	Brazil 2/4	—	4,096	4,096	1,024	1,024	—	256	4,096	4,096	2,048	64
	4	SNU/Hamster 85/4	64	1,024	1,024	—	64	—	64	2,048	2,048	512	32
	5	KHF 83/109	—	512	2,048	256	—	—	—	8,192	4,096	4,096	128
	6	80/39 # 211808	—	1,024	1,024	32	—	—	—	8,192	2,048	2,048	64

7	SR/11 #191811	—	1,024	1,024	128	32	—	—	4,096	4,096	4,096	256
8	B/1	—	1,024	1,024	128	32	—	—	8,192	2,048	2,048	32
9	Girard Point	—	1,024	1,024	128	32	—	—	8,192	4,096	1,024	64
10	Tchoupitoulas	—	1,024	1,024	32	32	—	—	8,192	2,048	2,048	64
11	Singapore R/36	64	4,096	4,096	64	64	64	—	4,096	8,192	2,048	128
12	Hong Kong R/14	—	65,536	1,024	4,096	1,024	—	—	4,096	8,192	4,096	256
13	Hong Kong R/35	—	65,536	512	4,096	1,024	—	—	4,096	16,384	4,096	256
14	Hong Kong R/40	—	65,536	1,024	4,096	4,096	—	—	4,096	16,384	4,096	256
15	Hong Kong R/90	—	65,536	1,024	4,096	1,024	—	—	8,192	16,384	4,096	256
16	Egypt R/12915	32	1,024	256	64	64	—	—	4,096	8,192	2,048	64
17	Egypt R/13120	—	4,096	4,096	64	64	—	—	4,096	2,048	1,024	32
18	SNUSD 84/34	—	256	256	—	64	64	—	4,096	2,048	1,024	64
19	Sri Lanka	16	4,096	256	16	16	—	—	4,096	4,096	1,024	64
20	Hubei/1	—	—	—	128	128	—	—	8,192	4,096	2,048	64
21	JTRN/82/17	—	—	—	256	—	—	—	8,192	1,024	1,024	64
22	TR/352	—	—	—	128	32	—	—	8,192	4,096	2,048	128
23	I/RN/82/3	—	—	—	—	—	—	—	8,192	1,024	1,024	64
Puumala virus 1	Hällnäs B	64	64	—	—	—	—	—	256	128	16,384	8,192
2	USSR Cl/18-20	—	—	—	—	—	—	—	64	64	4,096	4,096
Prospect Hill virus 1	Prospect Hill	—	—	—	—	—	—	—	256	256	2,048	1,024
Leaky virus 1	Leaky	16	16	n.d.	16	16	16	n.d.	n.d.	256	1,024	512

and Korea. The results of serologic relationships between hantaviruses using monoclonal antibody assays suggest that there are 7 serotypes as shown in Table 2.

Our results suggest that Nephropathia Epidemica (NE) in Finland is caused by either of 2 serotypes of Puumala virus; 1 is Puumala virus, the other virus has not yet been isolated. Evidences for this is that of 2 sera from NE patients that contained high FA antibody titers (4,096–16,384) against Puumala virus (Hällnäs B and U.S.S.R. C1/18–20 strains), one serum (NE/Fin 85–797) showed high antibody titers (512–4,096) to Hantaan and Seoul viruses but the other serum (NE/Fin 85–802) had low antibody titers (32–256) to Hantaan and Seoul viruses (Table 2).

Table 3 shows the serotypes of hantaviruses isolated from HFRS patients and from other vertebrate hosts. We suggest that it is better to use serotype names for hantaviruses, rather than using the name of the host of origin to classify viruses. First, because more than one hantavirus serotype may be isolated from a single host and second, because many species of animals are reservoir hosts of the same hantaviruses.

The recent findings of hantaan virus antigens in tissues of birds in the U.S.S.R. (E.A. Tkachenko, pers. comm., 1989) may have great impact on our understanding of both the epidemiology and the ecology of hantaviruses. It is

Table 3. Serotypes of hantavirus isolated from HFRS patients and animals in the world

Name of host	Serotypes of hantavirus isolated in Vero E6 cells
HFRS patient	Hantaan virus
	Seoul virus
Apodemus agrarius	Hantaan virus
	Maagi virus[a]
Urban rats	Seoul virus
	Seoul-like viruses[a]
Laboratory rats	Seoul virus
	Hantaan virus
Clethrionomys glareorus	Hantaan virus[b]
	Puumala virus
Microtus pennsylvanicus	Prospect Hill virus
Golden hamster	Seoul virus
Bandicota indica	Seoul virus
Mus musculus	Leaky virus
Suncus murinus	Thottapalyam virus

[a] Probable new serotype
[b] An isolate from U.S.S.R. but origin is not clear

possible that HFRS may become recognized as one of the important hemorrhagic diseases in many parts of the world in 1990s.

Acknowledgements

Given the limited space available, it is impossible to specify the names and institutions of the many investigators who willingly and enthusiastically collaborated with us to make these studies possible and successful. However, we would be remiss if we were to not emphasize that this has been a truly international partnership, one that has been both personally and professionally gratifying.

This work was supported in part by Grant no. DAMD 17-86-G-6011 from the U.S. Army Medical Research and Development Command, Frederick, Maryland.

References

1. Baek LJ, Yanagihara R, Gibbs CJ, Miyazaki M, Gajdusek DC (1988) Leaky virus, a new hantavirus isolated from *Mus musculus* in the United States. J Gen Virol 69: 3129–3132
2. Couland X, Chonaib E, Georges AJ (1987) First human case of hemorrhagic fever with renal syndrome in the Central African Republic. Trans R Soc Trop Med Hyg 81: 686
3. Gavrilovskaya IN, Apekina NS, Miasnikov YA (1983) Features of circulation of hemorrhagic fever with renal syndrome virus among small mammals in the European USSR. Arch Virol 75: 313–316
4. Hemorrhagic fever with renal syndrome: memorandum from a WHO meeting (1982) Bull WHO 61: 269–275
5. Karabatsos N (ed) (1985) International catalogue of arboviruses including certain other viruses of vertebrates, 3rd edn. American Society of Tropical Medicine and Hygiene, San Antonio, Texas
6. Lee HW (1988) Global update on distribution of hemorrhagic fever with renal syndrome and hantaviruses. Virus Info Exch Newslett 5: 82–84
7. Lee HW, Lee PW (1976) Korean hemorrhagic fever. I. Demonstration of causative antigen and antibodies. Kor J Intern Med 19: 371–394
8. Lee HW, Lee PW, Johnson KM (1978) Isolation of the etiologic agent of Korean hemorrhagic fever. J Infect Dis 137: 298–308
9. Lee HW, van der Groen G (1989) Hemorrhagic fever with renal syndrome. Prog Med Virol 36: 62–102
10. Lee PW, Gibbs CJ, Gajdusek CJ, Yanagihar R (1985) Serotypic classification of hantaviruses by indirect immunofluorescent antibody and plaque reduction neutralization tests. J Clin Microbiol 22: 940–944
11. Lee PW, Meegan JM, LeDuc JW, Tkachenko EA, Tvanov AP, Rezapkin GV, Drozdov SG, Kitamura T, Tsai TF, Dalrymple JM (1989) Serologic techniques for detection of Hantaan virus infection, related antigens and antibodies. In: Lee HW, Dalrymple JD (eds) Manual of hemorrhagic fever with renal syndrome. WHO Collaborating Center for Virus Reference and Research (HFRS), Institute for Viral Diseases, Korea University, Seoul, pp 75–106
12. Schmaljohn CS, Hasty SE, Dalrymple JM, LeDuc JW, Lee HW, von Bornsdorff CH, Brummer-Korvenkontio M, Vaheri A, Tsai TF, Goldgaber D, Lee PW (1985) Antigenic and genetic properties of viruses linked to hemorrhagic fever with renal syndrome. Science 227: 1041–1044
13. Tkachenko EA, Tvanov AP, Gonets MA (1983) Potential reservoir and vectors of hemorrhagic fever with renal syndrome in the USSR. Ann Soc Belg Med Trop 63: 267–269

14. van der Groen G, Leirs H, Verhagen R (1986) Polyhostal nature of Hantaan virus. In: Proceedings Second Symposium Recent Advances in Rodent Control, Kuwait, pp 197–207
15. Wah LT, Mangalam S, Lee HW (1987) Hemorrhagic fever with renal syndrome. Report on a case in Malaysia. J Infect Dis 156: 1035–1036

Authors' address: Dr. H. W. Lee, WHO Collaborating Center for Virus Reference and Research (hemorrhagic fever with renal syndrome), Institute for Viral Diseases, Korea University, 4, 2nd Street Myungyun-Dong Chongno-Ku, Seoul 110-702, Korea.

Arch Virol (1990) [Suppl 1]: 19–28

Diagnostic potential of a baculovirus-expressed nucleocapsid protein for hantaviruses

Cynthia A. Rossi[1], C. S. Schmaljohn[2], J. M. Meegan[3],
and J. W. LeDuc[1]

[1] Disease Assessment Division and [2] Virology Division,
U.S. Army Medical Research Institute of Infectious Diseases, Fort Detrick,
Frederick, Maryland, U.S.A.
[3] Communicable Diseases Division, World Health Organization, Geneva, Switzerland

Accepted January 16, 1990

Summary. A non-infectious recombinant, baculovirus-expressed protein, analogous to the nucleocapsid of Hantaan (HTN) virus, and standard Vero E-6 cell culture-prepared HTN antigen, were evaluated by enzyme-linked immunosorbent assays (ELISA) to detect IgG antibodies to various strains of hantaviruses. Rat and human sera previously found to have hantavirus-specific antibodies detectable by the plaque-reduction neutralization test (PRNT) were examined. Results obtained by the immunofluorescent antibody (IFA) test, ELISA, and PRNT were compared. Sera with IFA titers greater than 128 showed a high degree of correlation by both ELISA and PRNT. ELISA tests utilizing baculovirus-expressed protein were almost as sensitive as PRNT in detecting antibodies to HTN, Seoul, and Porogia viruses. The expressed nucleocapsid protein was found to be less sensitive in detecting antibody to Puumala virus, as has been observed with other HTN viral antigen-based tests. Our results suggest that the IgG ELISA with the baculovirus-expressed recombinant nucleocapsid protein is a useful alternative to the IFA for routine screening for antibodies to HTN virus. This test is sensitive, specific, easy to perform, rapid, and requires very small quantities of reagents. Furthermore, the recombinant antigen is non-infectious for humans and is, therefore, safe to prepare and use.

Introduction

Hantaan (HTN), Seoul (SEO), Porogia (POR), and Puumala (PUU) viruses are antigenically related members of the genus *Hantavirus* of the family *Bunyaviridae*; all are known to cause disease in humans [1, 13]. The general term, hemorrhagic fever with renal syndrome (HFRS), is used to describe these diseases, which range in severity from a subclinical to fatal disease, and are characterized by fever, headache, abdominal or back pain, renal insufficiency, and various degrees of hemorrhage [16]. Hantaviruses are maintained in nature by association with small rodents. Rodents are chronically infected, with virus which is shed in urine, saliva, and feces, but has no obvious pathological effects on them [6].

As with many other viral diseases, it is difficult on clinical grounds alone to diagnose hantaviral infections with certainty. The indirect immunofluorescent antibody (IFA) test currently is the easiest and most widely accepted method for serological detection of hantaviral antibodies. A recently developed enzyme-linked immunosorbent assay (ELISA) has afforded a rapid, simple, and less subjective test with which to detect antibodies. Both these tests require reagents that must be produced and inactivated under stringent biological-containment conditions. Production of HTN virus is hazardous and extremely time-consuming; yields of final product are insufficient for large-scale use. The development of a non-infectious, easily produced hantaviral antigen for use in diagnostic tests is of great importance. A recombinant baculovirus-expressed nucleocapsid (N) protein of HTN virus, such as that recently developed by Schmaljohn et al. [14] holds promise as such an antigen. To investigate the usefulness of the expressed N protein as a diagnostic antigen, reactivities of sera representing various hantaviral strains were compared by ELISA, using the recombinant antigen and a standard, inactivated antigen prepared in cell culture, by IFA, and by plaque-reduction neutralization test (PRNT).

Materials and methods

Viruses and cells

Prototype HTN (strain 76-118) virus isolated in 1976 [5], SEO (strain HR80-39) virus isolated in 1982 [7], PUU (strain 83-223L) virus isolated in 1984 [10], and POR virus isolated in 1986 [1] were propagated in Vero E-6 cells [2, 8]. Expression of the S genome segment of HTN virus by recombinant *Autographa californica* nuclear polyhedrosis virus and characterization of the expressed N protein have been described [14]. The recombinant baculovirus was propagated in *Spodoptera frugiperda* (SF-9) cells according to previously published methods [15].

Sera

Serum samples from 70 wild rats and 98 humans tested in our laboratory between 1983 and 1989 for the detection of antibody to hantavirus were used. These included sera from Greece, Yugoslavia, China, Japan, Sweden, Argentina, Korea, Philippines, Thailand, Brazil, Burma, Belgium, Taiwan, Hong Kong, Egypt, Sudan, and (various locations within) the United States. Also included were 90 serum samples collected from laboratory rats experimentally infected intramuscularly or by aerosol with hantaviruses [11].

These 258 sera were tested by PRNT and used as standards to compare the relative sensitivity and specificity of IFA and ELISA with either viral or expressed antigens. Positive sera included 28 with highest PRNT titers to HTN virus, 15 with highest titers to POR virus, 65 with highest titers to SEO virus, and 28 with highest titers to PUU virus. Of the latter, four reacted only with PUU virus in the PRNT, and not with HTN.

Plaque-reduction neutralization test

PRNT were performed according to previously published methods [3, 8, 12]. Briefly, PRNT were conducted in duplicate by mixing equal volumes of diluted serum with 200 plaque forming units of HTN, SEO, POR, or PUU virus, incubating them overnight at 4 °C, and then inoculating 0.2 ml into 25 cm^2 flasks containing Vero E-6 cell monolayers. A 1% agarose overlay was applied after 1 h of adsorption and flasks were stained with a neutral red-agarose solution after 10 to 14 days of incubation at 37 °C. Plaques were counted 24–48 h later. Neutralization titers were recorded as the reciprocal of the highest serum dilution neutralizing 50% of the plaque dose.

Immunofluorescent antibody test

Serial two-fold dilutions of sera were examined by IFA using spot slides containing Vero E-6 cells infected with HTN virus [3] (slides were prepared by Salk Institute, Swiftwater, PA). Fluorescein-labeled anti-human or anti-rat IgG heavy- and light-chain-specific, conjugate (Organon Teknika, Malvern, PA) was used at 1:30 dilution. Incubation periods consisted of 30 min each at 37 °C with two 15-min washes between steps. Titers were recorded as the highest serum dilution yielding characteristic cytoplasmic fluorescence.

Recombinant baculovirus-expressed antigen preparation for ELISA

Recombinant-infected SF-9 cells (MOI = 10) were harvested 2 days post-infection, centrifuged for 10 min at 3000 × g at 4 °C and resuspended to a concentration of 1 × 10^6 cells/ml in 0.01 M phosphate-buffered saline (PBS) (pH 7.4) with 1% sodium dodecyl sulfate. The cells were then sonicated continuously for 3 min with a cup sonicator (Tekmar Sonic Disruptor, Thomas Scientific, Swedesboro, NJ) at 50% maximum output while maintaining samples on ice. After a 15 min incubation period at room temperature, disrupted cells were again centrifuged. The supernatant then was collected and stored at 4 °C until used in ELISA.

Viral antigen preparation for ELISA

Ten days after the cells were infected with HTN virus supernatant fluid from Vero E-6 cell cultures was collected and concentrated 25-fold by precipitation with 8% (w/v) polyethylene glycol (MW 6000), buffered with 0.1 M Tris (pH 8.6), and inactivated with 0.4% beta-propiolactone (Oneal, Jones, and Feldman, St. Louis, MO) followed by cobalt irradiation with 2 × 10^6 Roentgens (~8.1 × 10^5 R/h in a Gammacell 220 Cobalt 60 irradiator, Radiochemical Company, Kanata, Ontario, Canada). This viral antigen was stored at −70 °C until used in ELISA. Safety tests were conducted by co-cultivating antigen with Vero E-6 cells for 50 days and periodic examination by IFA for characteristic cytoplasmic fluorescence. If no viral antigen was detected in cells by 50 days, the material was considered to be inactivated.

Enzyme-linked immunosorbent assay

The ELISA was done according to previously established methodology [9, 14] with slight modifications. The optimal dilutions of all reagents used in the ELISA were determined by checkerboard titrations with known positive and negative sera. Plates were washed three times in PBS, with 0.1% Tween 20 detergent between each step. Except for the initial overnight incubation, all incubations were for 1 h at 37 °C, and volumes in each well were 100 µl. Optical densities were determined spectrophotometrically at 414 nm, 30 min after addition of a chromogenic substrate, ABTS (2'-azino-di-[3-ethyl-benzthiazoline sulfonate]; Kirkegaard and Perry, Gaithersburg, MD) with a Multiscan microplate reader. Samples were considered positive for HTN-specific antibodies if the difference in optical densities of positive and negative control antigen wells was at least three standard deviations above the mean difference in optical density obtained from four negative control sera. All positive sera were titrated to endpoint.

Lysate prepared from the recombinant-infected SF-9 cells, or a negative control lysate consisting of uninfected SF-9 cells, was diluted 1:200 in PBS and allowed to bind to wells of polyvinyl microtiter plates (Dynatech Laboratories, Chantilly, VA) overnight at 4 °C. Serum samples were diluted in PBS with 0.1% Tween 20 and 5% fetal bovine serum (serum diluent), then added to both positive and negative antigens at 1:100 and diluted in serial two-fold dilutions to 1:1280 in serum diluent. Peroxidase-labeled, anti-rat IgG (Kirkegaard and Perry, Gaithersburg, MD; 1:1000 dilution) or anti-human IgG (Accurate Scientific Corp, Westbury, NY; 1:8000 dilution) conjugate diluted in serum diluent was used as detector antibody.

For comparison, a second ELISA utilized a 1:40 dilution, in serum diluent, of inactivated and concentrated HTN-infected cell culture-produced viral antigen, or a negative control antigen consisting of uninfected E6 cells, captured onto wells of a microtiter plate coated overnight (4 °C) with HTN-specific mouse monoclonal antibodies (hybridomas prepared by Dr. J. B. McCormick, Center For Disease Control, Atlanta, GA). Serum samples were added, followed by addition of a peroxidase-labeled detector antibody as above.

Results

Table 1 presents a comparison of sensitivity and specificity for each of the tests with the PRNT as a standard. The enzyme immunoassay with the cell culture-prepared viral antigen was the most sensitive (99%) and specific (98%), with only three false positive sera and two false negatives. The ELISA with the recombinant antigen was only slightly less sensitive (93%) and specific (98%), with two false positive sera and nine false negatives, while the IFA test was even less sensitive (88%) and specific (84%), with 19 false positive and 16 false negative reactions. Antisera to each of the hantaviral strains revealed that the viral antigen-based ELISA failed to identify only one SEO and one PUU virus-specific antiserum, while the recombinant antigen ELISA failed to identify one SEO and eight PUU virus-specific antisera (Table 2). The IFA was the least sensitive, unable to identify one POR and 15 PUU specific antisera.

Table 3 provides a summary of the disparity in test results, highlighting reactions seen with four sera containing neutralizing antibody only to PUU virus. Ten sera were positive by PRNT to both HTN and PUU, and reacted in the ELISA with either antigen, but were negative by IFA. Four sera were likewise positive to both HTN and PUU by PRNT, and were positive by IFA and ELISA with cell culture-prepared viral antigen, but were negative when the recombinant antigen was used. A single serum was positive to both viruses by PRNT, but negative in all other tests. Of four sera positive only to PUU virus by PRNT, one was positive in all other tests, one only by IFA and the cell-culture-antigen ELISA, and the remaining two were positive only with the ELISA with the cell-culture antigen.

Table 1. Comparison of sensitivity and specificity for Hantaan IFA test and ELISA with viral (EIA-VA) or a recombinant baculovirus-expressed (EIA-RA) Hantaan antigen, with hantavirus PRNT as a standard

Assay	Sensitivity %		Specificity %		False positive %		False negative %	
PRNT	100	(136[a])	100	(122)	0	(0)	0	(0)
IFA[b]	88	(120)	84	(103)	16	(19)	12	(16)
EIA-VA[c]	99	(134)	98	(119)	2	(3)	1	(2)
EIA-RA[c]	93	(127)	98	(120)	2	(2)	7	(9)

[a] Total PRNT positive sera tested = 136; total PRNT negative sera tested = 122; subsequent values in parentheses are number of serum samples with concordant or discordant results
[b] IFA positive ≥ 128
[c] EIA-VA and EIA-RA positive ≥ 100

Table 2. Comparison of reactivity of antisera specific for different hantaviral strains as determined by PRNT, with results obtained by IFA and ELISA with a viral (EIA-VA) or a baculovirus-expressed recombinant (EIA-RA) antigen

Antiserum to	Total no. sera tested	Number sera positive		
		EIA-VA[a]	EIA-RA[a]	IFA[b]
Hantaan	28	28	28	28
Porogia	15	15	15	14
Seoul	65	64	64	65
Puumala	28	27	20	13
Total	136	134	127	120

[a] EIA-VA and EIA-RA positive ≥ 100
[b] IFA positive ≥ 128

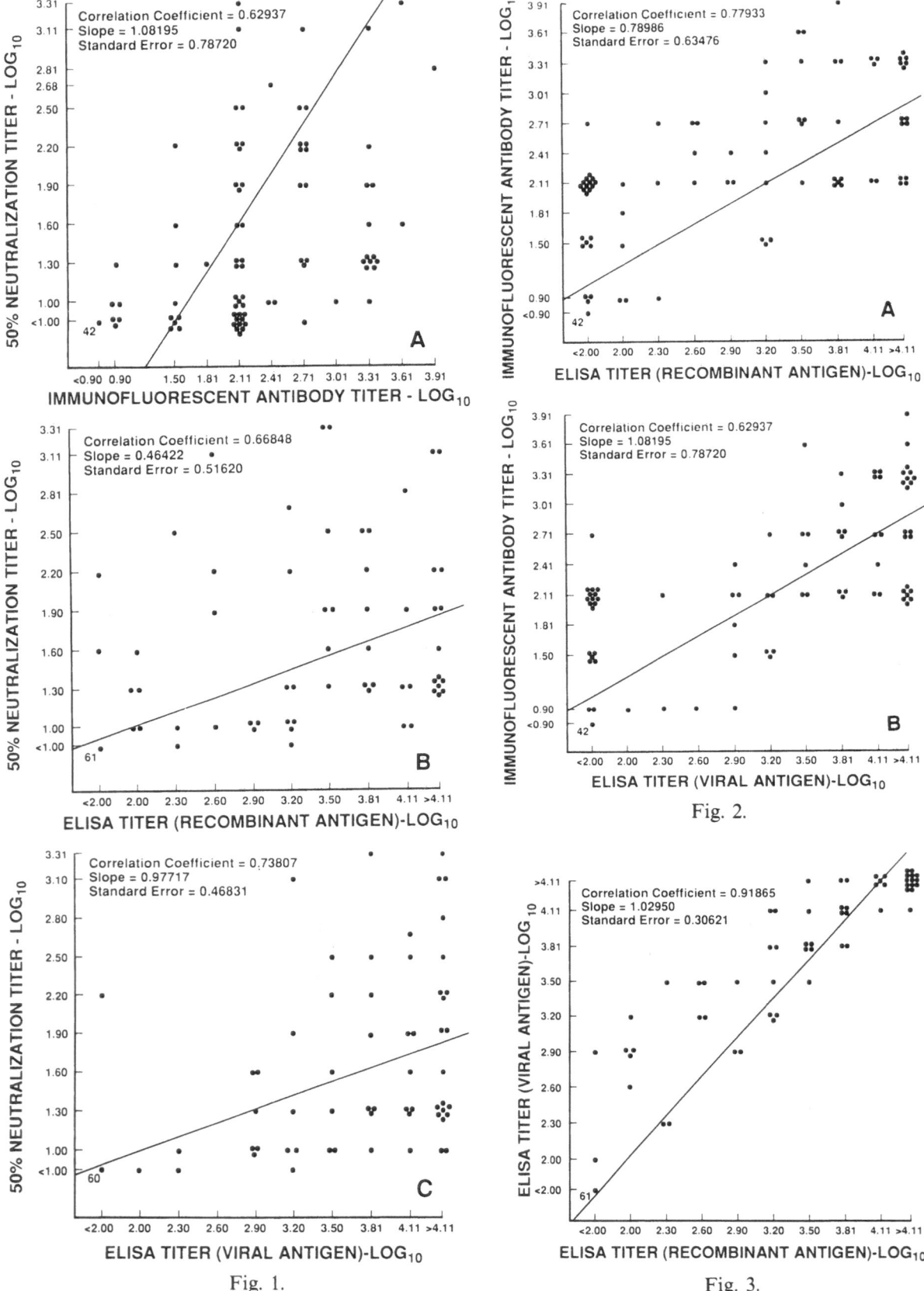

Fig. 1.

Fig. 2.

Fig. 3.

Table 3. Nephropathia Epidemica, Puumala-specific anti-sera showing a disparity in results when tested by PRNT, or ELISA with either a viral (EIA-VA) or a baculovirus-expressed recombinant (EIA-RA) antigen

No. of sera	PRNT[a]		IFA[b]	EIA-VA[c]	EIA-RA[c]
	HTN	PUU			
10	+	+	−	+	+
4	+	+	+	+	−
1	+	+	−	−	−
1	−	+	+	+	+
1	−	+	+	+	−
2	−	+	−	+	−

[a] Sera tested against both HTN and PUU viral strains. Sera considered PRNT-positive (+) if 50% neutralization titer ≥ 10
[b] IFA positive (+) ≥ 128
[c] EIA-VA and EIA-RA positive (+) ≥ 100

Sera from 70 rats and 51 humans (total $N = 121$) from the above collection were used to compare titers of antibody to HTN virus obtained by PRNT, IFA, and ELISA. Figure 1A shows a comparison of the titers obtained by the PRNT and IFA. The overall agreement between these tests was 89% (108/121) with a correlation coefficient of $r = 0.63$. A comparison of the titers obtained by the PRNT and baculovirus-expressed, recombinant antigen-based ELISA is shown in Fig. 1B. Overall agreement was 97% (117/121) with a correlation coefficient of $r = 0.67$. Figure 1C presents a comparison of the PRNT titers and viral antigen-based ELISA titers. The overall agreement was 97% (117/121) with a correlation coefficient of $r = 0.74$. A comparison of the titers obtained by the IFA test and baculovirus-expressed, recombinant antigen-based ELISA is shown in Fig. 2A. Overall agreement was 89% (108/121) with a correlation coefficient of $r = 0.78$. Figure 2B shows a comparison of the titers obtained by IFA and viral antigen-based ELISA. The overall agreement was 90% (109/121) with a correlation of $r = 0.81$. Figure 3 presents a comparison of the titers by ELISA

Fig. 1. Comparison of titers of antibody to Hantaan virus determined in 121 sera tested by PRNT and, **A** IFA; **B** a recombinant baculovirus-expressed antigen-based ELISA; **C** a viral antigen-based ELISA

Fig. 2. Comparison of titers of antibody to Hantaan virus determined in 121 sera tested by IFA and, **A** a recombinant baculovirus-expressed antigen-based ELISA; **B** a viral antigen-based ELISA

Fig. 3. Comparison of titers of antibody to Hantaan virus determined in 121 sera tested by ELISA with a recombinant baculovirus-expressed antigen and with a viral antigen

using two different antigen preparations. The overall agreement was 98% (119/121) with a correlation coefficient of $r = 0.92$.

Discussion

Four serological types currently are recognized within the genus *Hantavirus*: HTN, SEO, PUU, and Prospect Hill (PH) viruses [13]. Porogia virus is closely related to prototype HTN virus but is antigenically and ecologically distinct and may represent a different virus [1]. Seoul, PUU, PH, and POR viruses were each originally discovered because their infected rodent host's serum reacted by IFA with HTN-infected cells. Indeed, recognition of the broad cross-reactivity that anti-hantaviral antibodies exhibit when HTN virus-infected cells are used in IFA tests has made this test the most frequently used assay in epidemiological and clinical investigations. Nonetheless, the IFA often reacts nonspecifically to yield false positive or negative results; consequently, the PRNT has been most frequently relied on as a confirmatory test [1, 3, 4, 13]. The PRNT is quite time consuming, expensive, and technically demanding, and is not suitable for use in many laboratory settings. Clearly an alternative serological assay is needed.

The ELISA tests described here offer an attractive alternative to both IFA and PRNT. They are easy to perform in most laboratory settings and require small volumes of serum and reagents; this technology has been rapidly assimilated by our laboratory into many different routine laboratory applications. Both antigen preparations investigated offer acceptable sensitivity and specificity when examined with well-characterized, PRNT-confirmed sera.

The cell culture-produced viral antigen was slightly more sensitive than the recombinant antigen, but both antigens were equally specific. The lower sensitivity was almost exclusively associated with a failure to react with PUU-specific antisera. This was most notable with the recombinant antigen, where nine false negative reactions were seen. Because the recombinant antigen uses only gene products encoded by the S segment, while the PRNT measures at least products of the M segment and perhaps S segment as well, it may be possible to overcome this apparent deficiency by creating an expressed antigen that includes products of both the M and S segments of HTN, PUU, or both viruses. Efforts in this direction are in progress, but results are pending.

In spite of the minor limitations noted with the baculovirus-expressed recombinant antigen, the ELISA utilizing this antigen appears to be a useful alternative to IFA for routine screening of sera for antibodies to hantaviruses. This antigen is safe, easily produced in large quantities, and has sensitivity and specificity characteristics clearly superior to those of the IFA. Further, the ELISA is frequently used in many laboratories for determinations of antibody to other viruses, so that specific screening for antibodies to hantaviruses could be incorporated into routine testing without excessive complications or expenses.

Acknowledgements

In conducting the research described in this report, the investigator(s) adhered to the "Guide for the Care and Use of Laboratory Animals," as promulgated by the Committee on Care and Use of Laboratory Animals of the Institute of Laboratory Animal Resources, National Research Council. The facilities are fully accredited by the American Association for Accreditation of Laboratory Animal Care.

The views of the author(s) do not purport to reflect the positions of the Department of the Army or the Department of Defense.

Approved for public release, distribution unlimited.

References

1. Antoniadis A, Grekas D, Rossi CA, LeDuc JW (1987) Isolation of a hantavirus from a severely ill patient with hemorrhagic fever with renal syndrome in Greece. J Infect Dis 156: 1010–1013
2. French GR, Foulke RS, Brand OA, Eddy GA, Lee HW, Lee PW (1981) Korean hemorrhagic fever: propagation of the etiologic agent in a cell line of human origin. Science 211: 1046–1048
3. LeDuc JW, Smith GA, Johnson KM (1984) Hantaan-like viruses from domestic rats captured in the United States. Am J Trop Med Hyg 33: 992–998
4. LeDuc JW, Antoniadis A, Siamopoulos K (1986) Epidemiological investigations following an outbreak of hemorrhagic fever with renal syndrome in Greece. Am J Trop Med Hyg 35: 654–659
5. Lee HW, Lee PW, Johnson KM (1978) Isolation of the etiologic agent of Korean hemorrhagic fever. J Infect Dis 137: 298–308
6. Lee HW (1982) Korean hemorrhagic fever. Prog Med Virol 28: 96–113; 513–545
7. Lee HW, Baek LJ, Johnson KM (1982) Isolation of Hantaan virus, the etiologic agent of Korean hemorrhagic fever, from wild urban rats. J Infect Dis 146: 638–644
8. McCormick JB, Palmer EL, Sasso DR, Riley MP (1982) Morphological identification of the agent of Korean haemorrhagic fever (Hantaan virus) as a member of Bunyaviridae. Lancet 1: 765–767
9. Meegan JM, LeDuc JW (1989) Enzyme immunoassays. In: Lee HW, Dalrymple JM (eds) Manual of hemorrhagic fever with renal syndrome. WHO Collaborating Center for Virus Reference and Research (HFRS), Institute for Viral Diseases, Korea University, Seoul, pp 83–87
10. Niklasson B, LeDuc JW (1984) Isolation of the nephropathia agent in Sweden. Lancet 1: 1012–1013
11. Nuzum EO, Rossi CA, Stephenson EH, LeDuc JW (1988) Aerosol transmission of Hantaan and related viruses to laboratory rats. Am J Trop Med Hyg 38: 636–640
12. Schmaljohn CS, Hasty SE, Harrison SA, Dalrymple JM (1983) Characterization of Hantaan virions, the prototype virus of hemorrhagic fever with renal syndrome. J Infect Dis 148: 1005–1012
13. Schmaljohn CS, Hasty SE, Dalrymple JM, LeDuc JW, Lee HW, von Bonsdorff C-H, Brummer-Korvenkontio M, Vaheri A, Tsai TF, Regnery HL, Goldgaber D, Lee PW (1985) Antigenic and genetic properties of viruses linked to hemorrhagic fever with renal syndrome. Science 227: 1041–1044

14. Schmaljohn CS, Sugiyama K, Schmaljohn AL, Bishop DHL (1988) Baculovirus expression of the small genome segment of Hantaan virus and potential use of the expressed nucleocapsid protein as a diagnostic antigen. J Gen Virol 69: 777–786

15. Summers MD, Smith GE (1986) A manual of methods for baculovirus vectors and insect cell culture procedures. Department of Entomology, Texas A&M University, College Station, Texas

16. World Health Organization (1982) Report of the working group on hemorrhagic fever with renal syndrome. WPR/RPD/WG/82.16. WHO Regional Office for the Western Pacific, Tokyo

Authors' address: C. A. Rossi, Department of Epidemiology, Disease Assessment Division, USAMRIID, Fort Detrick, Frederick, MD 21701-5011, U.S.A.

Arch Virol (1990) [Suppl 1]: 29–33

Rapid serodiagnosis of hantavirus infections using high density particle agglutination

T. Tomiyama[1] and **H. W. Lee**[2]

[1] Laboratory of Microbiology and Immunology, University of Tokyo Branch Hospital, Tokyo, Japan
[2] The Institute for Viral Diseases, Korea University, Seoul, Korea

Accepted January 22, 1990

Summary. Antibody against hantaviruses was measured within forty minutes by a passive agglutination procedure using high density composite particles coated with purified Hantaan virus antigen. Antigen for the reaction was prepared from the brains of suckling rats infected with Hantaan virus, using ultracentrifugation, protamine and ethyl alcohol treatment. This method is more sensitive than the immunofluorescent antibody technique and the antigen reacted with antibodies to Hantaan, Seoul and Puumala viruses.

Introduction

The need for a rapid and simple serological test, such as passive agglutination, for hantaviruses (family *Bunyaviridae*, genus *Hantavirus*) is obvious and would be increasingly useful because of research on immunity, vaccination and epidemiology of diseases caused by these viruses, in addition to serodiagnosis of hemorrhagic fever with renal syndrome (HFRS) [2]. However, such tests as are available are scarcely used because of difficulties inherent in preparating suitable test antigens. This paper describes a new, rapid, and convenient and sensitive serological test for detecting antibodies against hantaviruses by means of high density composite particle agglutination (HDPA) [6], using highly purified Hantaan virus antigen, originally prepared for use as inactivated vaccine against HFRS [3]. Furthermore, because antigen coated HDP can be lyophilized, it can be preserved for a long period of time at room temperature.

Materials and methods

Carrier particles

High density composite particles [6] (HDP, Tokuyama Soda Co., Tokyo, Japan) were used as carrier. They have a silica core surrounded by a red dye layer and second silica layer covered the dyed layer. Furthermore, the particle surface is covered with functional groups designed to adsorb antigen. The density of the particles is 2.0 and their diameter is 1.8 µm.

Antigen

Hantaan virus, ROK84-105 strain [3], isolated directly in Vero E6 cell cultures in 1984 from blood of a HFRS patient was used in the experiments. The virus was passaged 7 times in suckling rat brains to increase titers and virus yield. The LD_{50} of strain ROK84-105 in suckling rats by intracerebral inoculation was $10^{9.5}$/ml. Supernatant fluid of 5% suckling rat brain suspension in phosphate buffered saline (PBS), pH 7.2 was inactivated with 0.05% formalin at 4 °C for 15 days. Purification of the inactivated virus suspension for use as antigen in HDPA was done according to the modified method used for preparation of Japanese encephalitis mouse brain vaccine [3]. Protein content of the purified antigen preparation was 43 µg/ml and antigen concentration of the preparation was 10,240 units/ml by ELISA test [5].

Preparation of antigen-coated HDP

For preparation of Hantaan virus antigen-coated HDP (Hanta-HDP), an equal volume of eight ELISA units/ml of Hanta-HDP antigen in PBS was added to the 0.5% HDP suspension in PBS in 1/60 mole, and incubated two hours at 20 °C, shaking each ten min. Then the HDP were washed with PBS twice, suspended in 0.1 the original volume of PBS containing 0.01% bovine serum albumin, 1% dextran, 1% sodium glutamate and 0.5% glycine as a stabilizer, and then lyophilized. Uninfected rat brains were also treated in same manner as the antigen control.

Serum

Seventeen sera from HFRS patients from Korea (K1–K8), Japan (J1–J4), Finland (F1–F4) and 2 antibody positive sera from healthy persons in the U.S.A., and 4 sera (K9–K12) from people who were inoculated with inactivated Hantaan virus vaccine in Korea were used for antibody determinations. Antibody negative serum from a healthy person in the U.S.A. was used as negative control serum. All serum samples from HFRS patients and from vaccinees from Korea had been shown in other tests to contain antibodies against Hantaan virus. Likewise, serum samples from HFRS patients in Japan and from healthy people in the U.S.A. were known to contain antibodies to Seoul virus and serum samples from Nephropathia Epidemica patients in Finland contained antibodies to Puumala virus.

Procedure of hantavirus HDPA test

A microtiter technique was used for HDPA tests for antibodies against Hantaan virus. The virus antigen- and normal antigen-coated HDP were suspended to a concentration of 0.5% with buffer. The procedure for titration of antibody against Hantaan virus is summarized in Table 1.

Immunofluorescent antibody technique (IFAT)

IFAT was carried out with Vero E6 cells infected with Hantaan virus as described previously [1]. IFA (IgG) titers were expressed as the reciprocal of the highest serum dilution giving specific fluorescence (1 + based on a subjective 0 to 4 + scale).

Table 1. Summarized procedure of hantavirus HDPA

	Well no.								
	1	2	3	4	5	6	7	8	9
Final dilution of serum	Control	40	80	160	320	640	1280	2560	5120
Diluent 25 µl	No	No	Yes	Yes	Yes	Yes	Yes	Yes	Yes
1:20 serum 25 µl	Yes	Yes	Yes	No	No	No	No	No	No
					Serial two-fold dilution				
Control-HDP 25 µl	Yes	No	No	No	No	No	No	No	No
Hanta-HDP 25 µl	No	Yes	Yes	Yes	Yes	Yes	Yes	Yes	Yes

Mix well and incubate at least 40 min at room temperature

ELISA assay

The test for demonstration of IgG antibodies against Hantaan virus was described previously [5]. The concentration of viral antigen in the preparation was determined by ELISA and one unit was expressed as the highest antigen dilution giving a specific reaction.

Results

Box titration of antigen-coated HDP and antibody in human serum was carried out as shown in Table 2. Positive agglutination patterns using antigen coated HDP were clearly demonstrable against antibodies after 40 min incubation at room temperature. The optimum antigen coating concentration was four to eight ELISA units/ml, while negative reactions were found with diluent. Alternatively, non-coated HDP antigen controls were always negative against both diluent and serum containing antibodies. Comparative antibody titers of sera from HFRS patients and from individuals vaccinated with Hantaan virus, as determined by IFAT, ELISA, and HDPA, are shown in Table 3. Hantaan virus antigen-coated HDP reacted with not only antibody to Hantaan virus (sera from Korea) but also with antibodies to Seoul virus (sera from Japan and U.S.A.) and with antibodies to Puumala virus (sera from Finland), as was shown by parallel IFAT and ELISA. It was also found that HDPA titers were about two to ten times higher than IFA titers but that ELISA antibody titers usually were higher than HDPA and IFAT.

Discussion

It has been recognized that HDP sensitizes more protein and lipid antigens on their surfaces, which results in higher sensitivity to antibody than similarly used erythrocytes or polystyrene latex particles [6]. As for hantavirus HDPA, we found that this test provides higher sensitivity than IFAT, which may depend on this binding property of HDP.

Table 2. Box titration of the Hantaan virus antigen-coated HDP and antibody positive serum

Antigen (ELISA U/ml)	Dilution of serum[a]							
	100	200	400	800	1600	3200	6400	Diluent
16	+ + +	+ + +	+ + +	+ + +	+ + +	+ + +	+	±
8	+ + +	+ + +	+ + +	+ + +	+ + +	+ + +	−	−
4	+ + +	+ + +	+ + +	+ + +	+ + +	+ + +	−	−
2	+ + +	+ + +	+ +		−	−	−	−
1	−	−	−	−	−	−	−	−
0	−	−	−	−	−	−	−	−

[a] Antibody titer of serum from this HFRS patient was 1:512 by IFAT

Table 3. Comparative antibody titers of sera from HFRS patients and vaccinees against Hantaan virus by HDPA, IFAT, and ELISA

Origin of country	Code no. of serum	Antibody titer to Hantaan virus by		
		HDPA	IFAT	ELISA
Japan	J-1	4,000	2,048	25,600
	J-2-1	320	256	3,200
	J-2-2	320	256	3,200
	J-3	4,000	512	25,600
	J-4	640	512	6,400
U.S.A.	US-1	1,000	64	800
	US-2	< 40	512	< 100
Finland	F-1	160	128	6,400
	F-2	40	32	3,200
	F-3	160	1,024	6,400
	F-4	320	1,024	6,400
Korea	K-1	16,000	1,024	25,600
	K-2	8,000	4,096	25,600
	K-3	16,000	2,048	25,600
	K-4	4,000	2,048	25,600
	K-5	1,600	1,024	6,400
	K-6	12,800	4,096	25,600
	K-7	12,800	4,096	6,400
	K-8	6,400	4,096	6,400
	K-9	3,200	256	6,400
	K-10	3,200	256	6,400
	K-11	200	256	800
	K-12	3,200	128	1,600
U.S.A.	Negative control	< 40	< 16	< 100

The highly purified Hantaan virus antigen as here applied reacted only with antibody to Seoul and Puumala viruses (genus *Hantavirus*) by HDPA test but not with antibody to other etiologic agents of viral hemorrhagic fevers, such as dengue, Rift Valley fever, Crimean-Congo hemorrhagic fever, Junin and Machupo hemorrhagic fevers, and Ebola, nor did it non-specifically agglutinate with sera from patients with leptospirosis and rickettsiosis [4].

As antigen-coated HDP can be lyophilized, this reaction is easily used for measurement of hantavirus antibody, without any technical complexity, within 1 h. The available serologic diagnostic tests for HFRS are IFAT, ELISA, plaque-reduction neutralization test, hemagglutination inhibition test and an immune adherence hemagglutination test [4], but these tests are complicated and time-consuming compared with hantavirus HDPA test. In our studies, only IgG antibodies in sera from HFRS patients and vaccinees were tested because hantavirus HDPA test cannot differentiate IgG and IgM antibodies.

It is expected that this test will be applied for clinical and epidemiological use, especially for rapid serodiagnosis of hantavirus infections among suspect HFRS patients at hospitals in areas endemic for HFRS.

Acknowledgements

This research was supported in part by Grant no. ICP/CDS011-A I.D. OCD/1/87 from WHO Western Pacific Region, Manila and DAMD17-86-G-6011 from the U.S. Army Medical Research and Development Command, Frederick, MD, U.S.A.

References

1. Lee HW, Lee PW, Johnson KM (1978) Isolation of the etiologic agent of Korean hemorrhagic fever. J Infect Dis 137: 298–308
2. Lee HW (1988) Global update on distribution of hantaviruses and hemorrhagic fever with renal syndrome. Virus Info Exch Newslett 5: 82–84
3. Lee HW, An CN (1988) Development of a vaccine against hemorrhagic fever with renal syndrome. J Kor Soc Virol 18: 143–148
4. Lee HW (1989) Hemorrhagic fever with renal syndrome in Korea. Rev Infect Dis 2: S864–S876
5. Meegan JM, LeDuc JW (1989) Enzyme immunoassays. In: Lee HW, Dalrymple JM (eds) Manual of hemorrhagic fever with renal syndrome. WHO Collaborating Centre for Virus Reference and Research (HFRS), Institute for Viral Diseases, Korea University, Seoul, pp 83–87
6. Mitani K, Une H (1988) Dyed inorganic composite particles and process for production thereof. U.S. Patent 4, 780, 422

Authors' address: Dr. Tetsuo Tomiyama, Laboratory of Microbiology and Immunology, University of Tokyo Branch Hospital. 3-28-6, Mejirodai, Bunkyo-ku, Tokyo 112, Japan.

Arch Virol (1990) [Suppl 1]: 35–47

Field trial of an inactivated vaccine against hemorrhagic fever with renal syndrome in humans

H. W. Lee[1], C. N. Ahn[2], J. W. Song[1], L. J. Baek[1], T. J. Seo[3], and S. C. Park[3]

[1] The Institute for Viral Diseases, Korea University, Seoul
[2] Mogam Biotechnology Research Institute, Kyunggi-do, and
[3] Korea University Hospital, Seoul, Korea

Accepted February 5, 1990

Summary. Hantaan virus was serially passaged in the brains of suckling rats. Vaccine inactivated with formalin was prepared from the suckling rat brains and the antigenic potency of the vaccine was determined by enzyme-linked immunosorbent assay (ELISA). Mice immunized with the vaccine were protected when challenged with live Hantaan virus. The antibody responses of 456 vaccinees given inactivated Hantaan virus vaccine against hemorrhagic fever with renal syndrome (HFRS) were then studied. Subcutaneous injection of the vaccine was better than intramuscular injection for production of antibodies in humans. Optimal immunogenic dose of the vaccine given subcutaneously to humans was 5,120 ELISA antigen units. Of 456 vaccinees, immunization of 336 with two doses of vaccine at one month interval resulted in 99% seroconversion by indirect immunofluorescent antibody tests. The vaccine was safe and only minimum side effects were observed. The efficacy of this vaccine against HFRS in the endemic areas of HFRS remains to be determined.

Introduction

Hemorrhagic fever with renal syndrome (HFRS) is a complex of acute viral hemorrhagic diseases affecting both humans and animals. These diseases are caused by certain viruses of the family *Bunyaviridae*, genus *Hantavirus* [6, 11]. There have been about 200,000 reported cases of HFRS with 3–8% fatality annually in Eurasia and accumulating evidence that hantavirus infections also occur in many parts of the world where HFRS is not known to exist [1, 2, 6, 7]. Since the discovery of Hantaan virus, the etiologic agent of Korean hemorrhagic fever [4, 5], in 1976, numerous Hantaan and Hantaan-related viruses have been isolated from field mice, urban rats, laboratory rats and other animals throughout the world. Hantavirus diseases became recognized as an important public health problem because HFRS was documented not only in rural and urban areas but also among laboratorians working with animals [2, 6]. Among hantaviruses, Hantaan, Seoul, and Puumala viruses are causative agents of HFRS and Hantaan and Seoul viruses cause severe symptoms with high mortality [6], therefore there is great need of vaccine against Hantaan and Seoul virus infections.

However, there is no effective vaccine or specific drug available to protect against HFRS and in 1984 WHO recommended developing an effective inactivated vaccine against HFRS as soon as possible. Recently, Lee and Ahn [3] and Yamanishi et al. [14] reported the development of inactivated vaccines against HFRS with Hantaan virus, isolated from an HFRS patient, and with Seoul virus, isolated from a rat tumor, respectively, and available evidence showed that these vaccines induced protective immunity in mice.

In this paper we report the results of a human field trial of inactivated Hantaan virus vaccine against HFRS, prepared in suckling rats brains, and evidence that this vaccine stimulates production of antibodies.

Materials and methods

Viruses

Hantaan virus, ROK84-105 strain, isolated from blood of an HFRS patient directly in Vero E6 cell cultures was used for vaccination strain [3] and prototype Hantaan virus, 76/118 strain [5], was used for challenge of vaccinated mice. The LD_{50} of strain ROK84-105 and strain 76/118 in suckling rats inoculated intracerebrally (IC) was $10^{9.8}$/ml and $10^{9.5}$/ml, respectively.

Vaccine preparation

Strain ROK84-105 was passaged 3 times in the brains of suckling ICR mice and 7 times in the brains of suckling SD rats and used as seed virus for vaccine preparation. Brains were harvested 6–8 days after virus inoculation, when suckling rats were paralyzed for preparation of vaccine. Supernatant fluids of 5% pooled suckling rats brain suspensions in phosphate-buffered saline (PBS) pH 7.2 were inactivated with 0.05% formalin at 4 °C for 15 days.

Purification and inactivation of virus suspensions were done according to a modification of the method used to prepare Japanese encephalitis virus mouse brain vaccine [14]. Sterility of the vaccine was checked in fluid thioglycolate medium. Residual live virus in vaccine was measured by inoculation of the vaccine both in Vero E6 cells and passaged 3 times, and in *Apodemus* mice IM and passaged 3 times for presence of virus in lungs of mice. After the sterility and safety of the inactivated vaccine were tested, one volume of vaccine preparation containing 10,240 ELISA units of antigen per ml was mixed with an equal volume containing 500 µg/ml of aluminum hydroxide gel. The adsorbed Hantaan virus vaccine was used for challenge of mice and for field trials in humans.

Hantaan virus antigen in the vaccine preparation

The concentration of viral antigen in the vaccine preparation was determined by antibody-bound enzyme-linked immunosorbent (ELISA) assay [8]. Briefly, convalescent serum from an HFRS patient containing high titer IgM antibody to Hantaan virus, 76/118 strain, was diluted with buffer solution to contain 16 ELISA units of IgM antibody and added to microtiter plates coated with goat anti-human IgM and kept 1 h at 37 °C. Then serially diluted antigen was added and the plates incubated for 1 h at 37 °C. Finally, rabbit anti-virus antibody and anti-rabbit antiserum labelled with peroxidase were added sequentially. The highest dilution of vaccine at which specific ELISA reaction was observed was regarded as 1 unit. The concentration of Hantaan viral antigen in the vaccine preparation was 10,240 ELISA units/ml. Total protein concentration of the purified vaccine was 27–50 µg/ml, as determined by Lowry's method [9].

Immunization of mice

Antibody negative normal *Apodemus agrarius*, 1–2 months old, were inoculated intramuscularly with 2,048 units of vaccine once or twice depending on antibody titers to Hantaan virus. The animals were bled by heart puncture 4 or 11 weeks after immunization and sera were collected for antibody tests.

Protection studies

Vaccinated mice were challenged IM with 10,000 suckling rat I.C. LD_{50} of prototype Hantaan virus or with one of a series of tenfold dilutions of that virus and killed 21 or 30 days later. The lungs and kidneys of immunized and non-immunized mice were tested for virus antigen by indirect immunofluorescent antibody technique (IFAT) and supernatants of tissue suspensions were inoculated into Vero E6 cells and passaged 3 times for demonstration of virus presence. Antibody titers were measured by IFAT, ELISA and neutralization test (NT) as described previously [5, 8].

Measurement of antibody to Hantaan virus

The available serologic tests for HFRS are IFAT, ELISA, plaque reduction neutralization test (PRNT), hemagglutination inhibition test (HIT) and immune adherence hemagglutination test (IAHT) [8]. IFAT and ELISA are simple, rapid, sensitive and serogroup-specific; HIT and IAHT are time-consuming and less sensitive: PRNT is sensitive, type specific and time consuming [2]. Therefore, simple and rapid IFAT and ELISA were employed for antibody measurement in this study and PRNT was used only for special purposes.

Volunteers

The vaccinees were volunteers and "informed consent" was obtained from all of them before vaccination. Age of the volunteers was between 20 to 50 and both sexes were included. The volunteers were civilian and their professions were laboratory workers, employees of a commercial manufacturing organization, professional golfers and employees of 4 golf clubs in Korea. IF and ELISA antibodies to Hantaan virus in sera from vaccinees were measured after vaccinations. General symptoms and local reactions of vaccinees after vaccinations were checked and recorded daily.

Results

Potency test in vivo

Mice were injected intramuscularly with 2,048 units of inactivated vaccine once or twice, with a four-week interval between injections, IF and ELISA antibodies were detected in all mice immunized with vaccine but only 2 out of 5 immunized mice had very low neutralizing (N) antibody titers as shown in Table 1.

Challenge of vaccinated mice with live virus

After inoculation of live Hantaan virus, virus could be isolated from the lungs of unimmunized mice but not from those of mice given vaccine. High titers of IF antibodies were detected in serum samples from unvaccinated but not from vaccinated mice 21 days after inoculation of live virus, as shown in Table 1.

All of 30 vaccinated mice contained IF antibodies (titer range 32–256), and none of 30 unvaccinated mice had antibodies to Hantaan virus at the time of challenge with the virus. Large amounts of virus in lungs and high titers of antibodies were detected in unimmunized mice and ID_{50} of the virus was $10^{6.8}/0.3$ ml; no virus was demonstrated ($ID_{50} < 10^{2.3}/0.3$ ml) in immunized mice, as shown in Table 2.

Route of vaccination in humans

To determine the appropriate route of vaccination in humans, various antigen doses of inactivated vaccine were injected into volunteers by IM or

Table 1. Challenge of Hantaan virus to vaccinated and non-vaccinated *Apodemus agrarius* mice by intramuscular inoculation of the live virus

Group of mice	Code no. of animal	Vacci-nation[a]	Antibody titer on day after vaccination				Chal-lenge[b]	IF antibody and growth of virus 20 days postinoc.	
			0		78			antibody titer	virus in lungs
			FA	N	FA	N			
Non-vaccinated normal	1	−	<16	<20	<16	<20	+	4,096	+ +
	2	−	<16	<20	<16	<20	+	1,024	+ +
	3	−	<16	<20	<16	<20	+	1,024	+ + +
	4	−	<16	<20	<16	<20	+	1,024	+ +
	5	−	<16	<20	<16	<20	+	2,048	+ + +
Vaccinated with inactivated vaccine	6	+	<16	<20	256	20	+	<16	−
	7	+	<16	<20	256	20	+	<16	−
	8	+	<16	<20	32	<20	+	<16	−
	9	+	<16	<20	32	<20	+	<16	−
	10	+	<16	<20	32	<20	+	<16	−

[a] Vaccination of Hantaan vaccine 2,048 units

[b] Challenge on day 78 with 10,000 LD_{50} live virus

subcutaneous (SC) routes, and IF and ELISA antibodies were measured 25–30 days after primary injection and 14 days after secondary injection of the vaccine. As shown in Table 3, seroconversion rates among vaccinees after primary SC and IM injections of 5,120 units of the vaccine was 96% and 69% and 100% and 84% after booster injection. Total number of vaccinees for the study of primary antibody response was 230 and for secondary antibody response was 150, respectively; the data indicate that SC route is better than IM route for vaccination.

Table 2. ID_{50} of challenge Hantaan virus in unvaccinated and vaccinated *Apodemus agrarius* mice

Group of mice	No. of antibody positive mice/no. of mice used	$ID_{50}/0.3$ ml[a] of virus in mice
Unvaccinated	0/30	$10^{6.8}$
Vaccinated with Hantaan virus vaccine (2,048 unit of antigen)	30/30[b]	$< 10^{2.3}$

[a] ID_{50} of Hantaan virus was measured 30 days after inoculation of serial log dilution of live virus into 5 *Apodemus* mice intramuscularly
[b] IF antibody titers of vaccinated mice were 32–512

Table 3. Comparison of seroconversion rate of vaccinees receiving formalin-inactivated suckling rat brain Hantaan virus vaccine by two different routes of injection

Route	Antigen dose (ELISA units)	One dose[a]		Two doses[b]	
		IF (IgG)	ELISA (IgM)	IF (IgG)	ELISA (IgM)
I.M.	2,560	2/26[c](8%)	8/26 (30%)	12/14 (86%)	6/14 (43%)
	5,120	29/42 (69%)	28/42 (67%)	27/32 (84%)	24/32 (75%)
	10,240	27/31 (87%)	20/31 (65%)	23/25 (92%)	23/25 (92%)
S.C.	2,560	18/38 (47%)	30/38 (79%)	16/18 (89%)	16/18 (89%)
	5,120	42/44 (96%)	38/44 (86%)	28/28 (100%)	26/28 (93%)
	10,240	45/49 (92%)	44/49 (90%)	33/33 (100%)	30/33 (91%)
Total no. of vaccinees		230		150	

[a] Test of primary antibody response 25–30 days after vaccination
[b] Test of secondary antibody response 14 days after injection of booster dose of 5,120 units of vaccine
[c] No. of antibody positive/no. of vaccinees

Seroconversion rate among vaccinees by IFAT and ELISA after injection of vaccine by different routes

Antibodies in serum samples from vaccinees after injection by different routes were measured by IFAT and ELISA. IF antibody test was more sensitive than ELISA in demonstrating low titer antibodies (IF antibody titers 32–64) against Hantaan virus after injection of more than 5,120 units of vaccine. Ninety-six percent of vaccinees were IF antibody positive after injection of one dose (5,120 units) of vaccine and 100% of vaccinees were IF antibody positive after receiving a booster dose (5,120 units) of vaccine SC, whereas 86% and 93% of vaccinees were ELISA antibody positive after receiving the same dose of primary and secondary injection of vaccines SC, respectively (Table 3).

Determination of primary immunogenic dose of inactivated vaccine in humans

Primary immunogenic dose of inactivated vaccine by SC inoculation in humans was determined by injecting various doses of vaccine (640–10,240 units of antigen) in 0.5–1.0 ml, as shown in Table 4. IF antibodies were measured 25–30 days later. Inoculation of either 5,120 units or 10,240 units of vaccine produced IF antibodies against Hantaan virus in 96% of vaccinees. Therefore, it was decided that 5,120 antigen units of vaccine is the optimal immunogenic dose for stimulating production of antibodies against Hantaan virus in humans.

Antibody response curves of vaccinees after injection of 5,120 antigen units of vaccine SC in humans

Figures 1 and 2 show IF and ELISA antibody responses of 83 vaccinees after injection of two doses of vaccine SC with a five-week interval between

Table 4. Determination of primary immunogenic dose of inactivated Hantaan virus vaccine by demonstration of IF antibodies in humans

Route of injection	Antigen dose (ELISA units)	No. seropositive/ no. vaccinees	
Subcutaneously	640	14/19	(73%)
	1,280	30/40	(75%)
	2,560	26/32	(81%)
	5,120	27/28	(96%)
	10,240	32/33	(96%)
Total no. of vaccinees		152	

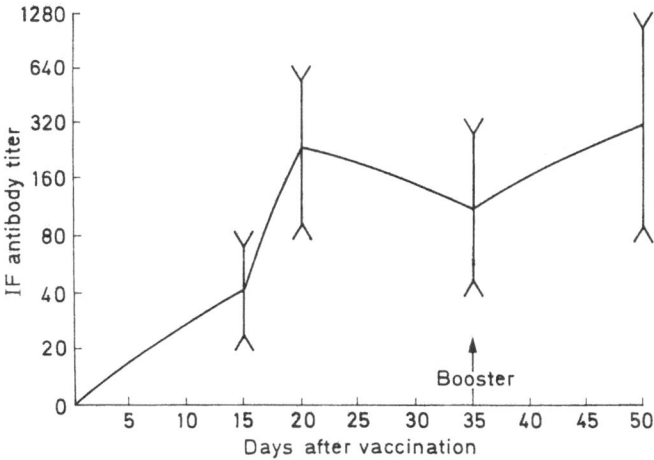

Fig. 1. Immunofluorescent IgG antibody response in vaccinees (n = 83) after inoculation with Hantaan virus vaccine SC in humans. Antigen dose: 5,120 units

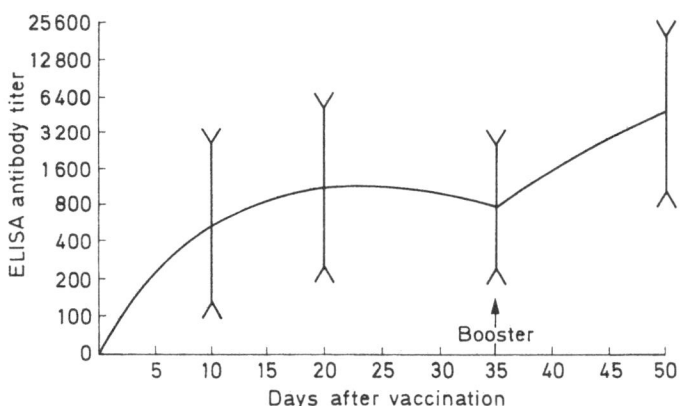

Fig. 2. ELISA IgM antibody response in vaccinees (n = 83) after inoculation with Hantaan virus vaccine SC in humans. Antigen dose: 5,120 units

immunizations. Primary IF antibody response in human reached a maximum three weeks after injection of vaccine, antibody titers were 80–640, and anamnestic antibody responses were observed after the volunteers received the booster injections. ELISA antibody responses were similar to those of IF antibody responses and titers ranged 200–6400. However, IFAT was more sensitive than ELISA for detecting antibodies against Hantaan virus after vaccination in humans, as shown in Table 3.

Seroconversion rate among vaccinees by sex

As shown in Table 5, seroconversion rates among vaccinees were compared by detecting IF and ELISA antibodies after vaccination. There was no

Table 5. Comparison of seroconversion rates of vaccinees by demonstration of IF and ELISA antibodies to Hantaan virus after subcutaneous injection of inactivated vaccine, by sex

Dose of vaccine (ELISA units)	Male				Female			
	One dose		Two doses		One dose		Two doses	
	IF	ELISA	IF	ELISA	IF	ELISA	IF	ELISA
5,120	30/33[a] (91%)	25/33 (76%)	29/29 (100%)	23/29 (79%)	70/76 (92%)	62/76 (82%)	68/68 (100%)	62/68 (91%)
Diluent	0/9 (0%)	0/9 (0%)	0/8 (0%)	0/8 (0%)	0/28 (0%)	0/28 (0%)	0/25 (0%)	0/25 (0%)

[a] No. of antibody positive/no. of vaccinees

difference in seroconversion rates among vaccinees by sex as determined by IFAT but there were slight differences as determined by ELISA. By ELISA, seroconversion rates among females were 82% and 91% (one-dose, two-doses) whereas among males the rates were 76% and 79% after one and two doses of vaccine. Sensitivity of IFAT is somewhat higher than ELISA for measuring antibody to this vaccine in humans.

Summary of seroconversion rates of all vaccinees by IFAT

The total number of vaccinees who received one dose of vaccine SC was 456; 406 (89%) of these had antibody against Hantaan virus. Of 336 vaccinees who received two doses of vaccine 333 (99%) had antibody (Table 6). None of 37 and 33 people who received only diluent once or twice developed antibodies. The data suggest that two injections of one immunogenic dose of vaccine at one month intervals by SC injection is an effective method of vaccinating humans against Hantaan virus.

Side effects in humans given inactivated vaccine

General symptoms and local reactions of 162 vaccinees who received one dose of vaccine SC are shown in Table 7.

A higher incidence of local reactions than general symptoms was observed. These reactions are compatible to these seen after administration of hepatitis B vaccine, probably because the adjuvant in both vaccine preparations is alum hydroxide gel. Main local reactions were itching, induration, and swelling but none had serious complaints. General symptoms and local reactions of 162 vaccinees who received 2 doses of vaccine at about one month intervals (Table 8) were about same as those in vaccinees

Table 6. Seroconversion rate of vaccinees by demonstration of IF antibodies to Hantaan virus after subcutaneous vaccination with inactivated vaccine

Group	Dose of vaccine (ELISA units)	Primary response[a]		Secondary response[b]	
		female	male	female	male
Vaccinee	5,120	227/254[c] (89%)	179/202 (88%)	205/206 (99%)	128/130 (98%)
Total		406/456[d] (89%)		333/336[e] (99%)	
Placebo	Diluent only	0/28 (0%)	0/9 (0%)	0/25 (0%)	0/8 (0%)
Total		0/37 (0%)		0/33 (0%)	

[a] Tested for antibodies 25–30 days after first dose
[b] Tested for antibodies 14 days after booster dose
[c] No. of antibody positive/no. of vaccinees
[d] Total no. of vaccinees given only one dose of vaccine
[e] Total no. of vaccinees given two doses of vaccine

Table 7. General symptoms and local reactions of 162 vaccinees after receiving one dose of 5,120 units of inactivated vaccine SC

Symptoms and reactions	No. of reactions	Disappearance of symptoms after vaccination (days)				
		1	2	3	4	7
Fatigue	1	1				
Nausea	2	2				
Myalgia	4	1	1	1		1
Redness	1					1
Pain	2	1	1			
Itching	11		1	1	5	4
Swelling	5	1	2	1	1	
Induration	8			3	1	4
Hyperpigmentation	3		1			2

who received only a single dose of vaccine. General symptoms and local reactions of 37 and 33 persons who received one and two injections of diluent with adjuvant SC are shown in Table 9; adverse reactions (swelling, itching, induration and hyperpigmentation) were essentially the same as those in vaccinees. The side effects of this vaccine seems to be primarily caused by adjuvant and it appears to be reasonably safe to use in humans.

Table 8. General symptoms and local reactions of 162 vaccinees after receiving a second dose of 5,120 units of inactivated vaccine SC

Symptoms and reactions	No. of reactions	Disappearance of symptoms after vaccination (days)					
		1	2	3	4	6	7
Dizziness	1			1			
Fatigue	2			2			
Myalgia	2		1	1			
Redness	1			1			
Pain	2	1	1				
Itching	8	4	1	2			1
Swelling	2	1		1			
Induration	6	2	1	1	1	1	
Hyperpigmentation	2			1	1		

Table 9. General symptoms and local reactions of 37 volunteers after first injection of vaccine diluent SC (placebo)

Symptoms and reactions	No. of reactions	Disappearance of symptoms after vaccination (days)			
		1	2	3	4
Dizziness	1	1			
Nausea	1		1		
Myalgia	1	1			
Redness	1	1			
Pain	1	1			
Itching	1			1	
Swelling	2		2		
Induration	2				2
Hyperpigmentation	1		1		

Discussion

Inactivated Hantaan virus vaccine, to be used to prevent HFRS, was prepared with virus harvested from suckling rat brains because the virus yield in Vero E6 and MRC-9 (human lung, ATCC CCL 212) cell cultures is not high and because in previous attempts to make such a vaccine, immunogenicity of Hantaan virus grown in cell cultures was very poor. We confirmed that the yield of virus and immunogenicity of virus from the brains of infected suckling rats were as satisfactory as virus harvested from the brains of infected suckling mice. Further, this is easy and economical to produce in large quantities from the brains of infected suckling rats than from the brains of infected suckling mice for vaccine production [3]. The strain of Hantaan virus adapted to suckling rats and used for vaccine preparation was isolated from the blood of an HFRS patient. Of several candidate strains of Hantaan virus isolated from HFRS patients, it had the greatest immunogenicity in mice and rats, in terms of its ability to rapidly produce high titers of IF, ELISA and neutralizing antibodies in laboratory animals. The vaccine seed virus killed suckling rats about 7 days after intracerebral inoculation and yielded $10^{9.8}$ LD_{50}/ml. Other advantages of using suckling rats for production of vaccine is that the weight of a suckling rat brain is 3 times greater than the weight of a suckling mouse brain and the number of offspring from a rat is 10–14, as compared to 8–11 from albino ICR mice. Finally, for production of a human vaccine, it is safer and more acceptable to use a virus isolated from a human patient than a virus isolated from rat malignant tumor cells [13].

The protein concentration of the final vaccine preparation that contains 10,240 units of Hantaan virus antigen is 27–50 µg/ml, which is relatively low, compared to the minimum requirement of protein content of Japanese encephalitis inactivated mouse brain vaccine (80 µg/ml) [10] and protein content of available commercial JE vaccines in Korea is 50–70 µg/ml. Therefore, the protein concentration of this vaccine is comparable to that of Japanese encephalitis (JE) vaccine, which has been used in millions of people for several decades in Asia. The protein concentration of the control vaccine (rat protein) is 27–50 µg/ml almost same as killed vaccine. In our limited experiment, antigen concentration and immunogenicity of the vaccine was higher than nucleocapsid protein of Hantaan virus obtained from baculovirus vector by genetic engineering (unpubl. data). *Apodemus* mice

Table 10. General symptoms and local reactions of 33 volunteers after secondary injection of vaccine diluent SC (placebo)

Symptoms and reactions	No. of reactions	Disappearance of symptoms after vaccination (days)				
		1	2	3	4	7
Dizziness	1	1				
Nausea	1		1			
Myalgia	1	1				
Pain	1	1				
Itching	3		3			
Swelling	2		2			
Induration	5			2	2	1
Hyperpigmentation	2		1	1		

immunized with our vaccine were completely protected against challenge with homotypic Hantaan virus (Tables 1 and 2) and against challenge with heterotypic Seoul virus (study in progress). It is noteworthy that immunized mice had high titers of IF antibodies and low titer of N antibodies in their sera and that immunized mice were completely protected against Hantaan virus infection.

It is remarkable that no anamnestic antibody response was observed in immunized mice and that IF antibodies disappeared after challenge with live virus. This phenomenon may have occurred after immunized mice with low level of antibody were challenged with excessive amounts of live virus antigen. Cellular immunity in vaccinated mice may play a very important role in protecting against Hantaan virus infection in vivo. Primary and secondary antibody responses of rats inoculated with vaccine were described previously [3] and N antibodies against Hantaan virus were demonstrated in sera from vaccinated rats that contain IF antibody titers > 256. IFAT was more sensitive than ELISA in demonstrating antibodies against Hantaan virus in vaccinated humans. However, we do not know the duration of persistence of antibodies in these vaccinees but IF and N antibodies persist at least 34 years after illness [6].

Aluminum hydroxide gel [12], used in this vaccine as an adjuvant, is permitted for use in other vaccines throughout the world. The human immunogenic dose of inactivated vaccine was determined for the first time by ELISA. Also we showed that 2 doses of vaccine by SC injection at one month intervals resulted in 100% seroconversion by IF but that PRN antibody titers were very low (10–20). In our limited study, 3 doses of vaccine injected SC at day 0, 28, and 60 produced high serum neutralizing antibodies against Hantaan virus in humans (study in progress). Recently, a human inactivated hantavirus vaccine was developed in North Korea from brains of infected suckling rats and hamsters with an antibody conversion rate of 78% as determined by IFAT. They also claim that the vaccine has been given to about 30,000 persons and efficacy tested and that protectivity is 80–100% (R. J. Kim, pers. comm., 1989).

The vaccine made with Hantaan virus and reported in our study was effective against Hantaan virus infection in mice and safe with minimum side effects in humans. The seroconversion rate of vaccinees after immunization with 2 doses was 100%; however, efficacy of this vaccine in endemic areas of HFRS remains to be studied.

Acknowledgements

This research was supported in part by Grant no. ICP/CDSO11-A.I.D. OCD/1/87 from WHO Western Pacific Region, Manila and a Grant from Green Cross Corporation Inc., Seoul.

References

1. LeDuc JW, Smith GA, Childs JE, Pinheiro FP, Maiztegui JI, Niklasson B, Antoniadis A, Robinson DM, Khin M, Shortridge KF, Wooster MT, Elwell MR, Ilbery PLT, Koech D, Rosa EST, Rosen L (1986) Global survey of antibody to Hantaan-related viruses among peridomestic rodents. Bull WHO 64: 139–144
2. Lee HW (1989) Hemorrhagic fever with renal syndrome in Korea. Rev Infect Dis 2: S864–S876
3. Lee HW, Ahn CN (1988) Development of a vaccine against hemorrhagic fever with renal syndrome. J Kor Soc Virol 18: 143–148
4. Lee HW, Lee PW (1976) Korean hemorrhagic fever. I. Demonstration of causative antigen and antibodies. Kor J Intern Med 19: 371–394.
5. Lee HW, Lee PW, Johnson KM (1978) Isolation of the etiologic agent of Korean hemorrhagic fever. J Infect Dis 137: 298–308.
6. Lee HW, van der Groen G (1989) Hemorrhagic fever with renal syndrome. Prog Med Virol 36: 62–102
7. Lee PW, Gibbs CJ, Gajdusek DC, Svedmyr A (1981) Antibody to Korean hemorrhagic fever virus in man in parts of the world where hemorrhagic fever with renal syndrome is not known. Lancet 1: 256–257
8. Lee PW, Meegan JM, LeDuc JW, Tkachenko EA, Tvanov AP, Rezapkin GV, Drozdov SG, Kitamura T, Tsai TF, Dalrymple JM (1989) Serologic techniques for detection of Hantaan virus infection, related antigens and antibodies. In: Lee HW, Dalrymple JM (eds) Manual of hemorrhagic fever with renal syndrome. WHO Collaborating Centre for Virus Reference and Research (HFRS), Institute for Viral Diseases, Korea University, Seoul, pp 75–106
9. Lowry OH, Rosenbrough NG, Farr AL, Randall RG (1951) Protein measurement with the folin phenol reagent. J Biol Chem 193: 265–269
10. Oya A (1987) New development of criteria on Japanese encephalitis vaccine requirements in Japan. WHO Japanese encephalitis and hemorrhagic fever with renal syndrome Bull 2: 11–13
11. Schmaljohn CS, Hasty SE, Dalrymple JM, LeDuc JW, Lee HW, von Bonsdorff CH, Brummer-Korvenkontio M, Vaheri A, Tsai TF, Regnery HL, Goldgaber D, Lee PW (1985) Antigenic and genetic properties of viruses linked to hemorrhagic fever with renal syndrome. Science 227: 1041–1044
12. Weeke B, Weeke E, Lowenstein H (1975) The adsorption of serum protein to aluminum hydroxide gel examined by means of quantitative immunoelectrophoresis. Scand J Immunol 4 [Suppl 2]: 149–154
13. Yamanishi K, Tomishita O, Tamura M (1988) Development of inactivated vaccine against virus causing hemorrhagic fever with renal syndrome. Vaccine 6: 278–282
14. National Institute of Preventive Medicine and Hygiene Japan (1976) Preparation of Japanese encephalitis vaccine. In: Yanagizawa G (ed) Vaccines in Japan, 2nd edn. Maruzen, Tokyo, pp 80–85

Author's address: Dr. Ho Wang Lee, The Institute for Viral Diseases, Korea University 4, 2nd Street, Myungyun-Dong, Chongno-Ku, Seoul 110-522, Korea.

Arch Virol (1990) [Suppl 1]: 49–56

Reservoirs and modes of spread of hemorrhagic fever with renal syndrome, a zoonotic nontransmissible human disease

M. P. Chumakov, **Irena N. Gavrilovskaya**, and **Y. A. Myasnikov**

Institute of Poliomyelitis and Viral Encephalitides, U.S.S.R. Academy of Medical Sciences, Moscow, U.S.S.R.

Accepted January 17, 1990

Summary. Data are presented on the distribution of zoonotic contagious non-transmissible (without transmission via arthropods) hemorrhagic fevers with renal syndrome on the basis of reported human diseases (33 countries), detection of antibody to these viruses (37 countries), or isolating a hantavirus and detecting antibody to it in small mammals (38 countries). Hantaviruses have been isolated from 53 species of mammals. The principal approaches to prophylaxis against this infection are elaborated.

*

Among numerous zoonotic virus diseases of humans three groups of infections are known:

1. *Arbovirus infections*, transmissive, with reservoirs among animals and virus transmission almost always by blood-sucking arthropods.
2. *Arenavirus zoonoses* (lymphocytic choriomeningitis, Lassa fever, Argentine hemorrhagic fever, Bolivian hemorrhagic fever and others) contracted by humans through contact with excreta of infected rodents, without participation of blood-sucking arthropods.
3. *Contagious infection* with aerosol, dust or alimentary virus spread from excreta of infected animals. Casually infected blood-sucking ectoparasites probably maintain the infection only in rodents; these insects usually are unable to transmit infection to humans. Hemorrhagic fever with renal syndrome (HFRS), a nontransmissive human disease, belongs to this group.

Hemorrhagic fever with renal syndrome is the name given to illnesses caused by a number of antigenically different viruses of the family *Bunyaviridae*, genus *Hantavirus*. The first scientific researchs of HFRS-like zoonotic-contagious diseases were described by various authors in the U.S.S.R., Sweden, Norway, Finland, and later in China, Japan, and elsewhere [1, 3, 4, 10, 12, 13, 18, 20, 21–23, 28]. These diseases were named Tula fever (1930), Scandinavian endemic nephropathy (1934), Far-Eastern hemorrhagic nephroso-nephritis (1940), epidemic hemorrhagic fever (China and Korea, 1940–1952), HFRS (1956) and Korean HF (1976) [14–17, 25].

Investigation of the possibility of HFRS virus transmission to humans by blood-sucking arthropods has been studied for more than 50 years. Japanese workers [10, 12] in 1940 and 1944 reported HFRS virus infection in gamasoid mites *Laelaps jittmari* collected from field mice in a focus of epidemic HF diseases in Manchuria, as well as HF virus perpetuation in field mice experimentally infected with a suspension of ground mites of this species. However, attempts by many workers (Soviet, Japanese, Chinese, American, and others) to obtain further evidence of the possible natural transmission of HFRS infection from arthropod vectors to humans failed completely.

Their epidemiological role in HFRS is assumed to consist of the maintenance of infection among murine rodents; it cannot be ruled out that HFRS-infected gamasoid mites in mouse nests dessicated and ground to dust, could become the source of human infection via aerosol. The observations of Japanese and Soviet workers confirmed that infection with HFRS virus occurs in several species of gamasoid mites, ectoparasites of murine rodents. However, the lack evidence for feeding of these mites on humans minimizes the possible importance of HFRS virus transmission to human via mites.

HFRS virus has reservoirs in many mammalian species and proven modes of horizontal and vertical transmission. HFRS is a zoonotic infection transmitted to man by contact with animal excreta without regular participation of blood-sucking arthropods [6, 19]. It is a serious public health problem in a number of countries, particularly in China and U.S.S.R. [7–9, 16], but also in Korea, and in Scandinavian and Balkan countries. Occasional cases of this infection have been observed in other countries of the world as well (Table 1).

The incidence of HFRS is increasing. In 1985 in European U.S.S.R., for example, it exceeded 11000 cases, and in China there were more than 100000. On other continents, thus far only occasional hantaviruses illnesses have been reported, and HFRS is not officially registered. Comparisons of the numbers of countries where human cases of HFRS have been reported, or antibodies were found in the human population, or hantaviruses were detected in animals, show great variation, even within an individual continent. This is probably due to different periods of observations, as well as reflecting the intensity of virological and serological effort. In recent years in the U.S.S.R. and other countries, effective methods have been developed for isolation of hantavirus strains and serologic diagnosis of hantavirus infec-

Table 1. Geographic distribution of hantavirus infections

	Europe	Asia	America	Africa	Australia and Oceania
No. of countries with HFRS diseases recorded	18	10	2	2	1
No. of countries with only antibody to HFRS demonstrated in humans	8	15	5	8	1
Maximum no. of HFRS cases/year	11,000 (USSR, 1985)	100,000 (China, 1985)	4?	1?	?
No. of countries with hantaviruses isolated from animals	12	12	3	7	4
No. of animal species with hantaviruses detected	24	32	8	8	3

tions and for identification of hantaviruses from animals, the natural reservoirs of this infection. At our institute, numerous HFRS virus strains were isolated from patients, from several vole species, from field mice, and from rats, by passage in Vero E6 cells or by passage in colonized bank voles (*Clethrionomys glareolus*) [6].

Studies of hantavirus reservoirs have demonstrated a virus carrier state in more than 50 species of small mammals belonging to several species of *Rodentia*, including members of families *Muridae* and *Cricetidae*, and species of the order *Insectivora*. Moreover, in intensive HFRS foci domestic cats and Siberian weasels (*Mustella sibirica*) and even hares (*Lepus mandshuricus*) and rabbits (*Aryctolagus cuniculus*) were shown to be hantavirus carriers or to have antibody to hantaviruses (Table 2). The most frequent carriers of HFRS virus, species that are epidemiologically important, belong to four rodent genera: *Apodemus*, *Clethrionomys*, *Rattus*, and *Microtus*. These animals usually are infected with distinct hantavirus serotypes, the serotype corresponding to the rodent species in which it is found. Animals spontaneously infected with hantavirus appear to remain asymptomatic but are capable of excreting virus into the environment essentially throughout their lives. In addition to horizontal transmission of hantaviruses between animals, vertical transmission also occurs, although it is rare in rodents, as documented

Table 2. Species of small mammals from which hantaviruses, virus antigen, or antibody to it have been found

Order	Family	Species[a]
Rodentia	Cricetidae	Clethrionomys glareolus (1, 2), C. rutilus (1, 2, 4, 6), C. rufocanus (1, 2, 4), C. gapperi (6), Microtus arvalis (1, 2, 3), M. rossiameridionalis (1), M. agrestis (1), M. oeconomus (1, 4), M. subterraneus (1, 3), M. majori (3), M. pennsylvanicus (6), M. californicus (6), M. fortis (4), Arvicola terrestris (3), Ondatra zibethica (1), Tscherskia albopictus (4), Cricetulus barabensis (4), Peromyscus maniculatus (6), P. truei (6), Meriones erythrourus (3)
	Muridae	Apodemus agrarius (1, 3, 4), A. peninsulae (4), A. sylvaticus (1, 3), A. flavicollis (1), Apodemus (sp.?) (1), Mus musculus (1, 4, 7, 8), Rattus norvegicus (1, 3, 4, 5, 6, 7, 8), R. rattus (1, 3, 4, 5, 6, 7, 8), Niviventer confucianus (4), Rattus (sp.?) (4), R. exulans (8), Bandicota indica (5), B. bengalensis (5)
	Praomidae	Praomys natalensis (7), P. erythroleucus (7), P. daltoni (7), Praomys (sp.?) (7)
	Steatomydae	Steatomys (sp.?) (7)
	Sciuridae	Tamias sibiricus (1, 4)
	Muscardinidae	Dryomys nitedula (3)
Insectivora	Talpidae	Talpa europaca (1)
	Soricidae	Sorex araneus (1, 2), S. caecutiens (1), S. minutus (1), Meomys fodiens (1), Crocidura suaveolens (1), C. russula (3), Anaurosorex squamipes (4), Suncus murinus (4)
Carnivora	Mustelidae	Mustela sibirica (4)
	Felidae	Felis domesticus (4)
Lagomorpha	Leporidae	Lepus mandshuricus (4), Oryctolagus cuniculus (4)

[a] Distribution areas and continents: Europe (1); Asia: Western Siberia (2), Transcaucasus (3), Far East (4), South-East (5); Americas (6); Africa (7); Australia and Oceania (8)

by virus isolation from embryos of infected animals. Table 2 presents a summary of proven small mammal carriers of hantaviruses.

At present, the causative agent of HFRS, its antigen, or antibody to it have been detected (according to the available information) in more than 50 species of small mammals. Besides, about twenty species and laboratory races of animals were shown experimentally to be susceptible to the virus.

Evidence for the presence of hantaviruses has been found in six continents, on Japanese islands, on Hawaii, and on the island of Madagascar. The greatest number of animal species found to carry these viruses was in Asia (32 species) and Europe (24 species), probably because human cases of HFRS have long been known in these areas and studies of animals have been most intensive. Eight species in Africa and in the Americas were found to be susceptible to the hantavirus infection, in Australia and Oceania only three.

In Asia virus distribution is irregular. While in western and middle Siberia five species have been found to be susceptible to hantavirus infection, infected animals of these species also have been found in Europe. That is, the foci of Asia may be considered to be discontinuous extensions from European foci. In the Far East, 16 small mammal species are susceptible to these viruses; about half of them are endemic to this territory. In Transcaucasus, hantaviruses were found in nine species, of which four are endemic. Finally, in south Asian countries only three species were found to be infected, all of which rats, bandicoots (*Bandicota indica*, *B. bengalensis*) are endemic. Thus far there is no convincing evidence of the existence of natural foci of hantaviruses in Central and Middle Asia, in India, or in the Middle East, but in these areas either very few animals have been examined or no observations made.

Hantaviruses have been found in animals of 37 countries [Europe 12, Asia (including the U.S.S.R.) 12, Africa 7, the Americas 3 and Australia and Oceania (including the state of Hawaii, U.S.A.) 3]. The presence of antibodies to hantaviruses in animal populations of other countries suggests that the list of the countries comprising the distribution of hantaviruses is far from complete.

These viruses are most wide spread in peridomestic animals—*Rattus norvegicus* (24 countries in all five continents), *Rattus rattus* (11 countries of five continents), and *Mus musculus* (5 countries in 3 continents). Among other main reservoirs of infection the highest prevalence is in *Clethrionomys glareolus* (9 countries of 2 continents), followed by *Apodemus agrarius* (4 countries in 2 continents) and *Clethrionomys rufocanus* (4 countries in 2 continents). Other affected species were found in only one or two countries.

The main sources of HFRS infection in Europe and Asia are *C. glareolus*, *A. agrarius*, and *R. norvegicus*. The role of each species in maintaining HFRS infection varies. In the European U.S.S.R. and Scandinavia, the most important species is *C. glareolus*, while *A. agrarius* and *A. peninsulae* predominate in Far Eastern foci of the U.S.S.R. and China. Thus far, there are comparatively few reports of hantaviruses in rats in the U.S.S.R. and more frequent reports of infections in rats in China, Japan, U.S.A. and elsewhere. Additional sources of infection, under certain conditions producing epizootics and consequently capable of infecting man, are represented by

A. peninsulae, *A. sylvaticus*, and *A. flavicollis* (Caucasus, Balkans, the Crimea), *C. rutilus*, *C. rufocanus*, and *R. rattus* (northern Europe, Far East).

Hantaviruses found in cats and rabbits by Chinese investigators belonged to the serotype Hantaan; these virus carriers seem to be a dead-end of the infection, as indicated by a very low prevalence of infection (less than 5%) and low virus titers in infected animals.

As for *Microtus arvalis* in the U.S.S.R. and in North America, Pennsylvanian and Californian voles (*Microtus*), despite high infection rates and high concentrations of virus, do not appear to play a great role in human infection and a hantavirus serotype from *Microtus* sp. thus far has not yet been found in humans, although antibody to this virus has been detected in humans.

The high susceptibility to hantaviruses of many species of small mammals is conductive to practically ubiquitous distribution of hantavirus infection in the world. Circulation of hantaviruses in a wide variety of animal hosts in natural foci of different regions appears to lead to considerable antigenic variability of the virus (more than 6 serotypes of the virus are known). In turn, antigenic variation may lend to wide variability of clinical manifestation of human infection: from latent to extremely severe forms, with a case fatality rate of 20%, for instance, in the Balkan areas where *A. flavicollis* is prevalent. These data attest to the necessity for further comparative studies

Table 3. Inactivated HFRS vaccine development in China and Korea, 1988–1989

Sources	Inactivation methods	Trials in
Chick embryo cells	Formalin (1:2000) 4 °C × 8 d	rabbits and humans
Golden hamster kidney cell culture (BHK-21)	Formalin (1:2000) 4 °C × 8 d	rabbits and humans
Suckling mouse brain	Formalin (1:2000) 4 °C × 8 d	rabbits
Suckling mouse brain	56 °C × 1 h	mice
Human diploid cells	?	rabbits
Vero E6	56 °C × 1 h	mice
Vero E6	Formalin (1:2000) 4 °C × 7 d	mice
Rat embryo lung cells	Formalin (1:2000) 4 °C × 8 d	mice rats, rabbits
Suckling mouse brain	60_{Co} (2.6×10 Rad)	rabbits and BALB/c mice
Mongolial gerbil kidney cells	Formalin (1:2000)	rabbits

of hantavirus serotypes circulating in nature so that we can select optimal candidates for vaccine production.

At present, control of HFRS is very complicated and methods primitive. Among relatively effective control measures, the most important is prevention of rodent penetration into the living or working premises of humans and regular control of the rodents in the premises and around them [27], carried out according to epidemiological need. In the future, a second important measure may be specific vaccination of high risk populations. At present, some countries are working to develop experimental technologies for the manufacture of vaccines against HFRS. Table 3 shows the main approaches and methods used in China and Korea (North and South) for the development and control of HFRS vaccines [5, 9, 11, 16, 18, 24, 27]. Given the importance of HFRS virus, we must be optimistic that one or more of these vaccines will be useful in public health practice.

References

1. Chumakov MP (1948) Etiology and epidemiology of hemorrhagic fever. Terapevt Ark (Moscow) 20: 68
2. Chumakov MP, Resnikov AI, Dzagurov SG, Leshinskaya EV, Glazunov SI, Dubniakova AM, Povalishina TP (1956) Hemorrhagic fever with renal syndrome in the Upper Volga region. Vop Virusol 4: 26–30
3. Chumakov MP (1965) On the problem of etiology and epidemiology of hemorrhagic fever with renal syndrome. In: Chumakov MP (ed) Endemic virus infections. Hemorrhagic fevers. Institute of Poliomyelitis, Moscow, pp 5–12
4. Churilov AV (1941) Clinical course of the so-called hemorrhagic nephroso-nephritis. Klin Med 19: 78–82
5. Dong JS, Jin WS, Lee HW (1989) A clinical study of inactivated vaccine against hemorrhagic fever with renal syndrome in volunteers. In: Lee HW (ed) First Int Conf on HFRS. Korea University, Seoul, pp 37
6. Gavrilovskaya IN, Apekina NS, Myasnikov JuA, Bernshtein AD, Ryltseva EV, Gorbachkova EA, Chumakov MP (1983) Features of circulation of HFRS virus among small mammals in the European USSR. Arch Virol 75: 313–316
7. Gajdusek DC (1952) Acute infectious hemorrhagic fevers in the U.S.S.R. First Annual Report of the Commission of Hemorrhagic Fever of the Armed Forces Epidemiological Board. Walter Reed Army Med C, Washington, DC
8. Gajdusek DC (1982) Rodent-borne viral nephropathy (hemorrhagic fever with renal syndrome; nephropathia epidemica). In: Report by the World Health Organization Working Group on Hemorrhagic Fevers with Renal Syndrome. WHO, Geneva, pp 1–3
9. Hsiang C, Zheng Z (1988) Identification of hemorrhagic fever with renal syndrome virus (HFRSV) in China. In: Proc Int Symp on Hemorrhagic Fever with Renal Syndrome. Hubei Medical University, Hubei, China, pp 2–23
10. Kasahara S, Kitano M (1943) Studies on pathogen of endemic hemorrhagic fever. Jpn J Pathol 33: 476–483
11. Kim DW, Park SC, In KH, Song GW, Back LG, Lee HW (1989) Clinical trial of hemorrhagic fever with renal syndrome vaccine in volunteers. In: Lee HW (ed) First Int Conf on HFRS. Korea University, Seoul, p 38
12. Kitano M (1944) Studies concerning epidemic hemorrhagic fever. J Jpn Infect Dis Soc 18: 303–317

13. Korchenov JG (1953) Some clinical aspects of so-called Tula fever. Voenno-Meditsinskii Zh (Moscow) 4: 26–99
14. Lee HW, Lee PW, Johnson KM (1978) Isolation of the etiological agent of Korean hemorrhagic fever. J Infect Dis 137: 298–308
15. Lee HW, Lee PW, Lähdevirta J, Brummer-Korvenkontio M (1979) Aetiological relation between Korean hemorrhagic fever and nephropathia epidemica. Lancet 1: 186–192
16. Lee HW, Chu YK (1988) Global update on Hantavirus infection and vaccines against hemorrhagic fever with renal syndrome (HFRS). In: Int Symp on Viral Diseases. Shanghai University, Shanghai, p 9
17. Lee HW, Dalrymple JM (eds) (1989) Manual of HFRS. WHO Collaborating Center for Virus Reference and Research (HFRS), Institute for Viral Diseases, Korea University, Seoul
18. Lee HW, Ahn CN, Baek LG, Song GW, Park SC, Seo TG, Kim DW (1989) Field trial of HFRS vaccine in man. In: Lee HW (ed) First Int Conf on HFRS. Korea University, Seoul, p 42
19. Myasnikov YuA, Bashkirev TA, Ozegova ZE, Goncharova MI, Bagan RN, Ivanova AA, Zaithseva AA, Loginov AJ (1980) Epidemiology HFRS in Middle Volga and in pre-Urals. In: Bashkirev TA (ed) Hemorrhagic fever with renal syndrome in Middle Volga and pre-Urals. Institute Pasteur, Leningrad, pp 40–57 (in Russian)
20. Myhrman G (1934) En njursijukdom med egenartad symptombild. Nord Med Tidskr 7: 793–794
21. Smorodintsev AA (1940) Results of the studies of the All-Union Institute of Experimental Medicine on encephalitis and nephroso-nephritis. Zh Microbiol Epidemiol Immunol (Moscow) 11: 88–89
22. Smorodintsev AA, Chudakov VG, Churilov AV (1959) Haemorrhagic nephroso-nephritis. Pergamon, London
23. Smorodintsev AA, Kazbintsev LJ, Chudakov VG (1964) Virus hemorrhagic fever. National Library of Medicine on behalf of the National Science Foundation, Washington, DC
24. Song G (1988) Prophylaxis of Hemorrhagic fever with renal syndrome (HFRS): development of inactivated cell culture vaccine against HFRS. In: Proc Int Symp on Hemorrhagic Fever with Renal Syndrome. Hubei Medical University, Hubei, China, pp 26–30
25. Sugyama K, Morikawa S, Matsuura Y, Tkachenko EA, Morata C, Komatsu T, Akao Y, Kitamura T (1987) Four serotypes of HFRS virus identified by polyclonal and monoclonal antibodies. J Gen Virol 68: 979–987
26. Vishniakov SV, Myasnikov YuA, Panina TV, Zhukova LD (1966) Development of a system of rat control measures in the forest foci of hemorrhagic fever with renal syndromes. Zh Microbiol Epidemiol Immunol (Moscow) 8: 12–17
27. Zhu Z (1988) Immunogenicity and protection of inactivated Epidemic hemorrhagic fever vaccine in rabbits. In: Int Symp on Viral Diseases, Shanghai University, Shanghai, China, p 6
28. Zetterholm SG (1934) Acute nephritis simulating acute abdomen. Svenska Läkartidningen 16: 425–429

Authors' address: M. P. Chumakov, Institute of Poliomyelitis and Viral Encephalitides, U.S.S.R. Academy of Medical Sciences, 142782 Moscow, U.S.S.R.

Arch Virol (1990) [Suppl 1]: 57–62

Pathogenesis of hemorrhagic fever with renal syndrome virus infection and mode of horizontal transmission of hantavirus in bank voles

Irena N. Gavrilovskaya, Natalia S. Apekina, Alla D. Bernshtein, Varvara T. Demina, Natalia M. Okulova, Y. A. Myasnikov, and M. P. Chumakov

Institute of Poliomyelitis and Viral Encephalitides, U.S.S.R. Academy of Medical Sciences, Moscow, U.S.S.R.

Accepted January 17, 1990

Summary. Inapparent, persisting hantavirus infection was demonstrated in bank voles (*Clethrionomys glareolus*) after experimental infection with strain Kazan 6 C.g. isolated in the U.S.S.R. Virus, viral antigen, antibodies, and the capacity for horizontal transmission of infection were demonstrable throughout the period of observation (13 months), the highest titers being observed 10–20 days postinfection. Direct correlation was detected between the intensity of horizontal transmission and the level of humoral immunity. The progeny of infected females were shown to be passively immune for 30–45 days after birth and to be relatively resistant to infection with strain Kazan 6 C.g. during that period.

Introduction

Bank voles (*Clethrionomys glareolus*) are known to be the main hosts of hantaviruses in European foci [1, 3, 6]. The epizootic processes in bank vole populations determine the pattern of epidemic processes in particular foci [6]. Comprehension of regularity of the epizootic process requires knowledge of the duration of the infectious process in bank voles, virus and viral antigen localization in their organs, and possible mode of virus transmission. Previously, studies at the population level showed that infection with hemorrhagic fever with renal syndrome (HFRS) viruses caused no deaths of bank voles and exerted no noticeable effect on their population condition [1]. Examination of naturally infected bank voles trapped in foci revealed marked viscerotropicity of the virus but could not reveal duration of HFRS virus infection. Therefore, we studied the pathogenesis of experimental hantavirus infection in bank voles to gain insight into the patterns of the epizootic process.

Materials and methods

Hantavirus infection was studied in two successive experiments in bank voles (100 and 162 animals, respectively) inoculated intramuscularly with 100 ID_{50} of strain Kazan 6 C.g., over a period of 13 months [7]. At 5, 10, 15, 20, and 30 days after infection, and then monthly, animals were bled from the retroorbital sinus and tested for antibody to the virus by indirect immunofluorescent antibody (IFA) technique [4]. Selectively, 5–8 animals from each observation interval were tested for antigen and infectious virus by ELISA and by bioassays in bank voles [5]; organs and tissues from these animals also were examined histologically [2]. The modes of transmission and the duration of horizontal transmission were studied by: (a) exposure for one day of 10 uninfected voles to 10 infected ones, individually, at intervals indicated above; (b) exposure of 5 uninfected voles to a litter of infected voles; (c) placing of cages containing uninfected voles at a distance of 0.1 to 1.5 m from cages with infected voles for 6–26 days after infection of the latter. Altogether, the exposure experiments involved 358 bank voles. The offsprings of 80 infected females (427 animals if three generations) were examined over time for the presence of antibody and, selectively (104 animals), for hantavirus antigen.

Results

Results of the experiments demonstrated an asymptomatic course of infection in bank voles. Morphologic changes were seen as endothelial hyperplasia in vascular walls and lymphohistiocytic infiltration of all organs [2]. Similar results had been obtained earlier in newborn white mice infected with Hantaan virus, strain 76-118 [8].

The infection was persistent (Fig. 1). The first signs were observed at 5–6 days (antigenemia, antigen in spleen, antigen and infectious virus in rectal tissue and feces, antibodies). Subsequently, virus and antigen were detected in lungs, blood, kidneys, liver, urinary bladder, salivary glands, thymus, brown fat, brain, and spinal cord. Antibodies, infectious virus and antigen were seen throughout the period of observation (13 months). The immune response began one week after infection (Fig. 2). Antibody titers varied considerably between individuals but three types of immune responses were seen: low (highest titers not exceeding 160; 30% of 162 animals tested), medium (titers not greater than 640; 57.6%), and high (titers above 640; 12.2%).

Most intensive virus replication and excretion into the environment was observed 14–20 days after infection. At that time virus antigen was found not only in the organs but also in excreta of most bank voles. All uninfected animals that had been in contact with the latter became infected (Fig. 3). The capacity for horizontal transmission declined markedly by the end of the first month, was absent during the third and fourth months and reappeared at 5–13 months, but its efficiency did not exceed 10–40%. The efficiency of horizontal transmission was observed to be directly associated with the level

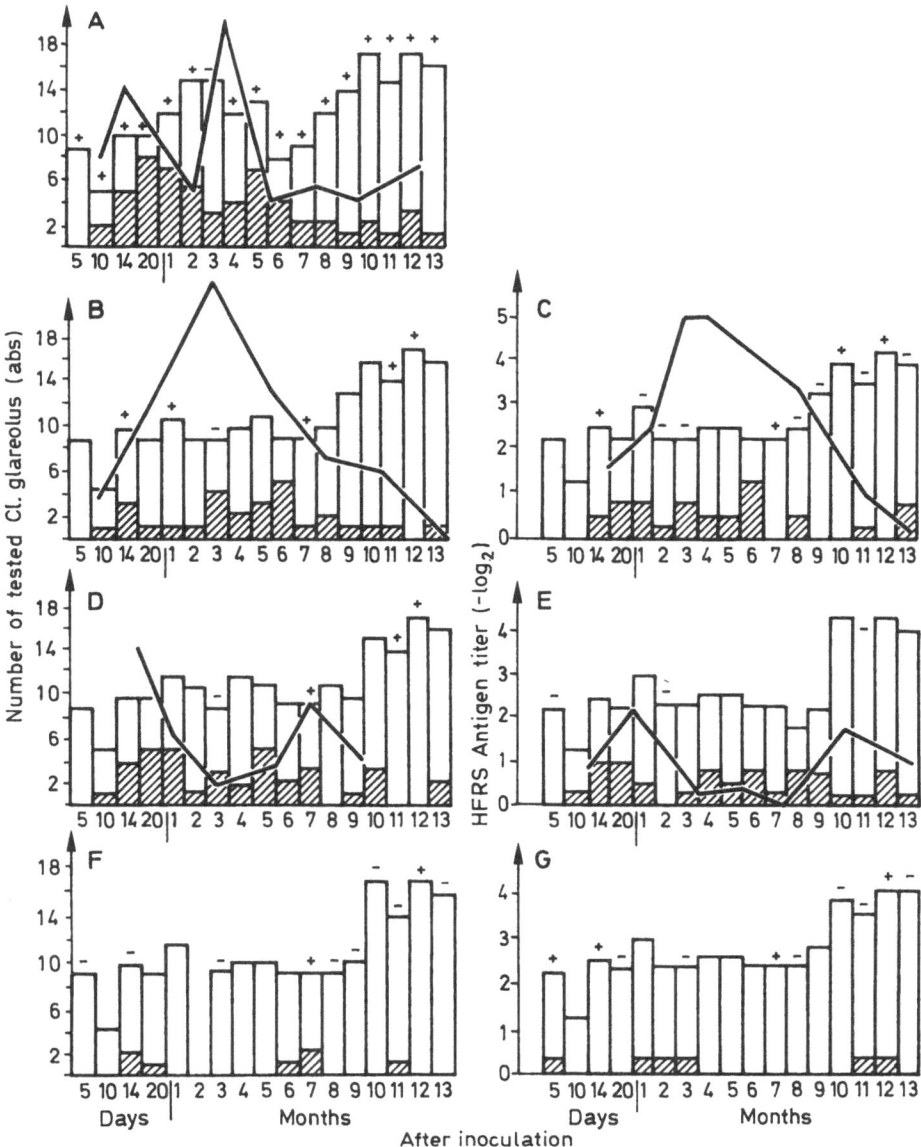

Fig. 1. HFRS virus and virus antigen persistence in experimentally infected bank voles. □ No. of bank voles tested; ▨ no. of animals with virus antigen in organs; **A** lungs; **B** brown fat; **C** salivary glands; **D** spleen; **E** liver; **F** kidney; **G** rectal tissue. Titers of antigen $-\log_2$. Presence of infectious virus in organs was detected by bioassay in bank voles. Virus positive (+), virus negative (−) samples. Data of two 13-month experiments are shown

of immune response (Association coefficient $K = 0.1497$, $p < 0.001$). Bank voles contracted the infection by direct contact with virus carriers, contact with their nest material, and contact with air in closed spaces at a distance of at least 1.5 m from the source of infection. The respiratory mode of infection seems to be as important for natural hosts of this virus as for humans. Air-dust transmission appears to depend mainly on virus excretion in feces and, to a lesser extent, in urine and saliva.

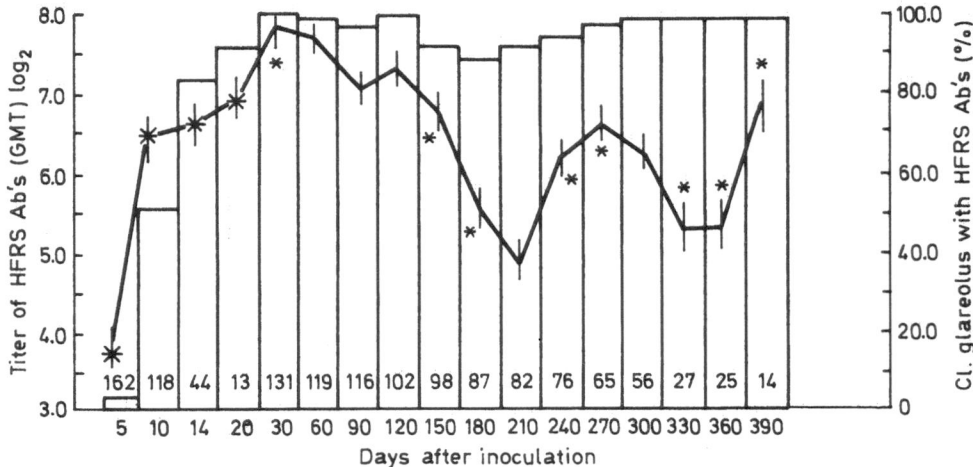

Fig. 2. Dynamics of immune response to HFRS virus in experimentally infected bank voles. Number of bank voles tested = 162. Fluorescent antibody titers (GMT in $\log_2$) in bank voles inoculated with 100 ID_{50} of strain Kazan 6 C.g. (—), frequency (%) of detection of antiviral antibody (□); presence of horizontal virus transmission (*)

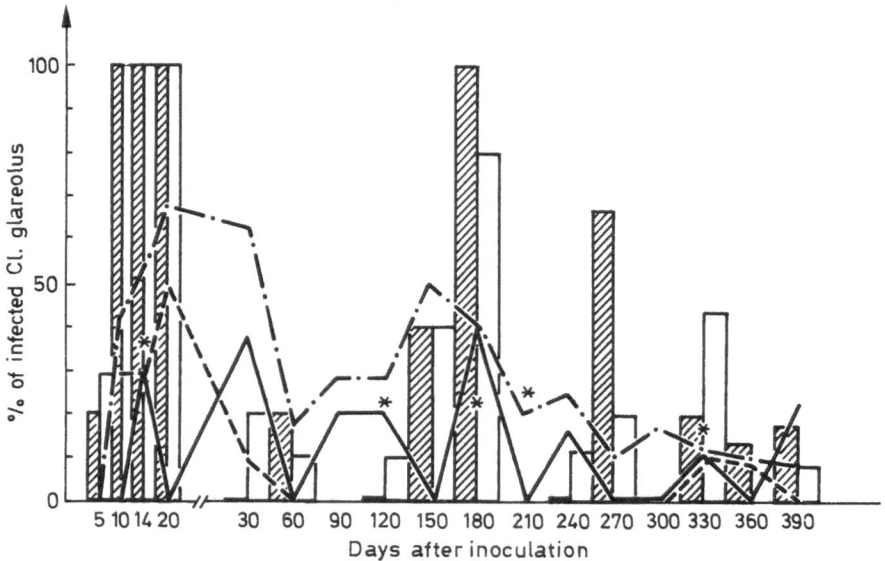

Fig. 3. Horizontal transmission of HFRS virus in bank voles. (i) Frequency (%) of detection of virus antigen in organs of bank voles inoculated with 100 ID_{50} of strain Kazan 6 C.g.: —·— lungs; —— salivary glands; – – – rectum, feces. (ii) Frequency (%) of detection of virus antigen in organs of bank voles exposed to infected voles in the same cage (□) or to their nest material (▨). Detection of virus antigen in kidneys (*)

Congenital immunity in the progeny of infected females was demonstrated both experimentally and in natural infections [10]. Passive antibody usually persisted for one month, in rare cases for 1.5 months (Fig. 4). In the first 20 days of life 78% of the progeny had postnatal passive immunity.

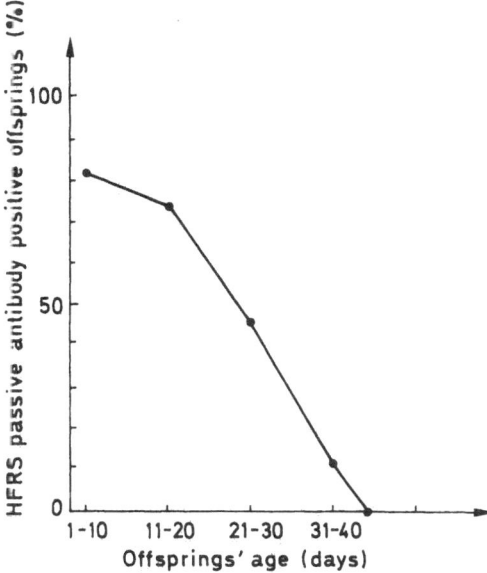

Fig. 4. Dynamics of postnatal passive immunity in bank vole offspring born of Kazan 6 C.g. virus-infected females

Table 1. Comparative titration[a] of hantavirus strain Kazan 6 C.g. in bank voles with (*A*) and without (*B*) passive immunity

Status	Presence of passive immunity	Contact with infected females for 1 month before inoculation	Geometric mean titer ($\log_2$) and no. positive/no. tested			Titer of virus in lungs ($\log_{10}$ ID_{50}/ml) after inoculation
			Antibody		Antigen after inoculation	
			before inoculation	after inoculation		
A born of infected females	yes	yes	4.9 ± 0.4	4.1 ± 0.9	0.9 ± 0.2	3.5 ± 0.8
			17/35	10/35	11/35	
B born of uninfected females	no	no	—	8.1 ± 0.5	4.1 ± 0.4	6.0 ± 0.5
			0/56	36/56	43/56	

[a] One month old bank voles were inoculated intramuscularly with 0.1–10000 ID_{50} contained in lung suspension of bank voles infected with hantavirus strain Kazan 6 C.g. and sacrificed 20 days later, at which time lung tissues were examined for viral antigen by ELISA and sera were examined for antibody by IFA

Despite longterm contact with infected females only two of 104 21-day-old animals had hantavirus antigen. Parallel titrations of hantavirus, strain Kazan 6 C.g. in 30-day-old progeny of infected and uninfected bank voles demonstrated that animals with passive immunity were more resistant to superinfection with this virus (Table 1).

Discussion

The model of bank vole–hantavirus (strain Kazan 6 C.g.) was used to study the inapparent virus persistence, most intensive in the first month after infection. These results agree with these reported by others [8, 9, 11] and may be universal for hantavirus infections in natural hosts. Our observations demonstrated an apparent life-long capacity of bank voles to transmit virus horizontally. Such transmission begins 5 days after inoculation, reaches highest levels 5–15 days later and subsequently only occurs periodically. The probability of horizontal transmission increases with the level of immune response, which may reflect the degree of intensity of the infectious process.

References

1. Bernshtein AD, Myasnikov YuA, Abashev VA (1980) Activity of HFRS foci and dynamics of main carrier population: an attempt of epidemiological prognosis. In: Bashkirev TA (ed) Hemorrhagic fever with renal syndrome in Middle Volga and pre-Urals. Institute Pasteur, Leningrad, pp 58–68 (in Russian)
2. Bogdanova SB, Gavrilovskaya IN, Boiko VA, Prokhorova IA, Linev MB, Apekina NS, Bernshtein AD, Chumakov MP (1987) Persisting infection induced by hemorrhagic fever with renal syndrome virus in Clethrionomys glareolus, natural host of the virus. Microbiol J 49: 99–106 (in Russian)
3. Brummer-Korvenkontio M, Henttonen H, Vaheri A (1982) Hemorrhagic fever with renal syndrome in Finland: ecology and virology of nephropathia epidemica. Scand J Infect Dis [Suppl] 36: 88–91
4. Coons AH (1958) Fluorescent antibody method. In: Danielli JE (ed) General cytological methods. Academic Press, New York, pp 399–422
5. Gavrilovskaya IN, Apekina NS, Gorbachkova EA, Chumakov MP (1981) Detection by enzyme linked immunosorbent assay of hemorrhagic fever with renal syndrome virus in lung tissue of rodents from European USSR. Lancet i: 1050
6. Gavrilovskaya IN, Apekina NS, Myasnikov YuA, Bernshtein AD, Ryltseva EV, Gorbachkova EA, Chumakov MP (1983) Features of circulation of hemorrhagic fever with renal syndrome (HFRS) virus among small mammals in the European USSR. Arch Virol 75: 313–316
7. Gavrilovskaya IN, Chumakov MP, Apekina NS, Ryltseva EV (1983) Adaptation to laboratory and wild animals of the hemorrhagic fever with renal syndrome virus present in the foci of the European USSR. Arch Virol 77: 87–90
8. Kurata T, Tsai TF, Bauer S, McCormick J (1983) Immunofluorescence studies of disseminated Hantaan virus infection of suckling mice. Infect Immun 41: 391–398
9. Lee HW, Lee PW, Baeck LJ, Song CK, Seong IW (1981) Intraspecific transmission of Hantaan virus, the etiologic agent of Korean hemorrhagic fever, in the rodent Apodemus agrarius. Am J Trop Med Hyg 30: 1106–1112
10. Verhagen R, Leirs H, Tkachenko E, VanDerGroen G (1986) Ecological and epidemiological data on Hantavirus in bank vole populations in Belgium. Arch Virol 91: 193–205
11. Yanagihara R, Amyx HL, Gajdusek DC (1985) Experimental infection with Puumala virus, the etiologic agent of Nephropathia Epidemica, in bank voles (Clethrionomys glareolus). J Virol 55: 34–38

Authors' address: I. N. Gavrilovskaya, Institute of Poliomyelitis and Viral Encephalitides, U.S.S.R. Academy of Medical Sciences, 142782 Moscow, U.S.S.R.

Arch Virol (1990) [Suppl 1]: 63–67

Hemorrhagic fever with renal syndrome in Bulgaria:
isolation of hantaviruses and epidemiologic considerations

S. Vasilenko, Irina Brodvarova, Yasen Topov, L. Shindarov, and **G. Kebedzhiev**

Institute of Communicable and Parasitic Diseases, Medical Academy, Sofia, Bulgaria

Accepted February 27, 1990

Summary. Epidemiological analysis of the incidence of hemorrhagic fever with renal syndrome, 1953–1988, showed that this infection is prevalent in many areas of Bulgaria. We developed a method ensuring 100% etiological diagnosis of the diseases by using albino mice for virus isolation and preparation of antigens for serological studies using a complement-fixation test. Experimental infection with hantaviruses was produced in newborn white mice. In these hosts, lesions were mostly observed in the brain, with considerable accumulation of virus detectable by electron microscopic and fluorescent antibody techniques. The virus of hemorrhagic fever with renal syndrome was successfully adapted to cell culture, in which it produced distinct cytopathic effects.

Introduction

Hemorrhagic fever with renal syndrome (HFRS) is caused by a hantavirus (family *Bunyaviridae*, genus *Hantavirus*). The disease is widespread geographically but occurs in many countries, including Bulgaria. In the latter it is widespread and highly lethal. Previously we studied the suitability of isolating the causative agent of HFRS (Hantaan virus) in mice and in cell cultures, obtaining encouraging results [5]. Our previous results proved the etiological relationship between HFRS cases occurring in Bulgaria, the European U.S.S.R., and Korea [1, 4].

Between 1953 and 1977, 389 patients in Bulgaria were shown to have HFRS; by 1989, 62 more cases had occurred. We found 74.4% of the cases in recognized foci; 83.8% of HFRS patients in Bulgaria resided in areas located more than 900 m above sea level. Location also correlated with patients: 73.4% were wood cutters or builders in alpine areas; 93.3% were males 20–40 years old. Natural foci of HFRS in alpine regions include an associated prevalence of bank voles (*Clethrionomys glareolus*) inhabiting biotopes at 1000–2000 m above sea level. In less hilly areas, reservoirs of infection also are in peridomestic rodents including house mice (*Mus musculus*), in addition to wild ones. In Bulgaria HFRS occurs mainly in two natural foci in mountainous regions, Rila–Rodopy, which includes the mountains Rila and Rodopy and the Balkans, which includes the mountains Stara Planina and Sredna Gora.

Serodiagnostic evaluations of acutely ill and convalescing patients from all over Bulgaria indicated that the highest prevalence of antibody to hantaviruses is in the Rila–Rodopy mountain region, where there are active foci of HFRS and highest morbidity in Bulgaria [1, 4].

Among 399 HFRS cases observed in Bulgaria between 1954 and 1988, 63 (15.8%) ended fatally. A significant rise in the incidence and lethality was observed in 1987–1988 when 11 of 30 patients died (36.7%).

Some cases were detected each year, varying from 1–3 to 30–40 or more. Usually they represent sporadic small epidemic outbreaks with 10–17 cases in one focus. The seasonal maximum was 54.6% during the warmest months of June–August and the lowest level (11.2%) during the cold months of November–January.

Of 214 selected persons tested for specific antibodies, 194 (90.7%) were found to be seropositive. The predominant part (70.2%) of the group of 131 seropositive persons diagnosed retrospectively on the basis of a single serum sample had antibodies against strains of the European virus (Udmurt, Kazan) only.

Analysis of 63 seropositive persons based on two and more serum samples showed the following: in 32 individuals (50.8%) antibody titres increased only against the European strains, in 26 persons (41.3%) the parallel increase of antibody against the European strain was higher than against Hantaan strain, and in 5 persons (7.9%) antibodies occurred only against the Hantaan strain. This paper presents the details of further results of the investigations of the etiology and ecology of HFRS in Bulgaria.

Materials and methods

For virus isolation attempts we used heparinized blood or 10% suspensions of organs (brain, liver, lungs, kidneys) from HFRS patients collected in the acute phase of their illnesses. Liver, lungs, and kidneys of rodents were tested for virus antigen by an indirect fluorescent antibody (IFA) technique. Serum samples for antibody tests were collected from patients in the acute- and convalescence-phases of their illnesses. Antigen was prepared from the Fojnica 2404 strain of Hantaan virus, isolated by A. Gligič in Yugoslavia from *Apodemus sylvaticus* [3]. This strain causes a lethal infection with considerable accumulation of virus in the brains and lungs of newborn albino mice within 9–10 days after inoculation. Therefore strain Fojnica 2404 proved to be very useful for preparation of antigen for complement-fixation tests.

Brains or lungs of newborn albino mice (NM) or newborn albino rats infected with this virus were frozen and thawed three times. A 10% suspension was prepared from the organs in borate-saline solution (pH 7.1) and clarified by low-speed centrifugation. This antigen was used as is or after inactivation with chloroform. Serum samples from patients convalescing from HFRS, sera from albino rats immunized with four doses (intraperitoneally) of non-inactivated antigen of Hantaan virus and rabbit anti-human FITC-conjugated gamma globulin were used for virus detection by the fluorescent antibody technique. Also, rhodamine-labeled albumin was used in IFA tests to minimize nonspecific fluorescence. Attempts were made to adapt the virus to human embryo lung diploid and Vero cells grown in MEM supplemented with 10% calf serum, 100 UI/ml penicillin and 10 μg/ml streptomycin.

NM were inoculated intracerebrally and subcutaneously with specimens from human patients or with suspensions of pooled brains and lungs of infected NM.

Supernatant fluid was removed from 24 h monolayers, the cells infected with specimens suspected to contain virus and, after 1 h adsorption, a serum-free culture medium was added. Complement-fixation tests were carried out according to the method of Clarke and Casals [2].

Results

We isolated strains from blood or tissues of 25 patients (9 fatal). All strains were isolated in NM and subsequently shown by IFA to be essentially identical to Hantaan virus strain Fojnica 2404 from Yugoslavia.

Table 1 presents the results of virological and serological examinations of 11 patients from north-western, south-western, and south-eastern Bulgaria. These results demonstrate the etiology of hantaviruses in HFRS in Bulgaria.

Table 2 presents summarized results of hantavirus IFA tests of rodents trapped in recognized foci of human infections. Most species were infected, including house mice. We noted a marked cytopathic effect in cell cultures inoculated with these isolates; monolayers showed general signs of degeneration, individual cells showed cytoplasmic granularity. On passage of isolates, the time of appearance of the cytopathic effect decreased until it stabilized at about 3–4 days.

In studies of experimental hantavirus infection of NM with strains isolated in Bulgaria, histopathological lesions were observed mainly in the brain. These lesions were encephalitic in nature: acute edema of the cerebral membranes, dilated blood vessels and perivascular hemorrhaging into the parenchyma, neuronal degeneration, small areas of encephalomalacia and areas of glial cell proliferation. In lungs, liver, kidneys, and heart, lesions were less significant: edemas, dilated blood vessels and leucocytic infiltrates, most marked in the lungs.

Electron microscopy of brains of infected NM revealed few virus-like particles, those observed being about 90–120 nm in diameter with electron-dense membranes and granular centers, some with an eccentrically located clear zone.

Table 1. Laboratory diagnosis of 11 cases of hemorrhagic fever with renal syndrome

Patient no.	District	Day of disease	Virus isolated	Titre of CF antibodies
1	Berkovitsa	12	yes	<8
		41		256
2	Etrpole	8	yes	8
		21		32
		47		32
3	Lovetsh	13	yes	<8
		40		16
4	Kiustendil	9	yes	8
		33		64
5	Sofia	5	yes	<8
		20		32
		58		64
6	Pasardgik	7	yes	8
		27		32
		>240		8
7	Sliven	11	yes	4
		84		32
8	Smolian	5	yes	4
		19		128
9	Smolian	5	yes	<8
		22		16
10	Smolian	22	yes	4
11	Smolian	22	yes	16

Table 2. Hantavirus antigen in organs of rodents trapped in a natural focus

Species	No. examined	No. positive	%
Apodemus flavicollis	41	14	34.1
A. sylvaticus	34	13	38.2
Mus musculus	10	5	50.0
Clethrionomys glareolus	3	2	
Rattus norvegicus	3	2	
R. rattus	1	0	
Glis glis	2	1	
Crocidura leucodon	1	1	
C. suaveolens	1	0	
Sorex araneus	1	1	
Total	97	39	40.2

Conclusion

We have isolated many strains of a hantavirus closely related to Hantaan virus. These were from humans and small mammals collected in endemic areas in Bulgaria.

Virus isolations were made in mice and in human embryonic and Vero cells cultures. Virus isolates were detected by cytopathic effects in cell cultures and identified by electron microscopy and by indirect immunofluorescence.

References

1. Gavrilovaskaya I, Vasilenko S, Chumakov M, Shindarov L, Katsarov G (1984) HFRS in Bulgaria: prevalence and serological evidence. Epidemiol Microbiol Infect Dis (Sofia) 4: 17–23
2. Clarke DH, Casals J (1958) Techniques for hemagglutination and hemagglutination-inhibition with arthropod-borne viruses. Am J Trop Med Hyg 7: 561–573
3. Glicič A, Frusič M, Obradovič M, Stojanovič R, Hlaca D, Gibbs CJ Jr, Yanagihara R, Calisher CH, Gajdusek DC (1989) Hemorrhagic fever with renal syndrome in Yugoslavia: antigenic characterization of hantaviruses isolated from *Apodemus flavicollis* and *Clethrionomys glareolus*. Am J Trop Med Hyg 41: 109–115.
4. Tchumakov M, Shindarov L, Gavrilovskaya I, Vasilenko S, Gorbachkova E, Katsarov G (1988) Seroepidemiology of HFRS in Bulgaria. Acta Virol 32: 261–266
5. Vasilenko S, Bradvarova I (1972) On the results of investigation of HFRS etiology in Bulgaria. In: Aktualnye problemy virusologii i profilaktiki virusnykh zabolevaniy. Abstracts 17th scientific session of the Institute of Poliomyelitis and Viral Encephalitis, Moscow, October 24–27, 1972, p 304

Authors' address: S. Vasilenko, Institute of Communicable and Parasitic Diseases, Medical Academy, 44a Boulevard Stoletov, BG-33 Sofia, Bulgaria.

Arch Virol (1990) [Suppl 1]: 69–80

Association of chronic renal disease, hypertension, and infection with a rat-borne hantavirus

G. E. Glass[1], J. E. Childs[1], A. J. Watson[2], and J. W. LeDuc[3]

[1] Department of Immunology and Infectious Diseases, The Johns Hopkins University
School of Hygiene and Public Health, and
[2] Division of Nephrology, The Johns Hopkins University School of Medicine,
Baltimore, Maryland
[3] Department of Epidemiology, U.S. Army Medical Research Institute
for Infectious Diseases, Frederick, Maryland, U.S.A.

Accepted April 4, 1990

Summary. We report an association between past infection with an indigenous rat-borne hantavirus and chronic renal disease, hypertension, and cerebrovascular accidents among individuals using the Johns Hopkins Medical Institution from January 1986 through October 1988. A sample population of 1148 patients receiving quantitative total urine protein tests was screened for IgM and IgG antibodies to three different hantaviruses. Fifteen seropositives (1.3%) were found, of which 12 resided in inner city Baltimore in areas where Norway rats infected with a hantavirus had been captured.

Comparisons of clinical histories for the 15 seropositive people and 73 age–sex matched seronegative controls demonstrated significantly higher rates of chronic renal disease (80% vs. 44%), and hypertensive renal disease (70% vs. 9%) among seropositive patients. Nearly all (14/15) seropositive individuals were clinically hypertensive, and they were nearly five times more likely to have suffered cerebrovascular accidents than seronegative persons. Only one acute illness, consistent with rat-borne hantaviral disease, was documented among these 15 seropositive individuals.

These data suggest that infection with rat-borne hantaviruses in inner city populations in the United States is associated with increased occurrence of chronic renal disease and hypertension.

Introduction

Hantaviruses (family *Bunyaviridae*), a recently isolated and characterized group of rodent-borne viruses, include the etiological agents of illnesses collectively referred to as hemorrhagic fever with renal syndrome (HFRS) [20, 27]. These viruses are suspected as being transmitted to humans by aerosol and fomites, producing diseases with a wide range of symptoms which typically include fever, headache, myalgia, and acute renal failure. More than 200,000 hospitalized cases of HFRS are estimated to occur annually, with the highest incidence in Asia, western Soviet Union and northern Europe [24, 27]. Case fatality rates range from < 1% to more than 15%, depending on the specific hantavirus responsible for infection [24]. Surviving patients typically recover without permanent sequelae [8]; however, Rubini et al. [18] reported residual renal function abnormalities and hypertension in some patients who contracted HFRS during the Korean War.

Currently, there are four well defined rodent–hantavirus associations; prototype Hantaan virus (HTN) and *Apodemus agrarius*, Puumala virus (PUU) and *Clethrionomys glareolus*, Seoul virus (SEO) (and other related rat-borne viruses) and *Rattus* spp., and Prospect Hill virus (PHV) with *Microtus pennsylvanicus* [27]. With the exception of PHV, each virus is responsible for a specific form of HFRS. HTN is associated with severe HFRS, PUU with nephropathia epidemica (NE), and SEO with mild epidemic hemorrhagic fever.

In the United States, Norway rats (*Rattus norvegicus*) are the major urban reservoir for Seoul-like hantaviruses, and prevalence of infection in this species can exceed 50% [3, 12]. While several studies have demonstrated serological evidence of past human infection with hantaviruses in the U.S., including infection with a rat associated virus, none have documented clinical illness [4]. This is curious in light of the known association between SEO and human disease in Korea and China [27]. Our goal was to look for acute, hospitalized cases of hantaviral disease in an area of the United States where human exposure to a rat-borne hantavirus was known to occur. Although only one case of possible acute hantaviral disease was documented, we observed associations between past exposure to a SEO related virus and hypertension and chronic renal disease. We conclude that hantaviral infection may represent a currently unrecognized cause of chronic disease among inner city residents of the United States.

Methods

Study population

All in-patients and out-patients using the Johns Hopkins Hospital from 15 January 1986 to 14 October 1988 receiving 24 h quantitative urine total protein tests and having blood drawn were considered for inclusion in the study. The population was drawn entirely from patients whose physicians were evaluating renal function. The majority of the sample population was selected from individuals excreting > 250 mg of protein/24 h, although two patients/week with urine total protein ≤ 150 mg/24 h were randomly selected from the same population and included. This criterion was selected because significant proteinuria is a consistent clinical finding in all recognized forms of HFRS regardless of the infecting virus [2, 9]. Patients were excluded from the study if they were currently undergoing chemotherapy for cancer or treatment for human immunodeficiency virus.

Serological techniques

Serum samples were obtained when patients enrolled in the study and examined by an IgG indirect enzyme-linked immunosorbent assay (ELISA) for antibodies to hantaviruses. The serological test is described in detail elsewhere [4]. Sera were considered suspect positive if their optical densities were greater than the mean plus three standard deviations of three known negative human sera from Baltimore residents that were included on each plate. Suspect positive sera and a random sample of negative sera were further examined by Western blot analyses using prototype Hantaan virus (HTN), and Baltimore rat virus (BRV; a local SEO virus isolate from a Norway rat [3]) as antigens. Sera producing a single band at approximately 50 kDa (corresponding to nucleocapsid antigen) were considered positive. To determine the specific hantavirus causing infection, each serum, was further examined by plaque reduction neutralization tests (PRNT) using prototype HTN, BRV and PHV [14]. The PRNT yields at least 4–16 fold differences in antibody titers between heterologous and homologous hantaviruses [5], allowing differentiation of persons infected with HTN while traveling overseas from those infected by an indigenous strain of hantavirus. All samples that were positive by ELISA and Western blot neutralized at least one of the hantaviruses. All seropositives were screened by an IgM capture ELISA to identify recent exposures and, potentially, acute cases of HFRS.

Clinical study

All persons reactive by IgG ELISA, Western blot, and PRNT were considered seropositive and previously exposed to a hantavirus. Their charts were reviewed for clinical and laboratory results and their histories were reviewed for illnesses consistent with HFRS [27].

Matched study

Seropositive patients were matched with patients from the remaining seronegative population. Seronegative patients were randomly selected and matched to seropositives for age (within 3 years), and sex. Five seronegatives were selected for each seropositive, with the exception of two positive females aged 87 and 88 for which only 8 negative patients could be identified. Medical charts were examined without prior knowledge of serological status by one of us (AJW) to determine clinical diagnoses. Variables recorded for each individual included race, age, occupation, home address and length of residency, travel history (if available), history of military service (if available), reason for admission, original admission ward, and medical history, including chronic disease history. Persons suffering from more than one chronic condition were included in each applicable category.

Diagnoses

Individuals with a history of systolic pressure greater than 150 mm, and/or diastolic pressure greater than 90 mm on more than three visits, were considered hypertensive. Those with impaired glucose tolerance, elevated blood sugar, especially if associated with retinopathy, or receiving appropriate medications were considered to have diabetes mellitus. The category of cerebrovascular accidents included individuals with strokes or ischemic attacks.

Chronic renal disease was identified based on a history of serum creatinine exceeding 1.4 mg/dl for more than 3 months.

Within the category of chronic renal disease, several potential causes were further identified. Diagnosis was made on the basis of previous medical histories, physical examinations, and previously obtained laboratory tests. Where available, biopsy information was obtained to provide a definitive diagnosis. In the absence of biopsy data, specific attention was paid to the presence of a family history of hypertension, the temporal relationship between the onset of hypertension and discovery of kidney disease, a history of diabetes mellitus, or any evidence suggesting glomerular process, e.g., nephrotic range proteinuria.

Individuals with chronic renal disease and at least a 10–15 year history of diabetes mellitus and/or associated retinopathy were considered to have diabetic nephropathy. Those with chronic renal disease in the setting of longstanding hypertension were diagnosed as having hypertensive renal disease. Individuals with a history of nephrotoxic drug ingestion and/or injection, followed by chronic renal disease, were categorized as drug induced. Obstructive renal disease was diagnosed for those with urological, radiological, or clinical evidence of obstruction, and autoimmune renal disease was diagnosed for individuals with positive laboratory findings for autoantibodies. In the absence of sufficient documentation, individuals that met the criterion of chronic renal disease were diagnosed as having disease of unknown origin.

Data analyses

All values are given as mean ± 1 standard deviation, except as noted. All frequency variables were tested as simple χ^2 tests for homogeneity, and odds ratios (O.R.) with 90% confidence interval (C.I.) were derived. Two-tailed tests were used in all comparisons. Statistical examination of primary diagnoses for chronic renal disease considered four categories; none, diabetes mellitus, hypertension, and other. Differences in cumulative distributions were examined using Smirnov two sample tests.

Results

Antibody prevalence and acute disease

Seventy-two percent of patients excreting ≥ 250 mg protein/24 h also had blood collected. Sera were not available from 446 patients with total urine protein > 250 mg/24 h. A total of 1,669 sera were obtained from 1,148 individuals who met the criteria for inclusion ($\bar{x}$ = 1.34 samples/person). Females predominated in the population (1.6:1), and averaged 41.4 ± 20.9 years of age. Men were significantly older (51.0 ± 19.2 years) than women (t = 8.25, 1146 degrees of freedom; p < 0.005). The median proteinuria for men (720 mg/24 h, 25–75% Quartiles 332.5–1614 mg/24 h) was significantly higher than for women (532 mg/24 h, 210–1425 mg/24 h).

Twenty samples ($\bar{x}$ = 1.33 bleeds/patient) from fifteen patients (1.31%) were seropositive for hantaviral antibodies by ELISA and confirmed by Western blot. All seropositive persons had markedly higher (≥ 10 fold) titers by PRNT to BRV than to HTN, and none neutralized PHV, implicating BRV in their infections. Sera from seronegative individuals (n = 93) were negative by Western blot and failed to neutralize any of the hantaviruses, indicating none of these individuals had been previously infected. Individuals with proteinuria (> 150 mg/24 h) were 2.7 times (0.5–14.7; O.R. ± 90% CI) more likely to be seropositive than individuals with normal protein levels. However, this difference was not statistically significant.

One seropositive individual was admitted with complaints of an acute illness consistent with rat-borne HFRS. This patient had a detectable IgM titer (800) to a hantavirus. The patient was a 72 year old African-American female who was admitted with a two week history of illness marked by rapid onset, later characterized by decreasing urinary output and edema. On admission, hepatomegaly was noted and she was in acute renal failure. She had an increased prothrombin time, and increased enzyme levels of alkaline phosphatase, aspartate aminotransferase and lactate dehydrogenase. However, her alanine aminotransferase was within a normal range. She was treated with aggressive diuretic therapy with little effect but her symptoms subsequently resolved. The patient was lost to followup and convalescent serum samples could not be obtained.

Most (10/15) seropositives, however, were admitted for complications from chronic diseases. Of the seropositives without chronic disease, one was admitted for severe preeclampsia, one for a psychiatric disorder, and three for complications from their inabilities to care for themselves. None of the seropositives reported histories of foreign travel. The median time of residence in Baltimore for 10 patients was 43 years. The duration of residence was not obtained for two Baltimore residents. Three individuals lived outside of Baltimore (one from Annapolis, MD, and two from New York, NY).

The seropositive individuals consisted of 10 women and 5 men, of whom 13 were African-American and 2 were Caucasian. The sex ratio among seropositive individuals was similar to the population as a whole ($\chi^2 = 0.16$, 1 df, n.s.). There was no significant difference in the ages of exposed men and women (Smirnov test; $T_1 = 0.5$, p = 0.20). Their ages ranged from 23 to 90 with a median age of 65 years.

Demographics of seropositives and seronegatives

The home addresses for 73 age and sex matched controls (seronegatives) was similar to that of the seropositives. Fifty-five resided in Baltimore, while 18 lived outside the city. Mapping of residences indicated that seropositives and control patients were drawn primarily from neighborhoods in eastern Baltimore (Fig. 1), where seropositive rats had been captured [3]. The proportion of Baltimore residents among seropositives and seronegatives, ($\chi^2 = 0.15$, 1 df; n.s.), and the length of their residencies (median 43.0 vs 38.5 years) did not differ significantly, nor did their occupational histories (Table 1). No seropositives and only one seronegative held a white collar position, and 57.4% of seronegatives and 42.9% of seropositives were either

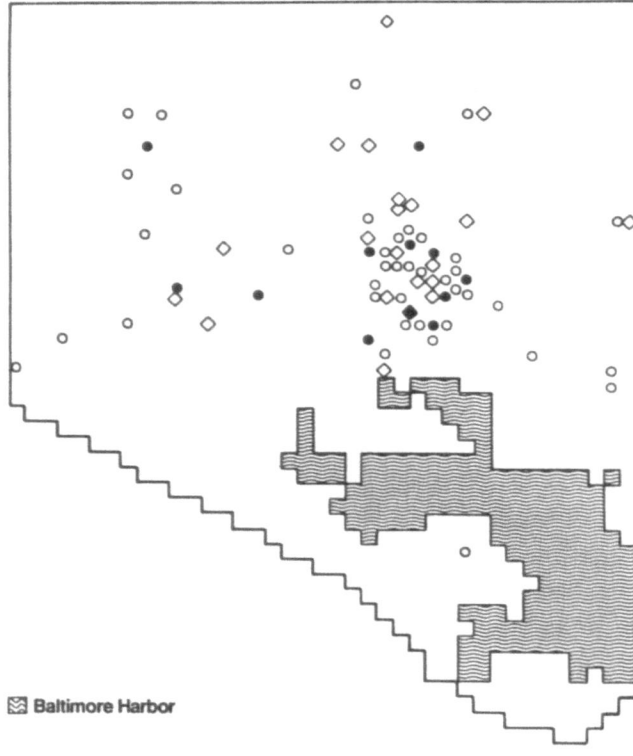

Fig. 1. Geographic distributions of residences for seronegative and seropositive individuals in Baltimore, MD. ●,◆ People with antibody to Baltimore rat virus (BRV), ○,◇ seronegatives. ○,● African-Americans, ◆,◇ others

unemployed or held unskilled/day-labor jobs. African-Americans represented 86.7% of the cases and 64.4% of the controls, a difference that was not statistically significant ($\chi^2 = 2.93$, 2 df, $0.1 < p < 0.5$).

Clinical study

Examination of patients' charts did not identify any consistent laboratory finding that differentiated cases from controls. Although seropositive patients tended to have higher values for systolic blood pressure, serum creatinine, and proteinuria levels, they were not significantly different (Table 1; $p > 0.10$). Diastolic blood pressure was higher among seronegatives; ($p < 0.10$), but did not reflect a lack of treatment in this group, as all hypertensive patients were medicated.

In contrast, seropositives represented a defined subgroup of patients in this population with regard to clinical histories of chronic disease. They had higher rates (80.0% vs. 43.8%) of chronic renal disease than seronegatives (O.R. = 5.1, 1.6–13.3), and there was a marked difference between seropositives and seronegatives in the presumed etiology of renal disease (Table 2). Seropositives had significantly higher rates of hypertensive renal disease (70% vs. 9.4% of those with assignable diagnoses), while diabetic nephropathy was the most common diagnosis in the control group (50% vs. 20%

Table 1. Epidemiological characteristics of seropositive and seronegative individuals examined for chronic disease. There were no significant differences between groups in any of these variables

Variable	Seropositive	Seronegative
Race		
African-American	13	47
Caucasian	2	24
other	0	2
Age		
male ($\bar{X} \pm$ sd)	58.2 ± 12.5	58.0 ± 12.0
female	66.7 ± 22.0	67.3 ± 21.2
Occupation		
unemployed	1	7
unskilled	5	28
blue collar	2	13
service oriented	3	12
white collar	0	1
retired	6[a]	35[a]
unknown	3	5
Residence		
Baltimore	12	55
non-Baltimore	3	18
Clinical		
blood pressure (mm/Hg)	$164/76 \pm 32/10$	$151/89 \pm 34/18$
serum creatinine (mg/dl)	2.69 ± 2.28	2.06 ± 1.96
proteinuria (mg/24 h)	2140 ± 3000	1660 ± 2440

[a] Includes individuals in other job categories

among cases). Other sources of chronic renal disease occurred at lower frequencies (6.2–12.5% and did not differ between groups). The differences in primary diagnoses among cases and controls were highly significant ($\chi^2 = 26.75$, 3 df, $p < 0.005$), and were primarily due to the overall higher prevalence of chronic renal disease among seropositive patients, and the differences in the frequency of hypertensive renal disease between the two groups (Table 2).

Fourteen seropositive individuals (93.3%) had hypertension. This prevalence was significantly higher (O.R. = 7.7, 1.2–23.5) than among the controls (64.4%) (Table 2). Seropositives also were 4.9 times (C.I. = 1.5–15.8) more likely to have suffered cerebrovascular accidents compared to controls. There was no difference in the prevalence of diabetes mellitus between the two groups (O.R. = 0.7, 0.3–1.9). Biopsies were available only for a minority (<20%) of patients, as the procedures were not generally performed on individuals with chronic renal failure. However, for all individuals, diagnoses

Table 2. Prevalences of clinically diagnosed chronic disease among individuals seropositive
or seronegative to Baltimore rat virus

Disease	Seropositive	Seronegative	Odds ratio	90% confidence intervals
Chronic renal disease	12	32	5.1	1.6–13.3
Hypertension	7[a]	3	20.4	5.1–60.5
Diabetes mellitus	2[b]	16	0.5	0.2–2.2
Drug-induced	1[b]	6	0.8	0.2–3.9
Obstructive	0	3	—	
Autoimmune	0	4	—	
Unknown	1[c]	0	—	
None	3	41	—	
Hypertension	14/15	47/73	7.7	1.2–23.5
Cerebrovascular accident	4/15	5/73	4.9	1.5–15.8
Diabetes mellitus	6/15	35/73	0.7	0.3–1.9

[a] Case deleted with chronic hypertension and chronic renal disease but requiring further testing

[b] IV drug user with diabetes mellitus included in drug-induced category

[c] Case with insufficient clinical data

were made by attending physicians, and none of our diagnoses differed from theirs.

Given the high rate of hypertensive disease among inner city African-Americans [12], their somewhat higher representation among the seropositive population was considered a possible source of confounding. However, these effects persisted after stratifying by race. Among African-Americans, prevalences of hypertension ($\chi^2 = 3.05$, 1 df, $p < 0.10$), chronic renal disease associated with hypertensive renal disease ($\chi^2 = 19.05$, 3 df, $p < 0.005$) and cerebrovascular accidents ($\chi^2 = 3.15$, 1 df, $p < 0.10$) remained significantly higher among seropositive individuals. The small sample size for seropositive Caucasians prevented a meaningful analysis.

Discussion

Hantaviruses are a recently described group of viruses. It is only since their isolation and tissue culture adaptation in the late 1970's that serological methods for identifying human infections were developed. With the advent of serological testing, the known geographic range of HFRS has spread to include southern and eastern Europe, Scotland, and Malaysia [1–3, 7, 23, 25], and infected rodents have been found worldwide [12].

The pathogenesis of hantaviral disease is unclear, although the basic lesion is endothelial cell damage or dysfunction [17], and viral antigen has been detected in vascular endothelia of experimentally infected animals [9] and fatal human cases [27]. Despite its name, HFRS presents with hemorrhagic manifestations in only 20–30% of cases. Rather, patients with moderate to severe illness experience an acute onset of fever, severe retro-orbital headache, blurred vision, photophobia, nausea, and abdominal, back, and flank pains. Laboratory findings during the febrile and oliguric stages include proteinuria, thrombocytopenia, leukocytosis with a left shift, elevated blood urea nitrogen and serum creatinine, and a slight transaminase elevation. Differential diagnoses of classic HFRS might include: leptospirosis, allergic interstitial nephritis, acute glomerulonephritis, rapidly progressive glomerulonephritis, and other causes of acute renal failure. However, at least 30% of clinical cases are mild, and may be misdiagnosed as influenza [13].

Although previous studies demonstrated human infection in the United States [4, 22, 26], most could not identify the probable source of the infecting virus and none identified associated illnesses [4]. Results presented here are the first to indicate that hantaviral infection in the United States, acquired by contact with domestic rodents or their excretion, may be associated with chronic human disease.

This population was drawn, primarily, from inner city, lower socio-economic status areas. Although most seropositive individuals resided in Baltimore, cases from New York and Annapolis indicate exposure is geographically widespread. There was no significant difference in exposure of the two sexes. This differs from other studies [27], in which males have higher infection rates and exposure is job-related. This suggests exposure to BRV may occur in non-occupational settings, such as in and around home. Childs and colleagues [3] have documented high rates of infection and abundant rat populations in most of these areas.

In our study, only one patient had a clinical history of an illness consistent with acute HFRS. This individual's illness was similar to HFRS described for rat-borne hantaviruses, especially in the marked liver involvement [2]. However, the lack of convalescent serum from this person precludes definitive diagnosis. The absence of additional cases may simply reflect incomplete medical histories because many charts did not extend more than three years before the most recent admission. None of the 14 IgM titers for the other seropositive individuals indicate recent infections. Alternatively, the infections may have been subclinical. Within endemic areas of Korea 1–4% of the population have been exposed to Hantaan virus but most infections did not produce disease [13].

Clearly, the data concerning chronic diseases are associational, and necessary precautions must be taken to consider alternate, confounding variables and hypotheses [15]. Matching of seropositives with seronegatives gave excellent agreement where it could be assessed. There were no significant differences in ages, sex ratios, geographic distribution of residencies, lengths of residence in Baltimore, occupational types, or uses of the hospital system (based on blood samples/person) for seropositive and seronegative individuals that would suggest other obvious confounding factors. In addition, stratification by race did not alter the association of seropositivity with chronic disease.

One potential explanation for the observed association is that individuals with pre-existing hypertensive disease have lifestyles that expose them to virus at increased frequency. In this situation chronic disease would predate exposure to a hantavirus. Our primary indicator of lifestyle, occupation, does not support this hypothesis. There were no differences in the types of work performed by seropositives and controls and seropositives showed no greater tendency to be incapacitated than controls. Among controls, 18/36 individuals under the age of 65 were retired, unemployed, or listed no occupation, while 3/7 cases were in this category. Similarly, the absence of any difference in the number of samples obtained from seropositive and seronegative patients suggests that seropositives were no more likely to have received hospital treatment than seronegative patients.

Another explanation is that seropositive patients may have had more complete histories obtained by their physicians. However, this seems unlikely, as all but one of the patients with chronic renal disease had been diagnosed by their physicians and histories emphasized evidence of hypertension and diabetes. A third possibility is that hypertensive individuals may be more susceptible to some renal pathology caused by hantaviruses. If so, hypertension may predispose individuals exposed to hantaviruses to develop renal disease.

Alternatively, hantaviral infection may precede and in some manner contribute to the onset of hypertension and hypertensive renal disease. Several uncontrolled studies of HFRS patients suggest that hypertension and chronic renal diseases may be a consequence of hantaviral infection in an unknown proportion of patients. Rubini and colleagues [18] noted that 2 to 5 years after apparent recovery from HFRS 7/13 patients had acquired hyposthenuria and 2/13 had developed hypertensive vascular disease. In addition, they reported one case of chronic glomerulonephritis and two cases of pyelonephritis among 31 cases after presumptive recovery. Lahdevirta's study of NE patients [10] also reported abnormal findings for blood pressure, sedimentation, and creatinine clearance rates among most of the 20 patients examined 1 to 6 years after infection. Cizman and colleagues' [6] recent report lends credence to these associations. Examination of patients two years after an outbreak of HFRS in Yugoslavia revealed that 18% had impaired tubular function, 9% were hypertensive, and 9% had chronic renal insufficiency. Our apparently higher rates of chronic diseases associated with infection may reflect study design. Whereas earlier studies followed patients who had HFRS, our study was of patients with high levels of proteinuria and prior exposure to a hantavirus. Population surveys need to be conducted to obtain comparable data.

The relationship between infection with a hantavirus and subsequent development of chronic renal disease in studies of HFRS patients requires further investigation. Studies are underway to examine the association of hypertension and hantaviral infection among inner city residents. This study suggests that despite the current absence of definitively recognized acute HFRS in the United States, human infection with indigenous hantaviruses is associated with chronic renal disease and hypertension; further studies on the sequential timing of infection with a hantavirus and the onset of hypertension and renal disease are needed to clarify the biology of this relationship.

Acknowledgements

We are indebted to Ray R. Arthur, Ph.D., Cynthia A. Rossi, M.A., and Muriel J. Tharrington, B.A. for technical assistance; to Patricia Charache, M.D. for access to specimens and medical data; and to Thomas P. Monath, M.D., Clarence J. Peters, M.D., Harbutune K. Armenian, Ph.D., and Paul Whelton, M.D. for reading earlier versions of this manuscript. This work was supported by contracts DAMD17-84-C-4015 and 17-87-C-7101 from the Department of Defense. The views of the authors do not purport to reflect the position of the U.S. Department of the Army, or the U.S. Department of Defense.

References

1. Antoniadis A, LeDuc JW, Daniel-Alexiou S (1987) Clinical and epidemiological aspects of hemorrhagic fever with renal syndrome (HFRS) in Greece. Eur J Epidemiol 3: 295–301
2. Chan YC, Wong TW, Yap EH, Tan HC, Lee HW, Chu YK, Lee PW (1987) Haemorrhagic fever with renal syndrome involving the liver. Med J Australia 147: 248–249
3. Childs JE, Korch GW, Glass GE, LeDuc JW, Shah KV (1987) Epizootiology of hantavirus infections in Baltimore: isolation of a virus from Norway rats, and characteristics of infected rat populations. Am J Epidemiol 126: 55–68
4. Childs JE, Glass GE, Korch GW, Arthur RR, Shah KV, Glasser D, Rossi C, LeDuc JW (1988) Evidence of human infection with a rat-associated hantavirus in Baltimore, Maryland. Am J Epi 127: 875–878
5. Childs JE, Glass GE, Korch GW, LeDuc JW (1988) The ecology and epizootiology of hantaviral infections in small mammal communities of Baltimore: a review and synthesis. Bull Soc Vector Ecol 13: 113–122
6. Čizman B, Furlan P, Kaplan-Pavlovčič S, Drinoveč J, Avšič T (1988) Follow-up of patients with HFRS. In: Proc Int Symp Hemorrhagic Fever with Renal Syndrome, Hubei, China, 31 October–2 November, 61: 150
7. Dournon E, Brion N, Gonzalez JP, McCormick JB (1983) Further case of haemorrhagic fever with renal syndrome in France. Lancet 2: 1419
8. Giles RB, Sheedy JA, Ekman CN, Froeb HF, Conley CC, Stockard JL, Cugell DW, Vester JW, Kiyasu RK, Entwistle G, Yoe RH (1954) The sequelae of epidemic hemorrhagic fever: with a note on causes of death. Am J Med 16: 629–638
9. Kurata J, Tasi TF, Bauer SP, McCormick JB (1983) Immunofluorescence studies of disseminated Hantaan virus infection of suckling mice. Infect Immun 41: 391–398
10. Lahdevirta J (1971) Nephropathia epidemica in Finland. A clinical histological and epidemiological study. Ann Clin Res 3 [Suppl 8]: 1–154

11. Lahdevirta J (1982) Clinical features of HFRS in Scandinavia as compared with East Asia. Scand J Infect Dis [Suppl 36]: 93–95

12. LeDuc JW, Smith GA, Childs JE, Pinheiro FP, Maiztegui JI, Niklasson B, Antonaidis A, Robinson DM, Khin M, Shortridge KF, Wooster MI, Elwell MR, Ilberg PLT, Koech D, Rosa EST, Rosen L (1986) Global survey of antibody to Hantaan-related viruses among peridomestic rodents. Bull WHO 64: 139–144

13. Lee HW (1988) Hantavirus infection in Asia. In: Nephrology II. Proc Xth Int Cong Nephrology 1987. Baillere Tindall, London, pp 816–831

14. Lee P-W, Amyx HL, Gajdusek DC, Yanagihara RT, Goldgaber D, Gibbs CJ Jr (1982) New haemorrhagic fever with renal syndrome-related virus in indigenous wild rodents in United States. Lancet 2: 1405

15. Lilienfeld AM, Lilienfeld DE (1980) Foundations of epidemiology, 2nd edn. Oxford University Press, New York, pp 191–225

16. Powell GM (1954) Hemorrhagic fever: a study of 300 cases. Medicine 33: 97–153

17. Rostand SG, Brown G, Kirk KA, Rutsky EA, Dustan HP (1989) Renal insufficiency in treated essential hypertension. N Engl J Med 320: 684–688

18. Rubini ME, Jablon S, McDowell ME (1960) Renal residuals of acute epidemic hemorrhagic fever. Arch Intern Med 106: 378–387

19. Schmaljohn CS, Hasty SE, Dalrymple JM, LeDuc JW, Lee HW, von Bonsdorff C-H, Brummer-Korvenkontio M, Vaheri A, Tsai TF, Regnery HL, Goldgaber D, Lee P-W (1985) Antigenic and genetic properties of viruses linked to hemorrhagic fever with renal syndrome. Science 227: 1041–1044

20. Tsai TF, Bauer SP, Sasso DR, Whitfield SG, McCormick JB, Caraway TC, McFarland L, Bradford H, Kurata T (1985) Serological and virological evidence of a Hantaan virus-related enzootic in the United States. J Infect Dis 152: 126–136

21. Van Ypersele de Strihou C, Vanderbroucke JM, Levy M, Dourn M, Cosyns C, Van der Groen JP, Desmyter J (1983) Diagnosis of epidemic and sporadic interstitial nephritis due to Hantaan-like virus in Belgium. Lancet 2: 1493

22. Van Ypersele de Strihou C, Mery JP (1988) Virus induced nephropathy: the hantavirus as an ubiquitous model. In: Nephrology II, Proc Xth Int Cong Nephrology 1987. Bailliere Tindall, London, pp 802–815

23. Walker E, Pinkerton IW, Lloyd G (1984) Scottish case of haemorrhagic fever with renal syndrome. Lancet 2: 982

24. Yanagihara R, Gajdusek DC, Gibbs CJ, Traub R (1984) Prospect Hill virus: serologic evidence for infections in mammalogists. N Engl J Med 310: 1325–1326

25. Yanagihara R, Gajdusek DC (1987) Hemorrhagic fever with renal syndrome: global epidemiology and ecology of hantavirus infections. Med Virol 6: 171–214

Authors' address: G. E. Glass: Department of Immunology and Infectious Diseases, The Johns Hopkins University School of Hygiene and Public Health, 615 N. Wolfe Street, Baltimore, MD 21205, U.S.A.

Arch Virol (1990) [Suppl 1]: 81–86

Prevalence of antibody to hantaviruses in humans and rodents in Italy. Provisional evidence of Hantaan-like virus infections in humans and Seoul-like virus infections in rodents

M. Nuti[1], Luisa Anna Ieradi[2], M. Cristaldi[2], and C. J. Gibbs, Jr.[3]

Departments of [1] Tropical Diseases and of [2] Animal and Human Biology, University
"La Sapienza", Rome, Italy
[3] National Institutes of Health, Bethesda, Maryland, U.S.A.

Accepted February 15, 1990

Summary. A serological investigation of 1583 healthy residents of several ecologic zones of central and northern Italy showed the presence of IgG antibodies to hantaviruses in 37 (2.3%) of them. No antibodies to hantaviruses were found in 158 individuals conceivably at risk to acquiring infection with these viruses. Two of 20 mammalogists and 3 of 51 patients on renal dialysis had low titer (16–32) antibody to hantaviruses. Of 257 others at possible risk, 9 of 192 farmers and 7 of 65 foresters had antibody to hantaviruses and more often had relatively higher titers than did the others with antibody.

Antibodies to hantaviruses were detected in 26 of 50 *Rattus norvegicus*, 3 of 17 *Rattus rattus*, and 6 of 31 *Mus musculus* captured in Rome. Titrations of rodent sera provided indications of infection with a Seoul-like virus, rather than with Hantaan or other hantaviruses. No antibodies to Hantaan, Seoul, or Puumala virus were found in 32 rodents trapped on the Pontine Islands near Rome.

Introduction

Until 1984, no information was available on the presence of hantaviruses in Italy. A preliminary seroepidemiological survey conducted in central Italy in 1985 indicated an antibody prevalence of 2.8% of 496 humans in the area; titers were low (< 32; M. Nuti, unpubl. data). We extended this investigation to include other areas of central and northern Italy, obtaining a similar pattern [7]. Other workers determined a 6% prevalence of antibodies to hantaviruses in inhabitants of western Sicily and 13% in renal dialysis patients [5]. Clinically-diagnosed hantavirus infections, unsubstantiated by laboratory tests, have been reported to occur in the area of Florence (M. Balducci, pers. comm., 1987).

We report the prevalence of antibodies to hantaviruses in healthy residents of central and northern Italy, in selected individuals at possible risk in Rome (central Italy) and in Cadore (northern Italy), and in wild rodents collected in Rome and on the Pontine Islands near Rome.

Materials and methods

Human serosurvey

Serum samples were collected at two sources: (a) Randomly collected sera, primarily from individuals ≥ 16 years of age, originally submitted to hospital laboratories for biochemical analyses. Samples were numbered so they could not be traced to the donor. The areas in which these people lived differ in terms of geographic and ecologic characteristics (urban areas, rural villages, coastal areas), population density, and occupational activities; (b) Serum samples from patients undergoing renal dialysis and individuals at risk of acquiring hantavirus infection and who voluntarily participated (mammalogists, rodent control personnel, garbage disposal workers, oarsmen, river police, farmers, lumberjacks, and foresters). Serum samples were labeled as to age and sex and were stored at −20°C until they were tested for antibody.

Rodent serosurvey

In Rome, rodents were captured at two sites during spring and summer 1985 and 1986. One site was at Tiberina Island, central Rome. The other was on the bank of the Tiber in a northern suburb of Rome. Rodents also were trapped in June, 1988 on three of the Pontine islands, located in the Mediterranean Sea between Rome and Naples. These islands are inhabited year around (Ponza), seasonally (Palmarola), or are practically deserted (Zannone). Rats and mice were captured using live-traps. Rats were anesthetized with chloroform and bled by cardiac puncture, mice were anesthetized and bled from the retroorbital sinus. Serum was separated from clot by centrifugation at 1500 rpm/15 min and were stored at −20°C until they were tested.

Antigens and serological techniques

A modification of a published indirect immunofluorescence test [4] was used to detect IgG antibodies against hantaviruses. Sera from rodents or humans initially were diluted 1:16 and 1:32 in phosphate buffered saline (PBS), pH 7.4. Vero E-6 cells were infected with one of the following hantaviruses and used to prepare spot slides: prototype Hantaan (HTN) virus (strain 76-118), Puumala (PUU) virus (strain Hällnäs B1), Prospect Hill (PH) virus (strain MP-40), and Seoul (SEO) virus (strain 80-39). Cells were used for preparation of spot slides when 30 to 50% were infected. Uninfected Vero E6 cells were included as negative controls. Human serum samples were considered positive if typical fluorescence was observed at 1:16 or greater dilution, whereas rodent serum samples were considered positive only at dilutions of 1:32 or greater.

Results

Healthy residents

Thirty-seven (2.3%) of 1583 serum samples from healthy individuals resident of central or northern Italy contained low titer (16–32) antibodies to HTN virus (Table 1). No medical history of hemorrhagic fever or renal disease was determined for these seropositive individuals. Of people with antibody, most were between 50 and 70 years of age; there was no significant difference in seroprevalence between males and females.

High risk subjects

Among 486 selected individuals at ostensibly high risk of acquiring infection or having had infections with hantaviruses in Rome, antibodies to HTN virus were found in 10.7% of foresters (three with titers as high as 128), 10% of mammalogists, 5.9% of renal dialysis patients, 4.7% of farmers, and none of the other 158 people (rodent control personnel, oarsmen, garbage disposal workers, and river police) tested. Because not all serum samples and not all spot slides were simultaneously available for testing, all samples were tested for antibody to HTN virus but not all samples were tested for antibody to the three other hantaviruses. When serum samples with antibody to HTN virus were titrated with spot slides containing HTN, PUU, PH, and SEO viruses, titers were four- to eight-fold higher to HTN than to the other viruses, with antibody to PUU virus next in titer rank.

Rodents

In Rome, *Rattus norvegicus* was the predominant rodent species captured and had the highest seroprevalence and antibody titers to SEO and HTN

Table 1. Antibody to Hantaan virus in 1583 humans in Italy

Status	No. with antibody/ no. tested	%
Healthy residents	37/1583	2.3
"High risk" residents		
Foresters (Cadore)	7/65	10.7
Mammalogists (Rome)	2/20	10.0
Dialysis patients (Rome)	3/51	5.9
Farmers (Cadore)	9/192	4.7
Trappers (Rome)	0/66	0
Oarsmen (Rome)	0/58	0
Garbage disposal workers (Rome)	0/21	0
River police (Rome)	0/13	0
Total	21/486	4.3

Table 2. Antibody to four hantaviruses in rodents collected in Rome

Species	Antibody to virus[a]			
	Hantaan	Seoul	Puumala	Prospect Hill
Rattus norvegicus	26/50 (52)	17/26 (65)	18/48 (37)	7/24 (29)
R. rattus	3/17 (18)	3/11 (27)	3/17 (18)	1/6 (17)
Mus musculus	6/31 (19)	1/11 (9)	3/31 (10)	3/20 (15)
Apodemus spp.	0/4	not tested	0/4	0/4

[a] Number with antibody/number tested (%)

viruses (Table 2). We found antibody to HTN virus in 52% of *R. norvegicus*, 17% in *R. rattus*, and 19% in *Mus musculus*. Antibody titers ranged from a maximum of 4096 to SEO virus to a maximum of 256 to HTN virus, with lower titers to other hantaviruses. Antibody rates were lower in other rodent species; antibodies were not detected in four *Apodemus sylvaticus*. Antibodies to HTN, SEO, or PUU viruses were not detected in 32 rodents (14 *Mus musculus* and 18 *Rattus* spp.) trapped on the Pontine islands.

Discussion

Although no cases of hemorrhagic fever with renal syndrome have been recorded in Italy, antibody to hantaviruses have been reported in people from central, northern [7], and southern parts of this country [5]. In addition, serologic evidence of HTN or related viruses in wild rodents has recently been demonstrated [7, 8].

Rodents are present in high numbers in Rome and constitute a public health problem as well as an esthetic one. Antibodies to hantaviruses have been detected in 14.6% of *R. norvegicus*, 9% of *R. rattus*, and 5.8% of *A. sylvaticus* in various regions of Italy and in 11 of 22 *Rattus* spp. in Tuscany (P. Verani, pers. comm., 1988). In addition, antibodies to hantaviruses have been found in 10 of 40 laboratory rats in Sicily [6].

Despite the high prevalence of hantavirus infection in rats captured in Rome, antibodies against HTN and related viruses were not detected in rodent control personnel and only 2 of 20 mammalogists had antibody to HTN virus (titer 32). This unexpected result prompted us to carry out an additional epidemiologic study of these workers to clarify this finding. Mammalogists, by the very definition of the term, would be expected to be in close contact with rodents [11], but the rodent control personnel tested in Italy did not have direct contact with animals because their work consisted of killing rodents with poisoned bait (M. Nuti, pers. obs.).

Serum samples from patients with renal disease frequently have been included in studies of hantavirus prevalence. However, the results of different surveys have ranged from negative [1] or very low prevalence (0.7% in Belgium [10]), to low (2.2% in U.S.A. [2]; 3.1% in Greece [3]; 4% in Hong Kong [9]) or high (39% in Germany; L. Zöller, pers. comm., 1986). Of the 51 dialysis patients in our study, antibody to HTN virus was found in 3 of 51, a value that does not differ significantly from the 5.6% positive found in a rural area north of Rome [8] but lower than the 13% reported for renal dialysis patients in Sicily [6].

At the outset of our studies, the high seroprevalence of hantavirus infections in wild rodents trapped in Rome led us to postulate than HTN virus infections might also be high in human residents of the city. However, the results reported here indicate that the rate of antibody is low even in individuals in ostensibly high risk groups. These results, in tandem with those of other studies in rural and forested areas of central and northern Italy (M. Nuti, unpubl. data) indicate that the risk of acquiring hantavirus infection in Italy is determined more by geography than by occupation. In fact, most of the people with antibody in our study were more than 50 years of age and live in rural areas. Similarly, in Tuscany, where clinically-suspected hemorrhagic fever with renal syndrome has occurred, HTN virus infections appear to be more frequent in rural areas, where antibody rates approximated 50% (P. Verani, pers. comm., 1988).

That the antibody prevalence rate to HTN virus in humans was higher than that for other hantaviruses is not surprising, in view of the fact that we used HTN virus for the original screening tests. However, the use of the same HTN virus antigen for screening rat serum samples should have similarly selected for antibody to this virus in rodents. In fact, when titrated for antibody to the four hantaviruses, titers in rats were higher to SEO virus than to HTN, PH or PUU virus. Whether this indicates a greater sensitivity of the SEO antigen in detecting hantavirus group-specific antigen or reveals infections with a SEO-like virus remains to be determined. Obviously, additional studies of hantaviruses in Italy are needed if we are to understand the complex epidemiology and clinical significance, if any, of human and rodent infections with viruses of this serogroup.

References

1. Brummer-Korvenkontio M, Vaheri A, Hovi T, von Bonsdorff C-H, Vuorimies J, Manni T, Pentinnen K, Oker-Blom N, Lahdevirta J (1980) Nephropathia epidemica: detection of antigen in bank voles and serologic diagnosis of human infection. J Infect Dis 141: 131–134
2. Gibbs CJ Jr, Takenaka A, Franko M, Gajdusek DC, Griffin MD, Childs J, Korch GE, Wartzok D (1982) Seroepidemiology of Hantaan virus. Lancet 2: 1406–1407
3. Lee HW, Antoniadis A (1981) Serological evidence for Korean haemorrhagic fever in Greece. Lancet 1: 832
4. Lee HW, Lee PW, Johnson KM (1978) Isolation of the etiologic agent of Korean hemorrhagic fever. J Infect Dis 137: 298–308

5. Mansueto S, Peters CJ, Tringali G, Rini GB, Vitale G, Quartarano P, Di Rosa S, Maggio AM (1987) Prime ricerche sulla presenza di anticorpi anti-virus Hantaan in Sicilia Occidentale. G Mal Infet Parassit 39: 99–101
6. Mansueto S, Rini GB, Vitale G, Tringali G, Di Rosa S, Pintagro C, Peters CJ, Maggio AM (1987) Febbre emorragica con sindrome renale e ratti di allevamento. Ricerca di anticorpi anti-virus Hantaan in uno stabulario. G Mal Infet Parassit 39: 102–103
7. Nuti M (1989) La febbre emorragica con sindrome renale in Italia. Oplitai 2: 23–34
8. Nuti M, Amaddeo D, Costa M, Cristaldi M, Ieradi LA, Montebarocci M (1989) Anticorpi contro Hantavirus e Leptospire in soggetti a rischio in Roma. Boll Inst Sieroter (Milan) 68: 284–288
9. Shortridge KF, Lee HW, LeDuc JW, Wong TW, Chau GW, Rosen L (1987) Serological evidence of Hantaan-related viruses in Hong Kong. Trans R Soc Trop Med Hyg 81: 400–402
10. van der Groen G, Piot P, Desmyter J, Colaert J, Muylle L, Tkachenko EA, Ivanov AP, Verhagen R, van Ypersele de Strihou C (1983) Seroepidemiology of Hantaan-related virus infections in Belgian population. Lancet 2: 1493
11. Yanagihara R, Gajdusek DC, Gibbs CJ Jr, Traub R (1984) Prospect Hill virus: serological evidence for infection in mammalogists. N Engl J Med 310: 1325

Authors' address: M. Nuti, Department of Tropical Diseases, Policlinico Umberto 1, I-00161 Rome, Italy.

Arch Virol (1990) [Suppl 1]: 87–94

Evidence of the presence of two hantaviruses in Slovenia, Yugoslavia

**Tatjana Avšič-Županc[1], M. Likar[1], Suzana Novaković[1], Borut Čižman[2],
Alenka Kraigher[3], G. van der Groen[4], R. Stojanović[5], Mirčeta Obradović[5],
Ana Gligič[6], and J. W. LeDuc[7]**

[1] Institute of Microbiology, Medical Faculty of Ljubljana, Ljubljana,
[2] Department of Nephrology, University Clinical Center, Ljubljana, and
[3] Institute of Public Health and Social Welfare, Ljubljana, Yugoslavia
[4] Institute of Tropical Medicine, Antwerp, Belgium
[5] Military Medical Academy, Belgrade, and
[6] Institute of Immunology and Virology, Belgrade, Yugoslavia
[7] U. S. Army Medical Research Institute of Infectious Diseases, Fort Detrick, Frederick,
Maryland, U.S.A.

Accepted March 27, 1990

Summary. Thirty-three cases of hemorrhagic fever with renal syndrome (HFRS) have been serologically confirmed in Slovenia during the last five years. The clinical picture varied from mild to severe, with a mortality rate of 3%. Serum samples from 240 patients suspected of having HFRS were tested by immunofluorescence using four different hantaviral antigens. Three reactivity patterns were observed. Using capture enzyme immunoassay, IgM was detected in 100% acute-phase serum samples of patients. The distribution of hantaviral infections in small mammals was examined in two natural foci of HFRS. Two hantaviruses were found in these mammals. In one area, *Clethrionomys glareolus* was the predominant species, and most of their sera reacted to highest titer with Puumala virus; mild illness was diagnosed in this area. In the second location, *Apodemus flavicollis* predominated, and most of their sera reacted to highest titer with Hantaan virus; severe illness was diagnosed in this area.

Introduction

Hemorrhagic fever with renal syndrome (HFRS) is an acute infectious disease with a variety of clinical manifestations [4, 21]. It is caused by viruses belonging to the family *Bunyaviridae*, genus *Hantavirus*, [15]. The prototype of this group is Hantaan virus, isolated in Korea in 1976 [9]. The severe form of HFRS is more often seen in the Far East [10]. In Europe, the mild form, known as nephropathia epidemica (NE) is found in Scandinavia [7] and western Europe [20]. Evidence for infection of wild, small mammals and of humans by hantaviruses has been reported nearly worldwide. Recognition of the number of small mammal reservoirs for hantaviruses has dramatically increased [19].

In Yugoslavia the first HFRS case was reported in 1952 [16], and several epidemics occurred between 1962 and 1986 [6]. The presence of HFRS in Slovenia, in northwestern Yugoslavia, was first reported in 1989 [1].

In this paper we report hantaviral antigen in lungs and antibodies to hantaviruses in sera of small mammals in three natural foci of HFRS in Slovenia. In addition, we present data on the prevalence of antibodies to these viruses in human sera, data indicating that at least two hantaviruses are present in Slovenia.

Materials and methods

Trapping procedure and locations

Small mammals were live-trapped in fields and forests surrounding villages in which HFRS cases had been diagnosed between 1985 and 1989. Study sites included Murska Sobota in northeast Slovenia, Novo mesto and Črnomelj in southern Slovenia, and a third location, Vipava, as an area where, to now, there is no evidence of hantaviral infections in humans (Fig. 1). Animals were identified by species and exaguinated in the field. Sera were diluted

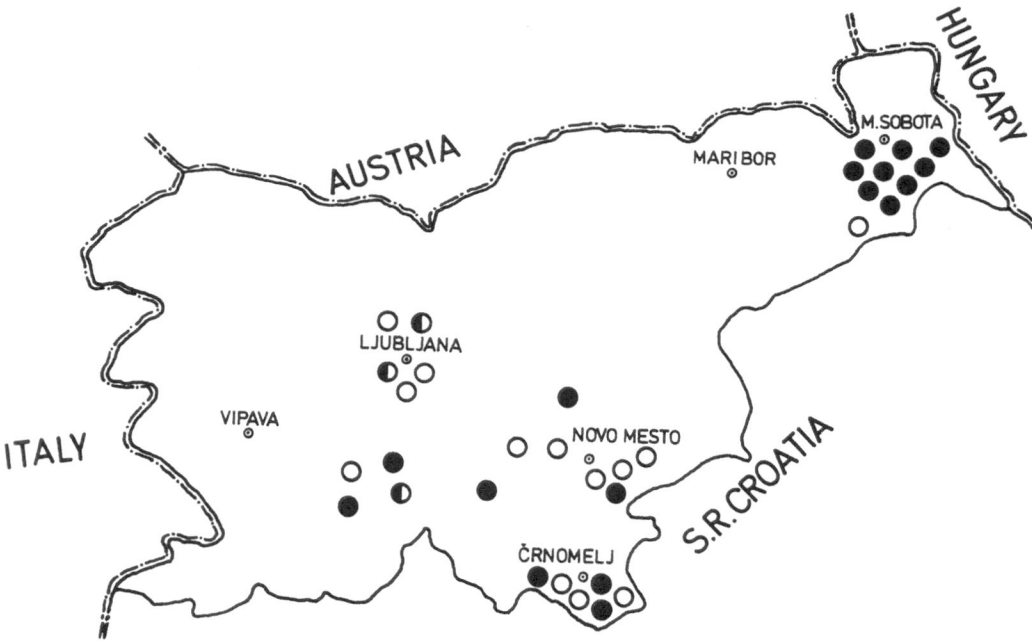

Fig. 1. Geographic distribution of 33 HFRS patients in Slovenia. ● Hantaan serotype,
○ Puumala serotype, ◖ Unidentified serotype

: 16 with phosphate buffer saline (PBS) pH 7.2 and stored at −20 °C until they were tested for antibody. Lung tissues were removed aseptically and stored in liquid nitrogen.

Trap sites in Murska Sobota consisted of corn and grape fields. Grasslands, sedges and bogs are characteristic for this part of Slovenia. The area of Novo mesto and Črnomelj is characterized by hilly-woodland area with rivers between.

Detection of hantavirus antigen

Cryostat-cut sections (4 µm thick) of lung tissue from small mammals were examined for the presence of hantaviral antigen by an indirect immunofluorescent antibody technique (IFA) [2]. Enzyme-linked immunosorbent assay (ELISA) for detection of hantavirus antigen in lung suspensions was also used [12].

Serological techniques

Human and animal sera were tested at a single 1:16 dilution by IFA using spot slides of Vero E-6 cells infected with hantaviruses as described previously [18]. Hantaviruses used in the study were: prototype Hantaan (strain 76-118), Puumala (strain Hällnäs B1), Prospect Hill (strain Prospect Hill 1) and Seoul (strain SR-11). Sera producing characteristic stippled fluorescence in the cytoplasma of infected cells were considered positive. All positive sera were diluted 2-fold and titrated for end-points; titers were expressed as the reciprocal of the highest dilution of serum giving typical focal cytoplasmic staining. Human IgM antibodies to Hantaan virus (HTN) and Puumala virus (PUU) were measured using an enzyme immunosorbent assay according to methods published previously [12–14]. An immunoblot method as described [1], was used for confirmation.

Results

Human serology

Between August 1985 and October 1989, 320 acute and/or convalescent sera were tested from 240 patients with clinical disease compatible with HFRS in Slovenia. Sera were stored at −20 °C until they were tested for antibody. Since then, 33 clinically diagnosed cases of HFRS have been serologically confirmed. Clinical pictures varied from mild to severe; the overall mortality rate has been 3% [3].

Sera from 240 patients suspected of having HFRS were examined by IFA using four hantaviruses and 33 cases of HFRS were serologically confirmed. Three reactivity patterns were observed. Table 1 presents nine examples of these patterns. Highest titers to HTN antigen and lower titers to PUU antigen were found in patients 1–3. Higher titers to PUU than to HTN virus were seen in patients 4–6. Sera of patients 7–9 reacted with almost equal (within two-fold) titers to all hantaviral antigens used. IFA positive human sera were confirmed by immunoblotting.

Patients from Novo mesto generally had more severe HFRS than patients residing elsewhere in Slovenia. Sera from two patients (Table 2, patients A and B) with severe disease collected at different time intervals and

Table 1. IFA antibody to four hantaviruses in serum samples from nine representative patients with HFRS in Slovenia

Patient	Reciprocal fluorescent antibody titer to				IB[b]	
	HTN[a]	PUU	SEO	PH	HTN	PUU
1	256	16	256	64	+[c]	NT[d]
2	1024	128	2048	512	+	+
3	2048	64	1024	16	+	−
4	16	2048	512	1024	−	+
5	64	2048	128	256	NT	+
6	256	2048	32	512	−	+
7	64	64	64	32	+	−
8	256	256	256	256	+	NT
9	2048	2048	1024	512	+	+

[a] HTN Hantaan
PUU Puumala
SEO Seoul
PH Prospect Hill
[b] *IB* Immunoblot
[c] +Positive reaction of a convalescent-phase patients serum with the electrophoretically blotted 50 kDa nucleoprotein of HTN or PUU virus on vitro cellulose strips. Serum dilution was 1/50
[d] *NT* Not tested

Table 2. IgM and IgG antibody to Hantaan and Puumala viruses in serum samples from two patients with HFRS in Murska Sobota and from six patients from Novo Mesto

Patient (days after onset)	Antibody titer			
	EIA (IgM)		IFA (IgG)	
	HTN	PUU	HTN	PUU
A (10)	>12800	200	512	128
(30)	6400	NT[a]	1024	128
(60)	800	NT	1024	32
B (6)	6400	<100	256	<16
(14)	>25000	400	1024	64
(90)	1600	<100	512	32
C (7)	200	6400	<16	1024
D (13)	800	3200	256	1024
E (21)	1600	3200	64	2048
F (14)	400	1600	32	2048
G (20)	100	400	128	512
H (14)	800	3200	64	2048

[a] *NT* Not tested

tested for hantaviral antibody, had significantly higher IgM and IgG titers to HTN virus than to PUU virus. Patients from Murska Sobota generally suffered a milder clinical picture and when their sera were tested by IFA their reactivity was higher to PUU antigen than to HTN. The same reactivity pattern was seen with IgM antibody (Table 2, patients C–H).

Small mammal survey

Of 93 animals collected in Murska Sobota, 12 (13%) had antibody by the IFA test. The highest positive rate was found in bank voles, *C. glareolus* (10/34). Hantaviral antigen was found in 18 (19%) animals (Table 3).

In Novo mesto and Črnomelj, animals were collected three times during 1987 and 1989. Seventeen (12%) of 146 animals from this region had IFA antibodies to hantaviruses (Table 3). Antigen prevalence in small mammals (24/146 = 15%) was slightly lower than in mammals from Murska Sobota (18/93 = 19%). *Apodemus flavicollis* was the predominant species infected with hantaviruses in Novo mesto and Crnomelj.

Although HFRS has yet to be diagnosed in residents of Vipava, seroprevalence (25%) and antigen prevalence (28.1%) of small mammals was

Table 3. Prevalence of hantaviral antibodies and antigens in small mammals captured at three locations in Slovenia

Species	Location								
	Murska Sobota			Novo Mesto/ Črnomelj			Vipava		
	T[a]	Ab[b]	Ag[c]	T	Ab	Ag	T	Ab	Ag
Rodents									
Apodemus flavicollis	21	2	4	75	12	16	6	1	4
A. sylvaticus	23	0	1	32	2	4	6	1	0
A. microps	2	0	0	1	0	0	0		
A. agrarius	2	0	0	0			7	4	2
Clethrionomys glareolus	34	10	11	6	0	1	0		
Mus musculus	3	0	2	14	3	3	13	2	3
M. agrastris panonicus	4	0	0	0			0		
Pitymys subterraneus	0			1	0	0	0		
Insectivores									
Crocidura suaveolens	1	0	0	10	0	0	0		
Sorex araneus	2	0	0	1	0	0	0		
Sorex spp.	1	0	0	0			0		
Total	93	12	18	146	17	24	32	8	9
% positive		13%	19%		12%	15%		25%	28%

[a] *T* Number of animals tested

[b] *Ab* Number with antibody

[c] *Ag* Number with antigen

high (Table 3). *Mus musculus* was the most abundant species in this region, but *A. agrarius* and *A. flavicollis* were the species most frequently found infected.

Discussion

Hemorrhagic fever with renal syndrome is a classical clinical syndrome, the etiology of which can be confirmed by virus isolations and antibody studies. Differences between clinical manifestations of this disease in different areas of the world are associated with the strain of hantavirus present. In Asia and the far eastern U.S.S.R. the disease is generally more severe than that observed in western Europe and Scandinavia. Korean hemorrhagic fever and epidemic hemorrhagic fever are severe forms of the disease and are associated with Hantaan virus; nephropathia epidemica, a milder form of the disease is associated with Puumala virus; a less severe form has been attributed to Seoul virus. In Yugoslavia, sporadic cases, small outbreaks, or extensive epidemics occur essentially every year in natural foci of HFRS [6]. Results of a previous study provide substantial evidence for the occurrence of HFRS in Slovenia [1].

Studies of antigen classification of hantaviruses has been done using a variety of methods [11, 17]. Cross-plaque reduction neutralization and IFA tests have permitted serotype classification of hantavirus infections in humans and rodents. Although antibody response to the four serotype (HTN, SEO, PUU, and PH) were sufficiently distinct to permit classification of hantavirus infection in most of our HFRS patients, results with some sera (Table 1) were not clear. Lee et al. [11] described similar observations with sera from HFRS patients from southern Yugoslavia. Based on the geographical distribution of the HFRS patients in Slovenia, it is clear there are at least two natural foci of this disease (Fig. 1).

Almost all patients from Murska Sobota had milder clinical course of the disease, similar to that described for NE in Scandinavia, for which the etiologic agent is PUU virus [7, 13]. More recently, HFRS due to Puumala or a closely related virus has been recognized in France, Belgium, Federal Republic of Germany, and southern Yugoslavia [5, 19].

In Novo mesto and Črnomelj rodents were captured around patients' homes and in the surrounding woods, and *A. flavicollis* and *A. sylvaticus* were the most abundant species captured. *A. flavicollis* was the dominant species infected with hantaviruses. Nine (64%) HFRS cases in this region of Slovenia showed a severe clinical picture with pronounced hemorrhagic manifestations. The one patient who died, did so on the fifth day after onset. When sera from patients with severe HFRS were tested for antibody to hantaviruses, a significant rise of titers was observed by ELISA (IgM) and IFA (IgG) test to Hantaan antigen.

A. flavicollis is widely distributed throughout the region where Balkan HFRS cases have been recorded. Our results support the presumption that this species is the principal maintenance host for a HTN-like virus, which causes severe forms of HFRS [5, 8].

In Vipava we found high antibody and antigen prevalence in *A. agrarius* and *A. flavicollis*. These data suggest that some inapparent infections with hantaviruses probably exists here also.

Our results provide evidence that at least two hantaviral serotypes circulate simultaneously in Slovenia. In geographic areas where more than one hantavirus circulates, it is important to distinguish between the two, because the prognosis for patients is quite different. Therefore, it is necessary to use diagnostic methods that provide adequately discriminatory results.

Acknowledgements

We gratefully acknowledge G. Hoofd (Institute of Tropical Medicine and Hygiene, Antwerpen, Belgium) for excellent technical assistance.

References

1. Avšič-Županc T, Čižman B, Gligič A, Hoofd G, van der Groen G (1989) Evidence for hantavirus disease in Slovenia, Yugoslavia. Acta Virol 33: 327–337
2. Chumakov MP, Gavrilovskaya IN, Boiko VA (1981) Detection of hemorrhagic fever with renal syndrome (HFRS) virus in the lungs of bank voles (*Clethrionomys rutilus*) trapped in HFRS foci in European part of the U.S.S.R. and serodiagnosis of this infection in man. Arch Virol 69: 295–300
3. Čižman B, Ferluga D, Kaplan-Pavlovcic S, Koselj M, Drinovec J, Avšič T (1989) Renal involvement in hantavirus disease. In: Amerio A, Caretelli P, Campese VM, Massry SG (eds) Drugs, systemic disease, and the kidney. Plenum, Washington, DC, pp 173–180
4. Desmyter J, van Ypersele de Stichou C, van der Groen G (1984) Hantavirus disease. Lancet 2: 158
5. Gligič A, Obradovič M, Stojanovič R, Hlaca D, Antonijevič B, Arnautovič D, Gaon J, Fursič M, Lee PW, Goldgaber D, Yanagihara R, Gibbs CJ Jr, Gajdusek DC, Svedmyr A (1989) Hemorrhagic fever with renal syndrome in Yugoslavia: detection of hantaviral antigen and antibody in wild caught rodents and serological diagnosis of human disease. Scand J Infect Dis 20: 261–266
6. Gligič A, Obradovič M, Stojanovič R, Vujosevič N, Ovcarič A, Frusič M, Gibbs CJ Jr, Calisher CH, Gajdusek DC (1989) Epidemic hemorrhagic fever with renal syndrome in Yugoslavia, 1986. Am J Trop Med Hyg 41: 102–108
7. Lahdevirta J (1971) Nephropathia Epidemica in Finland. A clinical, histological and epidemiological study. Ann Clin Res 3: 1–154
8. LeDuc JW, Antoniadis A, Siamopoulus K (1986) Epidemiological investigations following an outbreak of hemorrhagic fever with renal syndrome in Greece. Am J Trop Med Hyg 35: 654–659
9. Lee HW, Lee PW, Johnson KM (1978) Isolation of the etiologic agent of Korean hemorrhagic fever. J Infect Dis 137: 298–308
10. Lee HW (1982) Korean hemorrhagic fever. Prog Med Virol 28: 96–113
11. Lee PW, Gibbs CJ, Gajdusek CD, Yanagihara R (1985) Serotypic classification of hantaviruses by indirect immunofluorescent antibody and plaque reduction neutralization tests. J Clin Microb 22: 940–944

12. Meegan JM, LeDuc JW (1989) Enzyme immunoassays. In: Lee HW, Dalrymple JM (eds) Manual of hemorrhagic fever with renal syndrome. WHO Collaborating Center for Virus Reference and Research (HFRS), Institute for Viral Diseases, Korea University, Seoul, pp 83–87
13. Niklasson B, LeDuc JW (1987) Epidemiology of nephropathia epidemica in Sweden. J Infect Dis 155: 269–276
14. Niklasson B, Kjelsson T (1988) Detection of nephropathia epidemica (Puumala virus) specific immunoglobulin M by enzyme immunosorbent assay. J Clin Microbiol 26: 1519–1523
15. Schmaljohn CS, Hasty SE, Dalrymple JM, LeDuc JW, Lee HW, von Bonsdorff C-H, Brummer-Korvenkontio M, Vaheri A, Tsai TF, Regnery HL, Goldgaber D, Lee PW (1985) Antigenic and genetic properties of viruses linked to hemorrhagic fever with renal syndrome. Science 227: 1041–1044
16. Simic M, Miric V (1952) Uspela primena peritonealne dijalize kod jednog slucaja bubrezne insuficijencije. Vojnosanit Pregl 9: 285–290 (Summary in English)
17. Sugiyama K, Morikawa S, Matsura Y, Ikachenko EA, Morita C (1987) Four serotypes of hemorrhagic fever with renal syndrome viruses identified by polyclonal and mono-clonal antibodies. J Gen Virol 68: 979–987
18. van der Groen G, Piot P, Desmyter J (1983) Seroepidemiology of Hantaan related virus infection in Belgium populations. Lancet 2: 1943–1944
19. van der Groen G (1985) Haemorrhagic fever with renal syndrome: recent developments. Ann Soc Belg Med Trop 65: 121–156
20. van Ypersele de Strichou C, van der Groen G, Desmyter J (1986) Hantavirus nephropathy in western Europe. Ubiquity Nephrol 15: 143–171
21. Yanagihara R, Gajdusek DC (1988) Hemorrhagic fever with renal syndrome. A historical perspective and review of recent advances. In: Gear JHS (ed) Handbook of viral and rickettsial hemorrhagic fevers. CRC Press, Boca Raton, FL, pp 155–181

Authors' address: Dr. Tatjana Avšič-Županc, Institute of Microbiology, Medical Faculty of Ljubljana, Zaloska 4, YU-61105 Ljubljana, Yugoslavia.

Arch Virol (1990) [Suppl 1]: 95–100

Panhypopituitarism
in the acute stage of hemorrhagic fever with renal syndrome: a case report

J. S. Lee[1], J. S. Han[1], S. Kim[1], and K.-H. Chang[2]

Departments of [1] Internal Medicine and of [2] Radiology, College of Medicine, Seoul National University, Seoul, Korea

Accepted February 22, 1990

Summary. Necrosis and hemorrhage have been observed in the anterior lobe of the pituitary glands of humans who died in the acute stage of hemorrhage fever with renal syndrome (HFRS). This suggests that patients in the acute stage of this severe disease may have had pituitary apoplexy and subsequent residual pituitary hypofunction. We report an acute case of HFRS with panhypopituitarism, no response to a combined anterior pituitary stimulation test, and with pituitary hemorrhage detected by magnetic resonance imaging.

Introduction

Necrosis and hemorrhage in the anterior lobe of the pituitary gland has been observed in autopsy materials from humans in the acute stage of hemorrhagic fever with renal syndrome (HFRS) [6, 8]. Pituitary atrophy was observed by computerized axial tomography (CT) scan of the sella turcica in the late stage or after recovery from HFRS [5]. These findings suggest that patients who were in the acute stage of severe HFRS may have had pituitary apoplexy [3] and subsequent permanent residual pituitary hypofunction. However, reports of hypopituitarism following HFRS are rare, and for most of these patients complete endocrinologic or anatomical evaluations of the pituitary gland were not available [1, 7, 9, 10].

We report a case of acute stage HFRS with panhypopituitarism as determined by a combined anterior pituitary stimulation test and pituitary hemorrhage as determined by magnetic resonance imaging.

Report of a case

A 45-year-old Korean male had been well until 9 days prior to admission, when he developed sudden fever, chills and generalized myalgia, followed by nausea, vomiting and headache. He was admitted to a community hospital on the 5th day of illness. During admission, the patient complained of frequent loose stools, abdominal pain and headache. Transient loss of consciousness, low blood pressure and oliguria were observed on the 5th day of illness. He was transferred to a nearby university hospital and diagnosed as having HFRS and treated with hemodialysis for acute renal failure on the 8th day of illness. Abdominal pain with tenderness and distension, altered consciousness and hypotension developed. He was referred to Seoul National University Hospital on the 9th day of illness.

There was no history of tuberculosis, syphilis, diabetes mellitus, sarcoidosis, trauma, surgery or radiation therapy. On physical examination, he was drowsy and irritable. His blood pressure was 160/100 mm Hg, pulse 72/min and body temperature 36.5 °C. Puffy face and eyelid edema were observed. Petechiae on the axillae and a huge hematoma on the right groin were noted; melena was observed. The heart was normal, but crackling rales were heard on both lower lung fields. The abdomen was distended and abdominal tenderness was noted. Neurologic examination was negative. The 24 h urine output was 1,000 ml.

Laboratory findings were: Hemoglobin was 68 g/l, leukocytes $9.3 \times 10^9/l$ with 74% neutrophils, 23% lymphocytes and 3% monocytes, platelets $32 \times 10^9/l$. Prothrombin time was within normal range and partial thromboplastin time was 29 s. The urea nitrogen was 72.4 mmol/l, creatinine 1270 µmol/l, glucose 7.6 mmol/l, uric acid 1110 µmol/l, bilirubin 12 µmol/l, calcium 1.95 mmol/l, phosphorus 1.74 mmol/l, and protein 52 g/l (albumin 25 g/l). Serum sodium was 138 mmol/l, serum potassium 3.9 mmol/l, serum chloride 98 mmol/l, and blood carbon dioxide 19 mmol/l. The serum alanine and aspartate aminotransferase were 22 and 25 U/l, respectively. The urine contained protein (3+) but not glucose; the sediment contained many erythrocytes and 3–5 leucocytes per high-power field. An electrocardiogram showed a normal pattern and a chest X-ray revealed pulmonary congestion. Serologic tests for antibody to hantaviruses (indirect immunofluorescent antibody test) were positive with high titers (>1:1280) on the 9th and 14th day of illness. During his hospitalization, the patient received hemodialysis on 5 different occasions. In spite of diuresis, he complained of severe asthenia.

On the 10th day of illness, when the creatinine level was 1270 µmol/l, endocrinological studies were done. The basal 8 AM levels of hormonal studies were: T_3 uptake was 0.36, T_3 0.4 nmol/l, T_4 22 nmol/l, growth hormone (GH) below 1.0 µg/l, TSH 3.3 mU/l, cortisol 80 nmol/l and prolactin 2 µg/l. On the 20th day of illness, when the creatinine level was 710 µmol/l, the results of the hormonal assay at 8 AM were: T_3 uptake 0.18, T_3 0.4 nmol/l, T_4 29 nmol/l, GH below 1.0 µg/l, cortisol 80 nmol/l and prolactin 2 µg/l. To evaluate pituitary function, combined pituitary stimulation test, consisting of the simultaneous intravenous administration of 0.1 U/kg of regular insulin, 400 µg of thyrotropin releasing hormone (TRH) and 100 µg of luteinizing hormone releasing hormone (LH-RH), was performed on the 22nd and 26th day of illness. The results of these tests are shown in the Table 1. For morphologic evaluation of the pituitary gland, both sellar magnetic resonance (MR) imaging and postcontrast CT scan were performed on the 17th day of illness. The MR imaging showed a high-signal intensity in the anterior pituitary gland on T_1-weighted image, indicating hemorrhage (Table 2 and Fig. 1). The CT scan revealed a low density in the same area. On the two follow-up MR images, 20 days and 43 days after the first MR study, the high-intensity signal disappeared (Table 2).

The patient was discharged on the 40th day of illness with serum creatinine level of 190 µmol/l. After discharge, asthenia, weakness, dizziness continued and loss of body hair was observed. When he was readmitted to our hospital for further evaluation of pituitary

Table 1. Hormonal response to combined TRH, LH-RH, and insulin-induced hypoglycemia test

Day of illness	Phase	Urine volume (ml/day)	Serum creatinine (µmol/l)	Hormone[a]	Minutes after stimulation				
					0	30	60	90	120
22	late diuretic	4,300	540	Glucose (mmol/l)	3.8	0.5	4.2	1.8	5.1
				TSH (mU/l)	<0.05	<0.05	<0.05	—[b]	—
				PRL (µg/l)	2.9	3.2	2.6	—	—
				GH (µg/l)	<1.0	<1.0	<1.0	<1.0	<1.0
				FSH (IU/l)	1.3	—	2.9	1.6	1.7
				LH (IU/l)	3.0	5.4	5.9	—	—
				Cortisol (nmol/l)	50	90	100	90	—
26	late diuretic	6,500	310	Glucose (mmol/l)	4.2	1.0	1.3	9.1	3.9
				TSH (mU/l)	<0.05	<0.05	<0.05	—	—
				PRL (µg/l)	3.1	2.7	2.9	—	—
				GH (µg/l)	<1.0	—	<1.0	<1.0	<1.0
				FSH (IU/l)	1.3	—	2.9	1.6	1.7
				LH (IU/l)	3.0	5.4	5.9	—	—
				Cortisol (nmol/l)	30	—	90	100	90
80	recovery	4,000	130	Glucose (mmol/l)	3.8	1.5	4.0	4.2	6.6
				TSH (mU/l)	<0.05	0.1	0.1	—	—
				PRL (µg/l)	2.2	3.2	2.6	—	—
				GH (µg/l)	<1.0	—	<1.0	<1.0	<1.0
				FSH (IU/l)	4.4	—	4.6	5.8	2.9
				LH (IU/l)	5.3	7.1	6.9	—	—
				Cortisol (nmol/l)	170	200	180	230	—

[a] *TSH* thyroid stimulating hormone; *PRL* prolactin; *GH* human growth hormone; *FSH* follicle-stimulating hormone; *LH* luteinizing hormone

[b] — Not done

Table 2. Sequential changes of the anterior pituitary gland on sellar MR imaging and postcontrast CT scans

Day of illness	MR imaging	CT
17	Pituitary hemorrhage (high signal intensity on T_1-weighted image)	Resolving Pituitary hemorrhage
37	Disappearance of hemorrhage	Not done
60	Disappearance of hemorrhage	Not done
83	Pituitary atrophy	Pituitary atrophy

function because of persistent weakness, the combined pituitary stimulation test, sellar MR imaging and CT scan were performed. The result of the combined pituitary stimulation test showed no response to TRH, insulin-induced hypoglycemia or LH-RH. MR imaging and CT scan showed partially empty sella with pituitary atrophy (Table 2 and Fig. 1). After daily treatments with 5 mg of prednisolone and 0.125 mg of thyroxine, his symptoms improved.

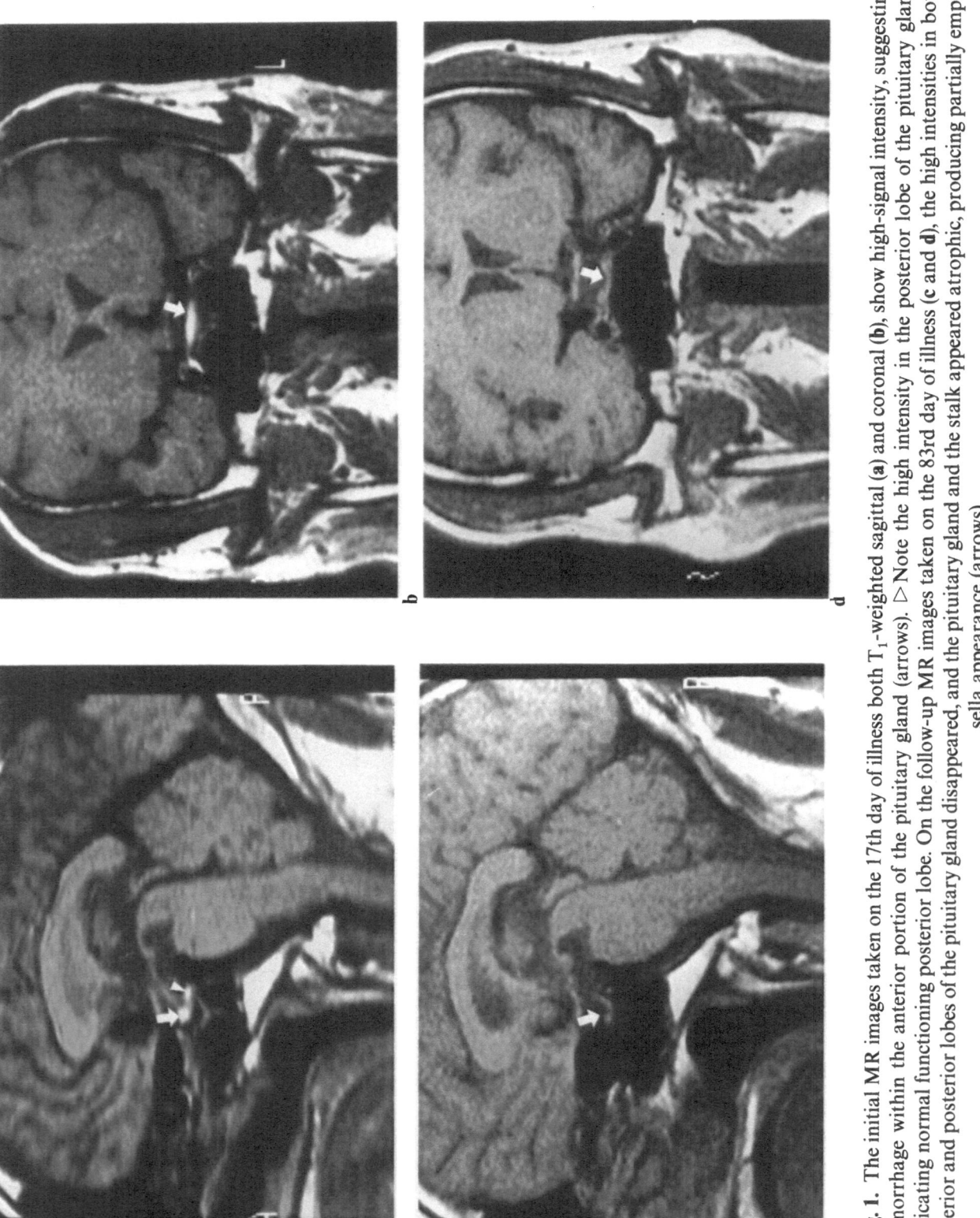

Fig. 1. The initial MR images taken on the 17th day of illness both T$_1$-weighted sagittal (**a**) and coronal (**b**), show high-signal intensity, suggesting hemorrhage within the anterior portion of the pituitary gland (arrows). △ Note the high intensity in the posterior lobe of the pituitary gland indicating normal functioning posterior lobe. On the follow-up MR images taken on the 83rd day of illness (**c** and **d**), the high intensities in both anterior and posterior lobes of the pituitary gland disappeared, and the pituitary gland and the stalk appeared atrophic, producing partially empty sella appearance (arrows)

Comments

Hemorrhagic fever with renal syndrome is an acute viral disease characterized by fever, circulatory disturbances, hemorrhagic manifestations and renal failure [4]. Most HFRS patients recover completely, except for those with rare neurologic sequelae [2, 4]. Hemorrhage and infarct-like necrosis in the pituitary gland are characteristic findings at autopsy in patients in the acute stage of this illness, although the presence and the extent of foci of necrosis in the anterior lobe of the pituitary gland depends to some extent on the stage of illness. Hemorrhage and necrosis in the anterior lobe of the pituitary gland develop in 50% to 100% of fatal cases when death occurred after the hypotensive phase [6, 8].

The mechanism of pituitary hemorrhage and necrosis in HFRS is not clear. There are hypotheses that pituitary necrosis may occur as the result of anoxemia from stasis, high viscosity, thrombosis due to disseminated intravascular coagulation and vascular collapse due to hypotension [2]. Despite this uncertainty, it is clear that severe destruction of the anterior pituitary gland will cause hypopituitarism.

Since Mayer [7] described the first case of clinical panhypopituitarism following HFRS, only a few more cases have been documented. According to previous reports, most hypopituitarism occurred after recovery from the acute phase of illness. The case presented here was unique in that hypofunction of the pituitary gland was confirmed by hormonal assay and found in the acute stage of this disease, and the morphologic abnormality of the pituitary gland was shown by sellar MR imaging and CT scan. The patient had low levels of pituitary secretory hormone and target gland hormones on the 19th day of illness and in follow-up studies. The patient was not responsive to the combined pituitary stimulation test.

The high intensity of the anterior pituitary gland by T_1-weighted MR imaging and the low density by CT scan suggest that the hemorrhage occurred at least 3 to 5 days before MR examination. This is because hemorrhage between 3 to 5 days and 8 weeks may appear as high-signal intensity by T_1-weighted image of MR due to methemoglobin in the hemorrhage, while hemorrhage of the hyperacute and chronic stage does not appear as high-signal intensity. On two follow-up MR images, the high-signal intensity of the pituitary gland observed on the initial MR imaging subsequently disappeared, suggesting absorption of the hemorrhage or changes in the chemical composition of the heme iron. By CT scan, the density of the hemorrhagic area is also variable, depending on the stage and location. The hemorrhage in resolution appears as iso- or low density by CT scan, which makes it difficult to differentiate between ischemic necrosis and hemorrhage.

In the late stage or after recovery from HFRS, the pituitary gland is not infrequently atrophied, as seen in the present case. The pituitary pathology in this case is considered similar to those described in previous reports which showed progressive pituitary atrophy by sequential follow-up sellar CT scans [5], stromal fibrosis in patients who died of HFRS in the late phase.

There were no other possible causes of pituitary failure in this patient. He had no past history of trauma, surgery, radiation therapy, other infectious and vascular diseases, and the sellar MR images and CT scans did not reveal tumors in the pituitary or parapituitary areas.

Obviously, HFRS is not the only cause of acute panhypopituitarism. Other disease entities, such as Sheehan's syndrome and viper snake bite, can which cause similar pathologic findings in the pituitary gland. However, when an acute panhypopituitarism of unknown origin occurs in the endemic areas of HFRS, the responsible physician should consider HFRS as a possible cause unless proven otherwise.

In conclusion, HFRS can induce hypopituitarism, which develops in the acute stage of infection. Physicians should be alerted to the possibility that their HFRS patients can develop hypopituitarism if signs of an endocrinological disorder begin to appear. If neglected, acute pituitary failure can be life-threatening.

References

1. Kim D (1965) Clinical analysis of 111 fatal cases of epidemic hemorrhagic fever. Am J Med 39: 218–220
2. Lee JS (1985) Korean hemorrhagic fever: hemorrhagic fever with renal syndrome. Inje Med J 6: 23–36 (in Korean)
3. Lee JS, Ahn C, Oh HY, Kim S, Lee M, Kim YI (1986) Panhypopituitarism and central diabetes insipidus as a complication of hemorrhagic fever with renal syndrome. Seoul J Med 27: 53–58
4. Lee M (1986) Korean hemorrhagic fever: hemorrhagic fever with renal syndrome, 2nd edn. Seoul National University Press, Seoul (in Korean)
5. Lim TH, Chang KH, Han MC, Chang YB, Lim SM, Yu YS, Chun YH, Lee JS (1986) Pituitary atrophy in Korean hemorrhagic fever. AJNR 7: 633–637
6. Lukes RJ (1954) Pathology of thirty-nine fatal cases of epidemic hemorrhagic fever. Am J Med 16: 639–650
7. Mayer CF (1952) Epidemic hemorrhagic fever of the far east or endemic hemorrhagic nephrosonephritis: morphology and pathogenesis. Lab Invest 1: 291–311
8. Steer A (1955) Pathology of hemorrhagic fever: a comparison of findings 1951 and 1952. Am J Pathol 31: 201–221
9. Whale GH, McKay DG (1955) Panhypopituitarism following epidemic hemorrhagic fever: pathologic findings. Ann Intern Med 43: 1320–1330
10. Zoeckler SJ, Orbinson JA (1955) Panhypopituitarism following epidemic hemorrhagic fever: clinical feature: report of a case. Ann Intern Med 43: 1316–1319

Authors' address: J. S. Lee, Department of Internal Medicine, College of Medicine, Seoul National University, 28 Yongon-dong Chongno-gu, Seoul 110-744, Korea.

Arch Virol (1990) [Suppl 1]: 101–108

Stabilized Ribavirin diphosphate analogs inhibit the vesicular stomatitis virus (Indiana) in vitro transcription reaction

M. J. Lachenmann, R. Fernandez-Larsson, and **J. L. Patterson**

Division of Infectious Diseases, Children's Hospital and Department of Microbiology and Molecular Genetics, Harvard Medical School, Boston, Massachusetts, U.S.A.

Accepted December 4, 1989

Summary. Ribavirin or Virazole (1-β-D-ribofuranosyl-1,2,4-triazole-3-car-boxamide) is a broad-spectrum antiviral agent whose molecular mode of action remains controversial. Results of earlier experiments indicated that ribavirin possesses a significant direct suppressive effect on the vesicular stomatitis virus (Indiana) (VSI) viral polymerase and that this inhibitory activity appears to block initiation of transcription. This direct effect can be mediated by the mono- and diphosphorylated forms of the drug as well as the triphosphorylated form. To confirm our previous results we examined the effects of analogs which stabilize as diphosphates on this VSI virus in vitro transcription reaction; all exhibited inhibitory activity to varying degrees. The results of these studies are discussed in relation to inhibition of primary transcription and its effect on decreased viral progeny production.

Introduction

In 1986 ribavirin was approved by the Food and Drug Administration as an aerosol for use in infants with serious infections due to respiratory syncytial virus. Ribavirin is and has been under clinical investigation against a variety of viral illnesses, including influenza, Lassa fever, hemorrhagic fever with renal syndrome, and human immunodeficiency syndrome. The drug possesses inhibitory activity against a broad spectrum of viral pathogens, including both DNA and RNA viruses [11], suggesting that it has a vast clinical potential that has yet to be realized.

Several hypotheses regarding the molecular mode of action of ribavirin have been proposed. One hypothesis is that the drug decreases intracellular pools of GTP, indirectly suppressing viral nucleic acid synthesis [10, 12]. Another proposes that ribavirin therapy of virus-infected cells results in synthesis of RNA with abnormal or absent 5' cap structures, which in turn leads to inefficient translation of viral transcripts [7]. A third states that the drug has a direct suppressive effect on viral polymerase activity. Specific RNA polymerases that have been shown to be affected are those associated with influenza virus [14], vesicular stomatitis virus (Indiana) [13] and, recently, La Crosse virus [2]. It has been somewhat problematic to determine experimentally the primary mechanism of action of ribavirin because none of these hypotheses are mutually exclusive, and indeed, ribavirin may act on many different levels. It is also important to note that ribavirin, unlike all other drugs whose structures resemble nucleoside analogs, has a modified base. Some antiviral agents, notably acyclovir, have modified sugars and generally are terminators of growing nucleic acid chains. Their mode of action was relatively simple to determine.

Most researchers have assumed that the structural similarity of ribavirin to guanosine was related to its functional similarity. Studies in which viral production was examined in tissue culture supported this hypothesis because the antiviral effect could be reversed by guanosine. However, work in our laboratory has indicated that the functional relationship of nucleosides to ribavirin may include all of the natural nucleoside triphosphates. Results of experiments which examined the effects of phosphorylated ribavirin compounds on an in vitro VSI polymerase assay indicated that the drug does indeed possess a significant direct suppressive effect on the viral polymerase [13]. In fact all three phosphorylated ribavirin molecules inhibited VSI virus transcription. The mono- and diphosphorylated forms of the drug possessed approximately two to three times the inhibitory activity of ribavirin 5'-triphosphate (RTP). Transcripts synthesized in the presence of all three forms of ribavirin were full length. Inhibition by ribavirin 5'-diphosphate (RDP) could be reversed by the addition of uridine, guanosine and cytidine 5'-triphosphates (UTP, GTP and CTP, respectively) while the addition of guanosine 5'-diphosphate (GDP) to the reaction did not reverse the inhibition. Nearest neighbor analysis showed that none of the phosphorylated forms were incorporated into the growing chain of RNA.

The activity of the mono- and diphosphorylated forms of the drug against VSI replicase was surprising. It has been shown previously that all nucleoside analogs inhibit viral polymerases in the triphosphorylated form [3, 6]. We used enzyme kinetics and product analysis to further investigate the mechanism of action of ribavirin. When analyzed by double-reciprocal plots both RDP and RTP gave similar patterns of inhibition, although the apparent Michaelis–Menton constants (Km) of the nucleoside triphosphates in the presence of the drugs are not the same [4]. Both RDP and RTP compete directly with all four of the nucleoside triphosphates, suggesting that they act on the viral polymerase in a similar fashion. Because there appears to be no incorporation of ribavirin into the growing viral nucleic acid chain and because the transcripts synthesized are full length even in the presence of both drugs, it was our hypothesis that RDP and RTP were blocking polymerase initiation at the 3'-terminus of the viral genome.

Given that a diphosphorylated form of ribavirin is an effective antiviral agent, we examined what effect compounds stabilizing as diphosphates would have on the VSI virus in vitro transcription reaction. The data reported here support our contention that diphosphorylated ribavirin can act as an antiviral agent.

Materials and methods

VSI virus was grown in Chinese hamster ovary cells and purified by sucrose gradient centrifugation. Structures of three ribavirin analogs (NARI #12,613, NARI #14,649 and NARI #14,650), gifts of Roland K. Robins and Daniel Smith (Nucleic Acids Research Institute, Costa Mesa, CA) are shown in Fig. 1. In vitro transcription reactions were as described by Banerjee [1]. The reaction mixture was composed of 100 mM NaCl, 50 mM Tris HCl (pH 8.0), 5 mM $MgCl_2$, 4 mM dithiothreitol, 0.05% Triton X-100, 10 U of placental RNase inhibitor (Boehringer Mannheim Biochemicals, Indianapolis, Ind.), 20 µCi of α-^{32}P UTP and 25 to 50 µg of VSI virions, in a total volume of 0.2 ml. All reactions were incubated and analyzed as described previously [4, 5].

Results

The extremely hygroscopic nature of the three analogs made accurate determinations of their weights difficult. UV spectra of the compounds dissolved in diethylpyrocarbonate-treated water were recorded so that the spectrum of a known concentration of RDP could be used as a standard for the calculation of their concentrations. However, a difference in extinction coefficients (e) was apparent when spectra were compared with empirical estimates of the concentrations.

It was determined that compound NARI #12,613 absorbed water slowest. Samples of this analog were weighed using a Mettler balance and UV spectra were recorded. All calculations were made on the assumption that no water was absorbed. Therefore, the reported concentrations represent an upper limit; actual values probably are somewhat lower. In our hands, the e_{220} of NARI #12,613 was estimated to be 11,000 M^{-1} cm, differing more than ten-fold from that of RDP which was e_{220} of 855 M^{-1} cm. The value for NARI #12,613 is slightly higher than that of any of the ribonucleosides, while the value for RDP is significantly lower [8]. From the similarities in the spectra and visual evidence, the e_{220} of the other analogs was estimated to be approximately the same as that of NARI #12,613.

Preliminary experiments to establish a concentration range over which the compounds displayed inhibitory effects were performed using the protocol for a standard polymerase assay, as described in Materials and methods, except that the concentrations of the nucleoside triphosphates remained fixed, while the concentrations of the RDP analogs were varied. As

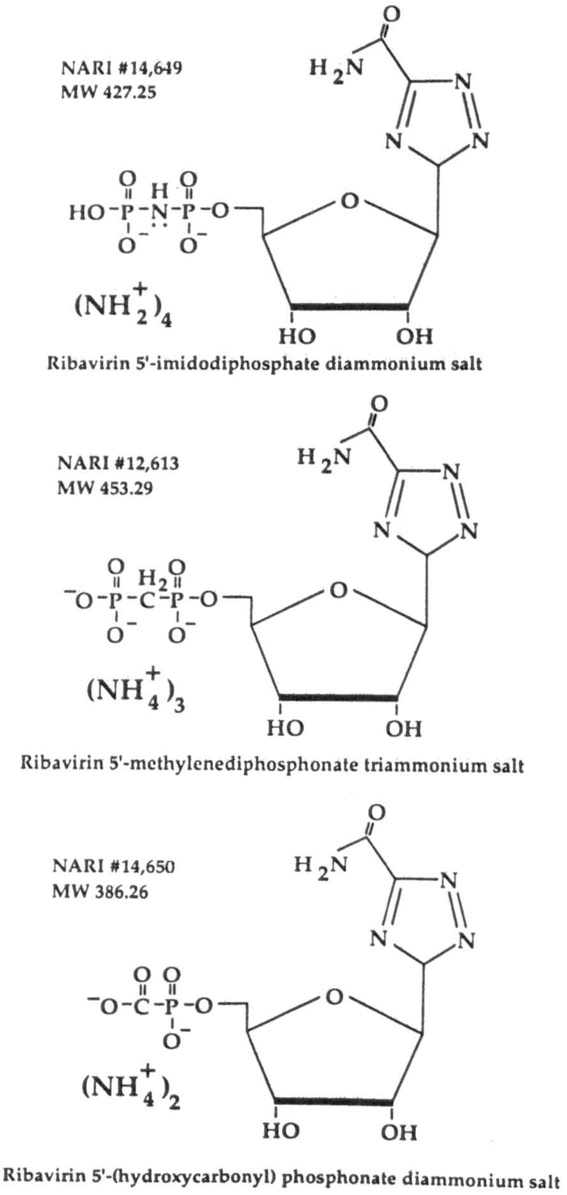

Fig. 1. Structures of compounds that stabilize as diphosphates

shown in Table 1, NARI #12,613 showed at least 80% inhibition at 5 mg/ml and was linear with respect to time to this concentration. Only #14,650 did not show linearity. At 2.5 mg/ml of #12,613 there was 51% inhibition. In comparison RTP produced approximately 55% inhibition at 0.15 mg/ml, a six fold difference. At 2.6 mg/ml of #14,649 and 2.1 mg/ml of #14,650 there was a 63% and a 15% inhibition, respectively. This is much less than that inhibition produced by RTP.

Table 1. Establishment of range of effective concentrations of three analogs

Compound	Stock concentration (mg/ml)	Dilution used in assay	Final concentration (in assay tube) (mg/ml)	% inhibition	
				60 min	90 min
RTP	2	—	0.15	55.6	44.5
12,613	336	5×	5.0	80.7	81.0
	336	10×	2.5	51.0	45.1
	336	50×	0.5	12.8	8.9
14,649	348	5×	5.2	91.0	89.1
	348	10×	2.6	63.5	60.0
	348	11×	2.4	47.7	48.6
	348	50×	0.52	11.9	11.9
14,650	279	5×	4.2	54.2	49.9
	279	10×	2.1	15.1	14.2
	279	50×	0.42	11.9	9.0

As noted, the concentrations of analogs represent upper limits, the actual inhibitory effects of these compounds probably being somewhat higher than reported; however it is unlikely that they approach the inhibition produced by RTP since there is a six-fold difference. The nonlinearity in the concentration dependence of #14,650 suggests a complex mode of inhibition.

At varying dilutions of each compound on the apparent Km and maximal velocity (Vmax) of the reaction in the presence of the inhibitors were calculated using linear regression analysis as described previously [4]. All the apparent Km values were raised to varying degrees as shown in Table 2. The apparent Vmax values were generally lower in the presence of the inhibitors, with the exception of NARI #12,613. The values of the apparent Vmax in the presence of #14,650 were relatively close to those of the uninhibited reaction.

Of the three compounds, apparently #14,649 is slightly more inhibitory than #12,613 and much more inhibitory than #14,650. The only chemical difference between #12,613 and #14,649 is the replacement of a methyl group for an amine group in #14,649. The reason for this slight difference in inhibition is unknown.

Discussion

Our results suggest that the compounds described here that stabilize as diphosphates can serve as inhibitors of the VSI virus polymerase. Therefore, at least some of the inhibitory effects of these compounds are related to the structural nature of the ribavirin base and can be mediated specifically as a diphosphate. These compounds are not as inhibitory as RDP, thus the

Table 2. Summary of kinetic experiments on RDP analogs

Compound	Graph method	Km, apparent (µM)		Vmax	
		no inhibitor	with inhibitor	no inhibitor	with inhibitor
RTP					
RTP	L-B[a]	211	306	479	279
NARI 613	L-B	163	636	485	585
#12,613					
(20 × Dil)					
	E-H[b]	134	289	468	435
NARI	L-B	108	433	588	294
#14.649					
(11 × Dil)					
	E-H	109	286	590	293
NARI	L-B	127	271	671	526
#14,650					
(5 × Dil)					
	E-H	126	213	670	498

[a] *L-B* Lineweaver–Burk (Double Reciprocal Plot)
[b] *E-H* Eadie–Hofstee Plot

possibility remains that a phosphorylating activity which converts RDP to RTP is partly responsible for the antiviral effect of RDP. However, as stated previously, we found that GDP could not reverse the effects of RDP, while GTP does [13]. Presumably GDP would triphosphorylate at least as easily as RDP. Also, there are no reports of nucleoside mono- or diphosphates actively playing a role in VSI virus in vitro transcription; only nucleoside triphosphates are effective in the reaction.

To date, all work regarding phosphorylating activities has centered on the phosphorylation of viral proteins, such as NS [9]. The major controversy surrounding phosphorylation seems to be whether the polymerase (L) protein provides the kinasing activity or if cellular kinases are packaged in the virion and phosphorylate proteins. One possibility that cannot be eliminated is that RDP may be phosphorylated to RTP and that this activity is inhibitory of the kinasing of viral proteins. However, there is no direct evidence that RDP or any diphosphate inhibits protein phosphorylation.

In any case, the RDP analogs reported here are not as effective at inhibiting the VSI in vitro polymerase reaction as RDP or RTP. The increased Km and relatively constant Vmax in the analogs #12,613 and #14,649 including RTP, indicate a more complex mechanism, which may be a combination of two or more different types of inhibition. The nonlinearity in the concentration dependence of #14,650 is interesting, although the reason for this effect is unknown. Possibilities include positive cooperativity between multiple inhibitor binding sites and different inhibitory sites with varying affinity for the analog. Proof of the actual nature of the mechanism awaits further experimentation.

These compounds, however, provide us with a clue as to the structural components of nucleoside analogs necessary for inhibiting RNA-dependent RNA polymerases. Work is in progress to determine the exact structural constituent that interferes with primary transcription of negative stranded RNA viruses.

Our data and those of others have suggested that ribavirin affects the initial steps of transcription by RNA polymerases and that this effect may be mediated by several phosphorylated forms. One reason the antiviral activity of ribavirin may be more difficult to assess than other commonly used antiviral agents is that inhibition of polymerases is without chain termination. However, if primary transcription is affected, direct linear effects of the drug on later stages of viral replication and assembly are difficult to ascertain, as any deregulation of the system at the mRNA level is presumably amplified at all levels of viral progeny production. These small effects on primary transcription may be enough to produce the entire antiviral activity seen, for example, with respect to diminished viral titers. The VSI virus in vitro transcription reaction could be developed as an effective screen for compounds that may interact with RNA-dependent RNA polymerases. This would be particularly important in examining drugs that cannot initially be screened efficiently in tissue culture because they are unable to cross a lipid membrane.

Ribavirin remains an important potential agent in the treatment of a broad spectrum of serious viral illnesses encountered worldwide and may serve in the future as a model for similar but more effective agents. Definition of its molecular activity is vital to the development of its use and the development of improved analogs.

Acknowledgements

This work was supported by US Army Grant DAMD17-87-C-7004. M. J. Lachenmann and R. Fernandez-Larsson were supported in part by NIH Training grants 5T326MO7306 and 5T326AIO7245, respectively.

References

1. Banerjee AK (1981) The in vitro mRNA transcription processes. In: DHL Bishop (ed) Rhabdoviruses, vol 2. CRC Press, Boca Raton, FL, pp 35–50
2. Cassidy LF, Patterson JL (1989) Mechanism of La Crosse virus inhibition by ribavirin. Antimicrob Agents Chemother 33: 2009–2011
3. Eriksson BE, Helgstrand E, Johnnson A, Larsson A, Misiorny JO, Noren L, Philipson K, Stenberg K, Stening G, Stridh S, Oberg B (1977) Inhibition of influenza virus ribonucleic acid polymerase by ribavirin triphosphate. Antimicrob Agents Chemother 11: 946–951

4. Fernandez-Larsson R, O'Connell K, Koumans E, Patterson JL (1989) Investigation of the vesicular stomatitis virus in vitro polymerase reaction using phosphorylated ribavirin. Antimicrob Agents Chemother 33: 1668–1673

5. Fernandez-Larsson R, Patterson JL (1989) Altered ATP function of vesicular stomatitis virus mutant detected by kinetic analysis of the transcriptase using phosphorylated ribavirin. J Gen Virol 70: 2791–2797

6. Furman PA, St. Clair MH, Spector T (1984) Acyclovir triphosphate is a suicide inactivator of the herpes simplex virus DNA polymerase. J Biol Chem 259: 9575–9579

7. Goswami BB, Borek E, Sharma OK (1979) The broad spectrum antiviral agent ribavirin inhibits capping of mRNA. Biochem Biophys Res Commun 89: 830–836.

8. Jaffe HH, Orchin M (1962) Theory and application of UV spectroscopy. Wiley, New York

9. Kingsford L, Emerson SU (1980) Transcriptional activities of different phosphorylated species of NS protein purified from vesicular stomatitis virions and cytoplasm of infected cells. J Virol 33: 1097–1105

10. Malinoski F, Stollar F (1981) Inhibitors of IMP dehydrogenase prevent Sindbis virus replication and reduce GTP levels in *Aedes albopictus* cells. Virology 110: 281–291

11. Sidwell RW, Huffman JN, Khard GP, Allen LB, Witkowski JT, Robins RK (1972) Broad spectrum antiviral activity of virazole: 1-β-D-ribofuranosyl-1,2,4-triazole-3-carboximide. Science 177: 705–706

12. Streeter DG, Witkowski JT, Khare GP, Sidwell RW, Bauer RJ, Robins RK, Simon LN (1973) Mechanism of action of 1-β-D-ribofuranosyl-1,2,4-triazole-3-carboxamide (Virazole), a new broad-spectrum antiviral agent. Proc Natl Acad Sci USA 70: 1174–1178

13. Toltzis P, O'Connell K, Patterson JL (1988) Effect of phosphorylated ribavirin on vesicular stomatitis virus transcription. Antimicrob Agents Chemother 32: 492–497

14. Wray SK, Gilbert BF, Knight V (1985) Effect of ribavirin triphosphate on primed generation and elongation during virus transcription in vitro. Antiviral Res 5: 39–48

Authors' address: J. L. Patterson, Division of Infectious Diseases, Children's Hospital, Boston, MA 02115, U.S.A.

Arch Virol (1990) [Suppl 1]: 109–117

Early events in infection with arenaviruses

Svetlana E. Glushakova[1], I. S. Lukashevich[1], A. E. Grinfeldt[2], Valentina A. Gotlib[2], and A. A. Lev[2]

[1] Department of Special Pathogens, Byelorussian Research Institute of Epidemiology and Microbiology, Minsk, U.S.S.R.
[2] Laboratory of Physical Chemistry of Cell Membranes, Institute of Cytology, U.S.S.R. Academy of Sciences, Leningrad, U.S.S.R.

Accepted February 7, 1990

Summary. Lysosomotropic agents, weak bases and ionophores, were used to study the mode of penetration of arenaviruses into cells. All drugs studied (ammonium chloride, chloroquine, amantadine, monensin) effectively inhibited replication of Pichinde, Mopeia, and Lassa viruses in BHK-21 and Vero cells. The experimental results indirectly indicate that endocytosis, using acid cellular endosomes, is a mode of entry of arenaviruses into cells. For prediction of arenavirus fusion protein we used an analysis of the elements of secondary and tertiary structure of precursor proteins of glycoproteins of five members of this virus family. Two regions of protein GP2 possessed properties that satisfied our criteria for selection of fusion peptides of enveloped viruses. One of the selected peptides was synthesized and model biophysical experiments demonstrated its ability to interact with lipid and to form single ion channels in the planar lipid bilayers, typical of fusion proteins of enveloped viruses. These data suggest that protein GP2 is the fusion protein of arenaviruses and that the fusion peptide is located in the immediate proximity to the N-end of this polypeptide.

Introduction

Early events in virus infection indicate the processes leading to virus genome entry into the cytoplasma or nucleus of the cell, where virus replication takes place [16]. Initiation of infection is a critical stage in this process, not only from the point of view of recognition of the mechanism of virus replication but also because this stage provides a target for intervening in virus replicative processes.

Two successive stages of virus entry are distinguished early in infection: (1) virus attachment to the cell surface and (2) penetration of virus nucleocapsid into the cytoplasm. With enveloped viruses, the latter process occurs by fusion of the virus and lipid membranes of the cell. This fusion is induced by virus envelope proteins. Fusion occurs either on the cell surface or within acid endocytic vacuoles. In the latter instance, transition of viral fusion protein from inactive to a functionally active form is under mildly acid pH conditions [15].

The early stages of arenavirus infection have, until now, been essentially unstudied [9, 23]. The purpose of this study was to elucidate the mode of arenavirus penetration into cells and to identify virus proteins and their fragments that function to carry virus genome into the cytoplasm, that is, possess membrane-fusing activity.

Materials and methods

Cells and viruses

Vero and BHK-21 cells were grown in Eagle's Minimal Essential Medium (MEM) containing 0.75 g/l NaHCO$_3$, 10 mM HEPES buffer, 2 mM glutamine, 100 μg/ml gentamycin and 10% fetal bovine serum. Lassa (Josiah) and Mopeia viruses were kindly provided by Dr. G. van der Groen (The Institute of Tropical Medicine, Antwerp, Belgium). Pichinde virus (AN 3729) was received through the courtesy of Dr. C. Pfau, Rensselaer Polytechnic Institute, Troy, New York, U.S.A. The viruses were propagated and titrated by plaque assay on Vero cells.

Effects of lysosomotropic drugs on arenavirus replication

The effect of lysosomotropic agents on replication of arenaviruses and on virus adsorption to cells was assessed according to methods published previously [9]. Briefly, BHK-21 or Vero cells were infected with viruses at a multiplicity of infection 0.1 PFU per cell. Ammonium chloride, amantadine, chloroquine and monensin were used as fresh solutions in MEM with 2% fetal bovine serum. All the drugs were added in culture medium at indicated time (see legends to figures). At 24–48 h post-infection viral replication was determined by plaque assay.

Analysis of the parameters of secondary and tertiary structure of glycoproteins

Defined amino acid sequences of precursor proteins (GPC) of two "Old World" arenaviruses (Lassa and Lymphocytic choriomeningitis virus) and two "New World" arenaviruses (Pichinde and Tacaribe) [1, 2, 6, 18, 19] were studied to determine their fusion proteins. The results of comparison of sequences according to the alignment principle are taken from the work of Auperin and McCormick [3]. The elements of secondary structure were analysed using commercially available computer programs from "PCGene": NOVOTNY [4, 20], SOAP [13, 14], AASCALE [11], RAOARGOS [17], and GARNIER [8]. Hydrophilicity profiles of GPC proteins were evaluated using SOAP computer program [13, 14].

Ability to interact with membranes and channel-forming activity
of the predicted Lassa virus peptide

Differential scanning calorimetry assay on dipalmitoylphosphatidylcholine liposomes and determination of channel-forming activity of the predicted peptide were performed by methods published previously [5, 10].

Results

Effect of lysosomotropic drugs on arenavirus replication

As shown in Figs. 1 and 2, all the lysosomotropic drugs under study inhibited replication of the three arenaviruses: Lassa, Pichinde, and Mopeia. This effect depended on the time of addition of the drug to the culture medium but was highest when the cells were pretreated with the agent, slightly less marked when the drug and the virus were added simultaneously, and greatly reduced when the drug was added after inoculation with the virus (Fig. 3). Mopeia virus was highly sensitive to lysosomotropic agents as shown by the fact that similar concentrations of ammonium chloride and monensin effected Mopeia virus 20–30 times as much as it effected Lassa and Pichinde viruses (Figs. 1c and 2d).

Having demonstrated inhibiting effects of these drugs on early stages of arenavirus infection, we obtained indirect evidence that these viruses enter the cells by receptor endocytosis [9].

Prediction of arenavirus fusion peptides

The search for fusion protein and fusion peptide of arenaviruses was done by analysis of amino acid sequences of the precursors (GPC) of envelope

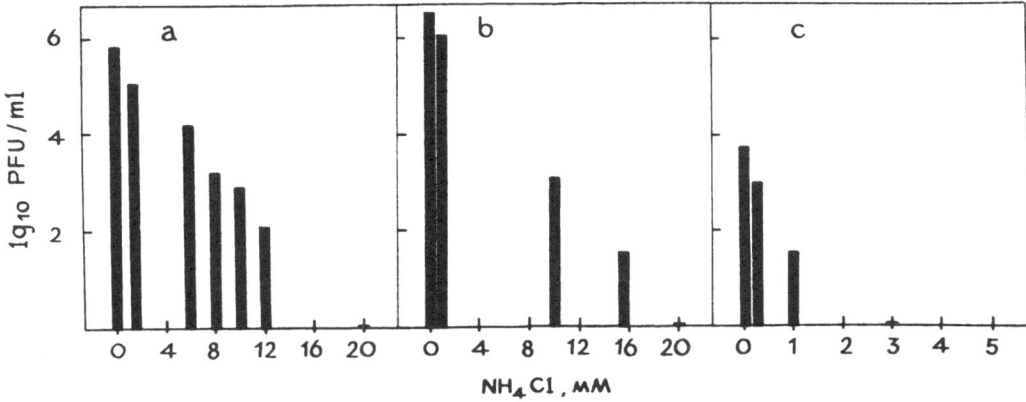

Fig. 1. Effect of ammonium chloride on arenavirus replication. **a** BHK-21 cells infected with Pichinde virus; **b, c** Vero cells infected with Lassa and Mopeia viruses respectively. The drug was present in culture medium 1 h before the virus inoculation, during the adsorption period and throughout the experiments

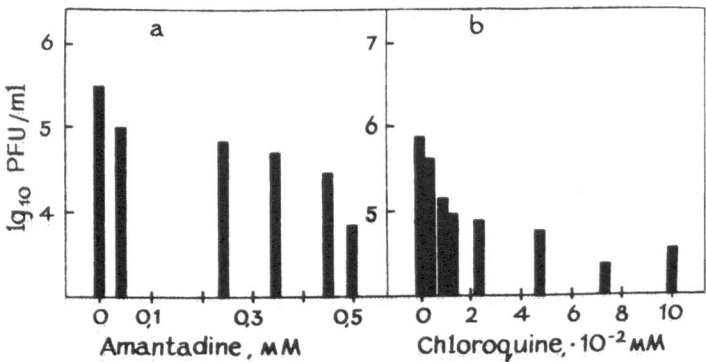

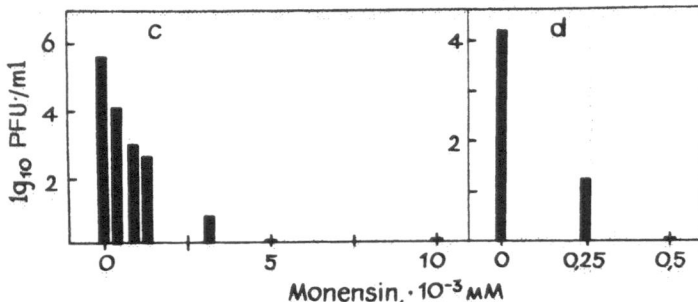

Fig. 2. Effect of other lysosomotropic drugs on arenavirus replication. **a–c** BHK-21 cells treated with the drugs and infected with Pichinde virus; **d** Vero cells treated with monensin and infected with Mopeia virus

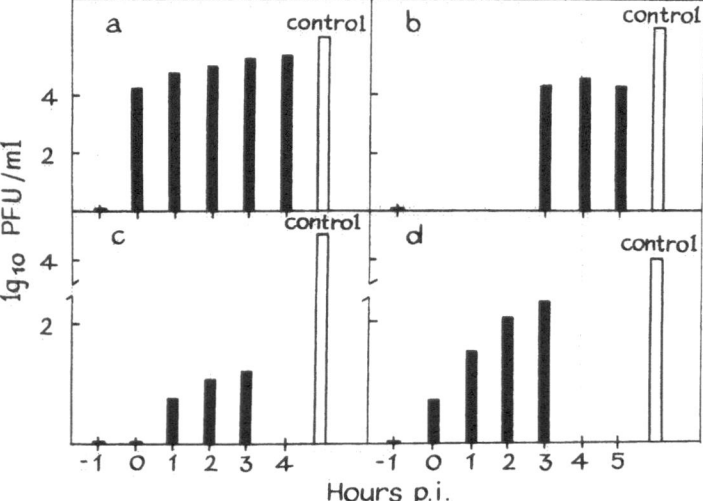

Fig. 3. Effect of time of addition of ammonium chloride (20 mM; **a–c**) and monensin (1×10^{-3} mM; **d**) on arenavirus replication. **a, d** BHK-21 cells infected with Pichinde virus; **b, c** Vero cells infected with Lassa and Mopeia viruses respectively

glycoproteins. As a result of the analysis, we chose regions of protein that had properties typical of fusion peptides of other enveloped viruses, namely: length about 20 amino acid residues, marked hydrophobicity, high degree of conservation within the virus family, remoteness from potential glycosylation sites, predominantly α-helical conformation, and location within the protein globule (for viruses entering cells by endocytosis). Figure 4 presents a linear map of Lassa virus GPC, which contains six long hydrophobic areas of the precursor protein that, by their physico-chemical properties, may be fusion peptide of this virus.

Two of these peptides, IV and V, by the sum of their properties are most probable fusion peptides. The sequence of peptide V of all arenaviruses is shown in Fig. 5. Identical amino acids are marked with dots, the positions with conserved amino acid substitutions are framed. Of those viruses analysed, peptide V has a maximal degree of homology and is conserved (more than 90%) among arenaviruses. Comparison of this peptide in Lassa and LCM (ARM and WE strains) viruses 23 of 24 amino acid residues were identical or have conserved substitutions. Among antigenically different viruses (Tacaribe and LCM), 22 amino acid residues are conservated, although in contrast to other arenaviruses, Tacaribe virus is believed to have not two but one envelope protein [6]. The hydrophobic index of this peptide is 0.554; 75% of the residues are in α-helical conformation; the peptide is probably located within the protein globule (GARNIER program) although it is located in immediate proximity to the N-end of GP2. Next to this peptide is a highly charged protein fragment, which is typical of certain proteins that interact with membranes [22].

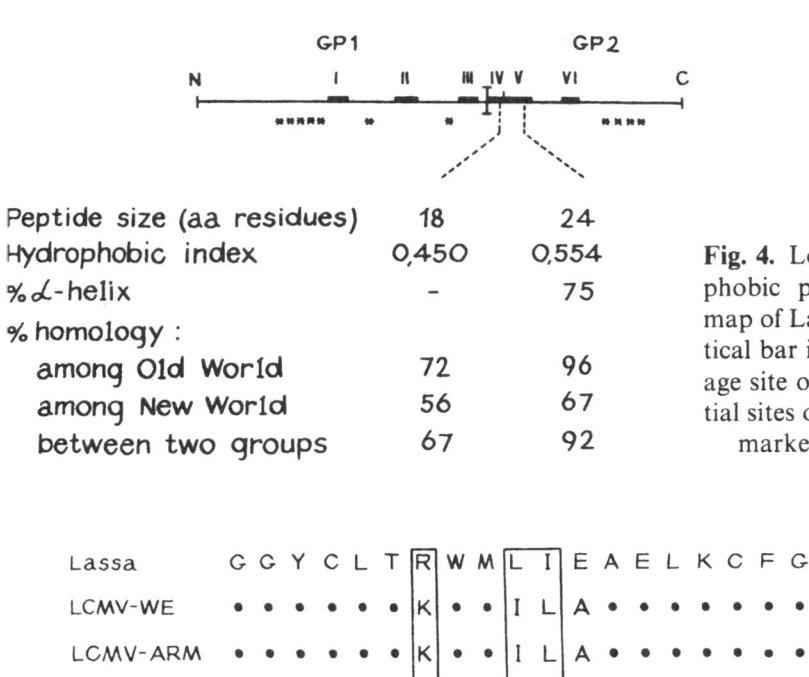

Peptide size (aa residues)	18	24
Hydrophobic index	0,450	0,554
% α-helix	–	75
% homology :		
among Old World	72	96
among New World	56	67
between two groups	67	92

Fig. 4. Location of hydrophobic peptides on GPC-map of Lassa virus. The vertical bar indicates the cleavage site of GPC, and potential sites of glycosylation are marked with asterisks

Fig. 5. Potential fusion peptides of arenaviruses. ● Identical amino acids; positions with conserved amino acid substitutions are framed

The other candidate for fusion peptides, peptide IV, is the N-end 18-member fragment of protein GP2. It meets many requirements for these functional regions of viral proteins. This peptide is less hydrophobic and conservative than peptide V is not α-helical, however, it is located at the N-end of polypeptide which is typical of many fusion peptides, and contains sequence Gly–X–Phe which is present in fusion peptides, for instance, paramyxoviruses and HTLV-I viruses [7]. Believing the inner peptide V to be a more probable fusion peptide, we cannot refute participation of peptide IV in the process of fusion of biological membranes.

The functional activity of peptide V in biophysical experiments on lipid bilayers and liposomes

It has been shown [12] that viruses of many families, interacting both with cell membranes and with artificial lipid bilayers, markedly alter the conductance of these substrates. It is assumed that fusion proteins may act as channel formers, altering ion penetration of the lipid bilayer.

In further experiments peptide V was synthesized on Beckman 990 Peptide Synthesizer by means of solid phase method and the synthesized peptide was tested in model experiments using lipid bilayers and liposomes. In differential scanning calorimetry assays using dipalmitoylphosphatidylcholine liposomes, peptide V exhibited well-pronounced changes in the lipid-phase transition (Fig. 6): reduced main phase transition temperature of the

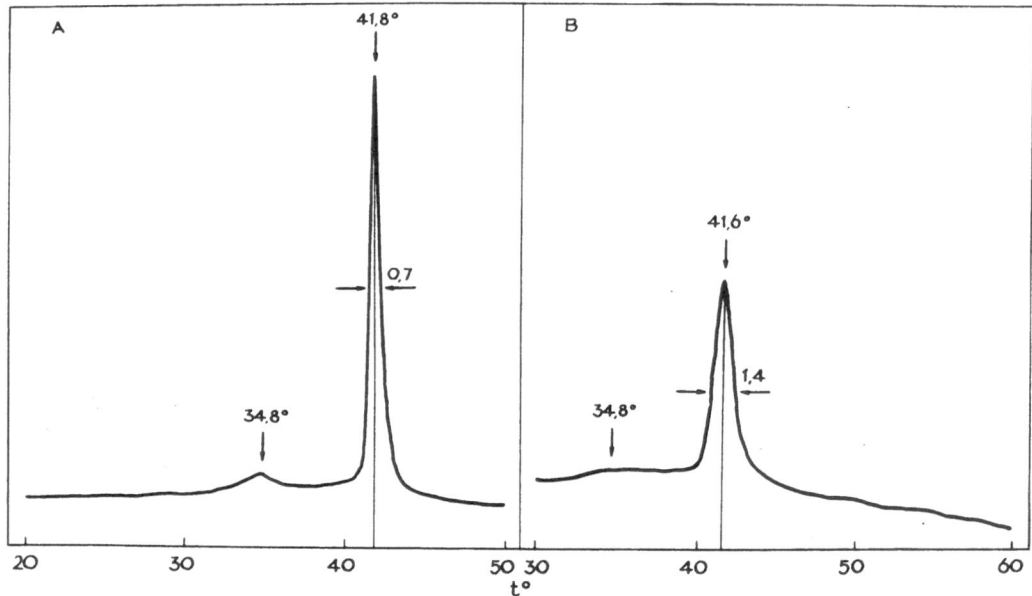

Fig. 6. Calorimetric assays of Lassa virus peptide–lipid bilayer interaction. **A** Control assay: dipalmitoylphosphatidylcholine $(3.5 \times 10^{-4}$ M)–water, multilamellar vesicles; **B** Lassa virus peptide V $(7.1 \times 10^{-5}$ M)–water, multilamellar vesicles

lipid (gel-liquid crystals) by 0.2 °C, changed the enthalpy at the main phase transition by 19%, and doubled the width of the transition peak (increase of its width at the half-height).

The peptide V showed marked channel-forming activity in Montall-Muller membranes (Fig. 7). In these experiments, bilayer membranes separated 0.1 M NaCl and KCl solutions (symmetrical electrolytic conditions). Under the conditions of the study, distinct single channels were obtained with three peaks of conductance (1.25, 3.75, and 5.42 pS) in NaCl solution, indicating possible dependence of the "size" of single ionic channels on the degree of aggregation of peptide molecules. The maximum values of conductance of single channels in NaCl were 10 pS, in KCl about 19 pS. The maximal length of single channels under two electrolytic conditions was 20 s. The observed channels showed marked cation–anion selectivity ($t_+ \geq 0.85$) and potassium–sodium selectivity (G/KCl:G/NaCl was 2.0–4.5). This characteristic is similar to that of selectivity of gramicidin, a well-studied natural channel-forming agent.

Discussion

Two types of lysosomotropic agents were used in the study. The first included ammonium chloride, amantadine, and chloroquine, substances which

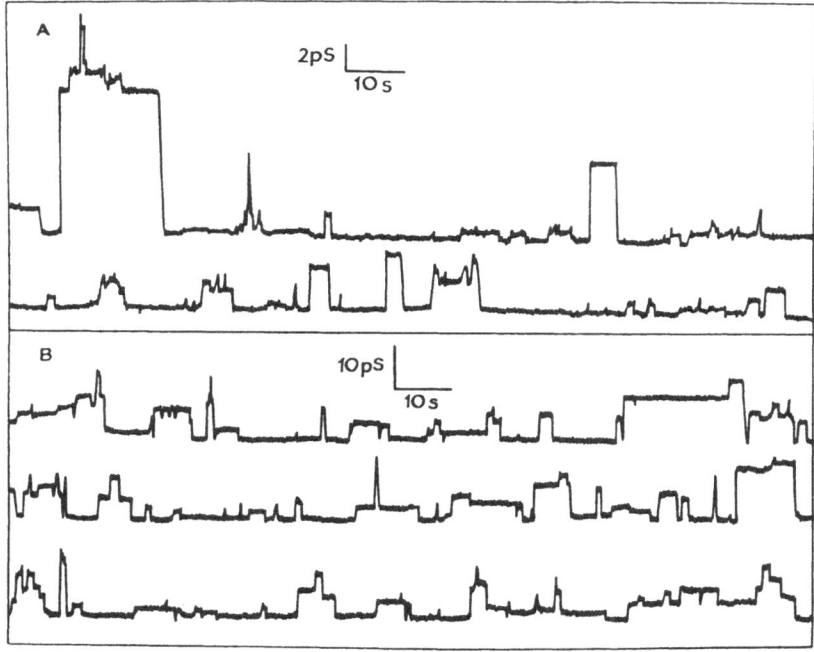

Fig. 7. Channel-forming activity of Lassa virus peptide in planar lipid bilayers. **A** Asolectin, 0.1 M NaCl, pH 5.1, 27 °C; Lassa virus peptide V, 2×10^{-7} M. **B** Asolectin, 0.1 M KCl, pH 7.4, 27 °C; Lassa virus peptide V, 2×10^{-7} M.

penetrate into acid cell vacuoles and neutralize their contents by binding protons; the second, monensin (ionophore), eliminates pH gradients cell vacuoles. Thus, all the substances used neutralize the mildly acid medium of vesicles and, if they contain virus entering the cells by endocytosis, block the stage of membrane fusion required for virus nucleocapsid release into the cytoplasm; thereafter, and as a consequence, progression of virus synthesis is prevented. If the virus induces membrane fusion on the cell surface and does not use endosomes for entry, early stages of infection are not sensitive to effects of lysosomotropic agents [21]. All lysosomotropic drugs studied effectively inhibited replication of arenaviruses in cells (Figs. 1 and 2). The inhibitory effect depended on the time addition of the drug and was most marked when cells were pretreated with the drug 1 h before virus inoculation (Fig. 3). These lysosomotropic agents had no virucidal effect and no effect on virus adsorption on cells. Having demonstrated inhibiting effects of these drugs on the early stages of arenavirus infection, we obtained indirect evidence that these viruses enter the cells by receptor endocytosis [9].

Analyses of amino acid sequences of arenavirus glycoproteins by means of computer programs and on the basis of our criteria for selection of fusion peptides of enveloped viruses, revealed two peptides in the most conserved protein (GP2) of arenaviruses. These may be fusion peptides of these viruses. The hydrophobic fragment (peptide V) of the GP2 is typical of fusion peptides of other enveloped viruses. In model biophysical experiments (Figs. 6 and 7) we showed that synthesized peptide V had the ability to interact with lipids and markedly change the conductance of lipid membranes, a characteristic typical of fusion proteins of enveloped viruses.

From these results it may be assumed that the most conserved protein (GP2) of arenaviruses is the fusion protein, that the fusion peptide is located in immediate proximity to its N-end, and that this peptide is an analogue of the internal fusion peptides, as it is in certain other viruses [7]. It is reasonable to assume that knowledge of the mechanisms of early stages of viral infection will help improve both anti-viral immunological and chemotherapeutical aspects of strategies, which is particularly important for control of arenaviruses·highly pathogenic for humans.

References

1. Auperin DD, Romanovski V, Galinski M, Bishop DHL (1984) Sequence studies of Pichinde arenavirus S RNA indicate a novel coding strategy, an ambisense viral S RNA. J Virol 52: 897–904
2. Auperin DD, Sasso DR, McCormick JB (1986) Nucleotide sequence of the glycoprotein gene and intergenic region of the Lassa virus S genome RNA. Virology 154: 155–167
3. Auperin DD, McCormick JB (1989) Nucleotide sequence of the Lassa virus (Josiah strain) S genome RNA and amino acid sequence comparison of the N and GPC proteins to other arenaviruses. Virology 168: 421–425
4. Chou PY, Fasman GD (1978) Prediction of the secondary structure of proteins from their amino acid sequences. Adv Enzymol 47: 45–148

5. Eband RM, Stertevant GM (1981) A calorimetric study of peptide–phospholipid interactions: the glucagon–dimyristoylphosphatidylcholine complex. Biochemistry 20: 4603–4606

6. Franze-Fernandez MT, Zetina C, Japalucci S, Lucere MA, Bouisso C, Lopez R, Rey O, Daheli M, Cohen GN, Zakin MM (1987) Molecular structure and early events in the replication of Tacaribe arenavirus S RNA. Virus Res 7: 309–324

7. Gallaher WR (1987) Detection of a fusion peptide sequence in the transmembrane protein of human immunodeficiency virus. Cell 50: 327–328

8. Garnier J, Osquthorpe DJ, Robson B (1978) Analysis of the accuracy and implications of simple methods for predicting the secondary structure of globular proteins. J Mol Biol 120: 97–120

9. Glushakova SE, Lukashevich IS (1989) Early events in arenavirus replication are sensitive to lysosomotropic compounds. Arch Virol 104: 157–161

10. Hladky SB, Haydon DA (1972) Ion transfer across lipid membranes in the presence of gramicidin A. I. Studies of the unit conductance channel. Biochim Biophys Acta 274: 294–312

11. Jones DD (1975) Amino acid properties and side-chain orientation in proteins: a cross correlation approach. J Theor Biol 50: 167–183

12. Kempf G, Michel MR, Kohler U, Koblet H (1987) Can viral envelope proteins act as or induce proton channels? Biosci Rep 7: 761–769

13. Klein P, Kanehisa M, DeLisi C (1985) The detection and classification of membrane-spanning proteins. Biochim Biophys Acta 815: 468–476

14. Kyte J, Doolittle RF (1982) A simple method for displaying the hydropathic character of a protein. J Mol Biol 157: 105–132

15. Marsh M, Helenius A, Matlin K, Simons K (1983) Binding, endocytosis and degradation of enveloped animal viruses. Methods Enzymol 98: 260–266

16. Marsh M (1984) The entry of enveloped viruses into cells by endocytosis. Biochem J 218: 1–10

17. Rao JKM, Argos P (1986) A conformational preference parameter to predict helices in integral membrane proteins. Biochim Biophys Acta 869: 197–214

18. Romanowski V, Bishop DHL (1985) Conserved sequences and coding of two strains of lymphocytic choriomeningitis virus (WE and ARM) and Pichinde arenavirus. Virus Res 2: 35–51

19. Romanowski V, Matsuura Y, Bishop DHL (1985) Complete sequence of the S RNA of lymphocytic choriomeningitis virus (WE strain) compared to that of Pichinde arenavirus. Virus Res 3: 101–114

20. Rose GD, Roy S (1980) Hydrophobic basis of packing in globular proteins. Proc Natl Acad Sci USA 77: 4643–4647

21. Spear PG (1987) Virus-induced cell fusion. In: Sowers AE (ed) Cell fusion. Plenum, New York, pp 3–32

22. Terwilliger TC, Weissman L, Eisenberg D (1982) The structure of melittin in the form I crystals and its implication for melittin's lytic and surface activities. Biophys J 37: 353–361

23. Welsh RM, Trowbridge RS, Kowalski JB, O'Connell CM, Pfau CJ (1971) Amantadine hydrochloride inhibition of early and late stages of lymphocytic choriomeningitis virus-cell interactions. Virology 45: 679–686

Authors' address: I. S. Lukashevich, Department of Special Pathogens, Byelorussian Research Institute of Epidemiology and Microbiology, Nogin str. 3, 220050 Minsk, U.S.S.R.

Arch Virol (1990) [Suppl 1]: 119–124

A study of the NS3 nonstructural protein
of tick-borne encephalitis virus
using monoclonal antibodies against the virus

**A. V. Timofeev[1], Alla A. Kushch[2], M. F. Vorovitch[1], S. M. Tugizov[2],
L. B. Elbert[1], and D. K. Lvov[2]**

[1] Institute of Poliomyelitis and Viral Encephalitides, Moscow
[2] D.I. Ivanovsky Institute of Virology, Moscow, U.S.S.R.

Accepted January 9, 1990

Summary. We obtained two monoclonal antibodies against native NS3 nonstructural protein of tick-borne encephalitis (TBE) virus, strain Sofyin. Using these monoclonal antibodies, we were able to determine that NS3 is a stable protein with preferential perinuclear localization. Both monoclonal antibodies precipitate NS3 proteins of TBE strains Sofyin and Absettarov and of Langat virus but do not precipitate analogous proteins of other flaviviruses (Powassan, yellow fever, Japanese encephalitis). Taking into account the work of others, we suggest that there are at least two distinct epitopes on NS3 protein. A putative functional model of NS3 is discussed.

Introduction

The NS3 protein is one of the major virus-specific components of cells infected with flaviviruses. This protein can be detected early in infection, the maximum rate of synthesis greatly depending on the multiplicity of infection [9]. By comparing genetic maps of flavivirus NS3 proteins we conclude that: (1) NS3 is one of the most conserved virus-specific proteins and is highly hydrophilic; the variable domain is only on the C-terminus and (2) in most flaviviruses NS3 has two conserved cysteines, in positions 1666 and 2068, and except for NS3 of tick-borne encephalitis (TBE) virus, has no potential sites for glycosylation [2, 6, 13; A.G. Pletnev, pers. comm.].

The cleavage site at the N-terminus of NS3 includes a pair of basic amino acids (Lys–Arg or Arg–Arg) in positions -2 and -1 relative to the point of cleavage, followed by an amino acid with a short side chain (Gly or Ser). Although no cellular proteases with such specificity have been found, the existence of a virus-specific protease has been predicted [10]. The analogous protease may act on the C-terminus of NS3. Processing of the NS3 protein takes place in the cytoplasm [2]. Until now, the function of NS3 has not been clear. Recently, two groups independently predicted the dual nature of this protein: as a helicase and as a virus-specific protease [1, 3, 4].

One of the main tools for studying the structural and functional features of proteins is monoclonal antibodies (Mabs). We therefore prepared two Mabs against native NS3 nonstructural protein of TBE virus and studied their type-specificity and certain features of the NS3 protein. The results of these studies and a hypothesis of the function of NS3 are presented in this paper.

Materials and methods

Cells and viruses

Porcine embryo kidney (PE) cells were propagated in Medium 199 supplemented with 10% bovine serum. Flaviviruses used were TBE virus (strains Sofyin and Absettarov), Langat, Powassan, Japanese encephalitis, and yellow fever (17D vaccine strain).

Preparation of radioactive cell lysates

Monolayers of PE cells were infected with 0.1–1.0% clarified suspensions of brains of flavivirus-infected suckling mice at an MOI of about 1–5 plaque-forming units/cell. At 20 h after infection, the medium was replaced with Earle's balanced salt solution supplemented with 1% newborn bovine serum containing ^{14}C-amino acid at 30 µCi/ml. At the end of the desired labeling period (20 h for long labeling experiments and 1 h for pulse-labeling) the medium containing radioactive amino acids was replaced either with Medium 199 containing 0.25% newborn bovine serum for different time intervals (chase experiments) or with radioimmunoprecipitation assay buffer (0.13 M NaCl, 10 mM Tris HCl, pH 8.0; 1 mM EDTA; 1% Nonidet P-40; 500 units/ml aprotonin). After homogenization and clarification, the sample was stored at −70 °C.

Cell fractionation procedure

After radioactive labeling, PE cells were collected and fractionated as described by Grun and Brinton [6]. Briefly, after homogenization the disrupted cells were pelleted by centrifugation at 1,000 × **g** for 5 min. The supernatant fluid was separated by high-speed centrifugation (105,000 × **g** for 1.5 h) into cytoplasmic and plasma membrane fractions. After additional homogenization and centrifugation (10,000 × **g** for 10 min), the pellet from the first centrifugation was separated from nuclei and large cell debris. After this procedure, the supernatant contained fragmented outer nuclear-associated membranes. Each fraction was equilibrated by volume and stored at −70 °C.

Preparation of monoclonal antibodies

Balb/c mice were intraperitoneally inoculated five times at weekly intervals with 0.2 ml volumes of a 10% clarified suspension of brains from suckling mice infected with TBE virus (strain Sofyin). For the first three immunizations antigen was inactivated with formalin and mixed with complete Freund's adjuvant. The last two immunizations were with live virus-containing material and without adjuvant. Spleen cells were fused with NS-O mouse myeloma cells in a proportion of 5:1 under standard conditions. Methods of incubation of fused cells in selective medium and detection of antibody-producing hybridoma cells by immunofluorescence were done as described earlier [7]. Immunoglobulins from culture fluid of clones positive by immunofluorescence were concentrated and partially purified using 50% ammonium sulfate precipitation. Ascitic fluids were obtained by intraperitoneal inoculation of mice with 10^7 hybrid cells. Before inoculation, mice were treated with Pristane (Sigma Chem. Co., St. Louis, Mo., U.S.A.). Immunoglobulin types of the Mabs were determined using culture fluids dialyzed against phosphate buffered saline (PBS), pH 7.4, and tested in Ouchterlony gel diffusion against antisera specific for each immunoglobulin isotype (Sigma Chem. Co., St. Louis, Mo., U.S.A.).

Passive protection test

Groups of 10 mice were each passively immunized with undiluted ascitic fluids containing Mabs. Each mouse received a dose of 0.2 ml subcutaneously and was then challenged 24 h later by intraperitoneal inoculation of 100 LD_{50} of TBE virus. They were then observed for 18 days for signs and symptoms of illness.

Enzyme immunoassay (ELISA)

Microtiter plates (Flow Labs. Ltd., Ayrshire, Scotland, U.K.) were coated by overnight incubation with horse serum containing antibodies to TBE virus polypeptides (Research Institute for Vaccines and Sera, Tomsk, U.S.S.R.) diluted in PBS. After three washes with PBSw (PBS containing 2% newborn bovine serum and 0.05% Tween-20) we added lysates of infected or mock-infected cells or TBE virus purified by high-speed centrifugation in PBS containing 2% newborn bovine serum and 0.5% Tween-20 and incubated them for 1 h at 37 °C. After four more wash cycles, Mabs from culture fluids were added and incubated 1 h at 37 °C. The wells were washed four times more and goat anti-mouse IgG peroxidase conjugate (Pharmacia Chem. Co., Uppsala, Sweden) was added, the plates incubated for 1 h at 37 °C, and the plates again washed five times. Finally, substrate (2 mg/ml o-phenylenedi-amine in phosphate-citrate buffer, pH 5.0, and 5 µl/ml 30% peroxide) was added and, after 30 min incubation at room temperature, the reaction was stopped by adding 2 N H_2SO_4. The absorbance at 492 nm was measured by photometer (Miniscan, Flow Labs. Ltd., Ayrshire, Scotland, U.K.).

Radioimmunoprecipitation (RIPA) assay

This was done by a modification of the procedure described by Stephenson et al. [14]. Aliquots (200 µl) of lysates of virus-infected PE cells or fractions of radiolabeled PE cells were incubated overnight at 4 °C with immunoglobulins in RIPA buffer. The reaction mixture was then incubated for 2 h at 4 °C with 50 µl of a saturated (0.3 g/ml) suspension of Protein A-Sepharose 4B (Pharmacia Chem. Co., Uppsala, Sweden) with gentle agitation. The gel was washed 5 times each with 1 ml of RIPA buffer. The sample was then eluted from the gel with 50 µl of sample buffer, boiled for 5 min and stored at −20 °C or used for polyacrylamide gel electrophoresis (PAGE) analyses. In control experiments we used the same procedure with lysates of mock-infected cells and normal mouse serum.

Analytical methods

Electrophoresis was done with 10% PAGE gels containing sodium dodecyl sulfate (SDS-PAGE), as described by O'Farrel [12]. After electrophoresis the gels were enhanced, dried and autoradiographed at −70 °C for various time intervals on X-ray film.

Results

Analysis of the monoclonal antibodies

Monoclonal antibodies positive by immunofluorescence were tested by ELISA using purified TBE virus or crude lysates of PE cells infected with TBE virus. Among others, we found two clones (N-1 and N-6) which reacted by ELISA only with crude homogenate in infected cells and which did not react by ELISA when purified virus was used as antigen. Immunoglobulin subclass of Mabs produced by those clones was IgG3.

To determine antigenic specificity, both Mabs were tested by an immunoprecipitation technique using crude extracts of radiolabeled TBE virus-infected PE cells. Immunoprecipitates were analyzed using SDS-PAGE electrophoresis. Both Mabs precipitated NS3 nonstructural protein of TBE virus. Mock-infected cells and normal mouse serum did not precipitate.

Type-specificity of Mabs against NS3

To study the type-specificity of Mabs, PE cells were infected with certain of TBE antigenic complex viruses (TBE strains Sofyin and Absettarov, Langat virus, and Powassan virus) and with other flaviviruses (yellow fever, Japanese encephalitis). The infected cells were labeled with ^{14}C-amino acid mixture. RIPA, using crude extracts of these cells, showed that Mabs N-1 and N-6 detected only NS3 of Sofyin, Absettarov, and Langat and did not precipitate NS3 of Powassan, yellow fever, or Japanese encephalitis viruses.   -

Stability of NS3

PE cells infected with TBE virus were pulse-labeled with the ^{14}C-amino acid mixture and chased for various time intervals. The lysates of these preparations were analyzed by immunoprecipitation with Mab N-6 and SDS-PAGE electrophoresed. The radiograph of the gel revealed that radioactive label could be detected in precipitated NS3 even after 20 h of chase. Therefore, we conclude that NS3 protein is relatively stable.

Localization of NS3 in infected cells

We separated TBE virus-infected PE cells into three fractions: (1) nuclear-associated membranes, (2) plasma membranes, and (3) cytoplasm. Each fraction was equilibrated by volume and, after immunoprecipitation with Mab N-6, was analyzed as described above. We found that NS3 is localized mainly in the fraction containing nuclear-associated membranes and in the cytoplasm but was not detected in the fraction containing the plasma membrane.

Discussion

We obtained two Mabs against native NS3 nonstructural protein of TBE virus; both were IgG3. In spite of high conservation of this protein among flaviviruses, these Mabs precipitated NS3 of TBE virus strains Sofyin and Absettarov and Langat virus but did not precipitate NS3 of Powassan, yellow fever or Japanese encephalitis viruses. We know of only one publication concerning Mabs against a flavivirus NS3. Gould et al. reported a lack of specificity of Mab against NS3 of Japanese encephalitis virus when this Mab was used to study NS3 of other flaviviruses [5]; that Mab also detected cellular protein of nuclear origin. Taking into account both the results of Gould et al. and our own, we propose the existence of at least two distinct epitopes on NS3: the first, conserved among flaviviruses, may be antigenically similar to one of the cellular proteins (cellular helicase?) and the second, a variable epitope, may be located within the regulatory region of NS3, the existence of which has been predicted to occur on the C-terminus of this protein [2, 4, 7].

Information about intracellular localization of NS3 is partially contradictory. Using immunofluorescence tests, some investigators have found NS3 distributed uniformly throughout the cytoplasm of infected cells, while others found preferential perinuclear localization [5, 11]. Our results confirm the possibility that NS3 is localized in the perinuclear region of infected cells (nuclear-associated membrane fraction) and in the cytoplasm as well. Contradictory to the immunogold study of Ng and Corner [11], who found limited amounts of NS3 on the surface of infected cells, we did not detect NS3 in the fraction containing plasma membranes. We tested Mabs N-1 and N-6 in passive protection tests but failed to detect reactivity, even when we used undiluted immune ascitic fluids. Ng and Hong have recently shown that NS3 protein is localized in close connection with cytoskeleton in the perinuclear region [10].

Hypothesis of the function of NS3

We suggest a putative functional model for the NS3 nonstructural protein. According to our studies and the studies of others, NS3 protein has two functions: as a protease, and as a helicase acting on double-stranded viral RNA during replication. NS3 is a highly hydrophilic protein. Nevertheless, it is found in quantity in the fraction containing nuclear-associated membrane. NS4a protein, which follows NS3 in the gene order sequence (5'-C-preM[M]-E-NS1-NS2a-NS2b-NS3-NS4a-NS4b-NS5-3') is a highly hydrophobic protein closely connected by C-terminus with membranes of the endoplasmic reticulum. Substrates for NS3 as a virus protease are localized near the membranes of the endoplasmic reticulum. According to these observations, we presume that NS4a protein plays a role in the functioning of

NS3. Although closely connected with NS4a, NS3 acts as a protease, taking part in virus polyprotein processing. After NS3 separates from NS4a it joins the polymerase complex. Attempts to verify this hypothesis are in progress.

References

1. Bazan JF, Fletterick RJ (1989) Detection of a trypsin-like serine protease domain in flaviviruses and pestiviruses. Virology 171: 637–639
2. Coia G, Perker MD, Speight G, Byrne ME, Westaway EG (1988) Nucleotide and complete amino acid sequences of Kunjin virus: definitive gene order and characteristics of the virus-specified protein. J Gen Virol 69: 1–21
3. Gorbalenya AE, Koonin EV, Donchenko AP, Blinov VM (1989) Two related super-families of putative helicases involved in replication, recombination, repair and expression of DNA and RNA genomes. Nucleic Acids Res 17: 4713–4730
4. Gorbalenya AE, Donchenko AP, Koonin EV, Blinov VM (1989) N-terminal domains of putative helicases of flavi- and pestiviruses may be serine proteases. Nucleic Acids Res 17: 3879–3897
5. Gould EA, Chanar AC, Buckley A, Clegg CS (1983) Monoclonal immunoglobulin M antibody to Japanese encephalitis virus that can react with a nuclear antigen in mammalian cells. Infect Immun 41: 774–779
6. Grun JB, Brinton MA (1986) Characterization of West Nile virus RNA-dependent RNA polymerase and cellular terminal adenylyl and uridylyl transferases in cell-free extracts. J Virol 60: 1113–1124
7. Hahn YS, Galler R, Hunkapiller T, Dalrymple JM, Strauss SH, Strauss EG (1988) Nucleotide sequence of dengue 2 RNA and comparison of the encoded proteins with those of other flaviviruses. Virology 162: 167–180
8. Kushch AA, Novak M, Melnikova YeF, Gaidamovich SY, Gresikova M, Sekeyova M, Novokhatsky AS, Mikheeva TG, Sveshnikova NA, Borecky LS, Zhdanov VM (1986) Preparation and characterization of hybridomas secreting monoclonal antibodies to tick-borne encephalitis virus. Acta Virol 30: 199–205
9. Lyapustin VN, Gritsun TS, Lashkevich VA (1986) Effect of multiplicity of infection on synthesis of the tick-borne encephalitis virus specific proteins during a single repro-duction cycle. Acta Virol 30: 289–293
10. Ng ML, Hong SS (1989) Flavivirus infection: essential ultrastructural changes and association of Kunjin virus NS3 protein with microtubules. Arch Virol 106: 103–120
11. Ng ML, Corner LC (1989) Detection of some dengue-2 virus antigens in infected cells using immunomicroscopy. Arch Virol 104: 197–208
12. O'Farrel PH (1975) High-resolution two-dimensional electrophoresis of proteins. J Biol Chem 250: 4007–4021
13. Rice CM, Lenches EM, Eddy SR, Shin SJ, Sheets RL, Strauss JH (1985) Nucleotide sequence of yellow fever virus: implications for flavivirus gene expression and evolution. Science 229: 726–733
14. Stephenson JR, Grooks AJ, Lee JM (1987) The synthesis of immunogenic polypeptides encoded by tick-borne encephalitis virus. J Gen Virol 68: 1307–1316

Authors' address: A. V. Timofeev Institute of Poliomyelitis and Viral Encephalitides, U.S.S.R. Academy of Medical Sciences, Moscow 142782, U.S.S.R.

Arch Virol (1990) [Suppl 1]: 125–135

The envelope protein E of tick-borne encephalitis virus and other flaviviruses: structure, functions and evolutionary relationships

F. X. Heinz, C. W. Mandl, F. Guirakhoo, Heidi Holzmann, W. Tuma, and **C. Kunz**

Institute of Virology, University of Vienna, Vienna, Austria

Accepted January 24, 1990

Summary. A structural model of the TBE virus E protein contains information on the arrangement of the polypeptide chain into distinct protein domains corresponding to antigenic domains defined by MAbs. The model is consistent with data obtained from other flaviviruses as well and it is reasonable to assume a common structural organization of all flavivirus E proteins. Indirect evidence exists that certain distinct sequence elements may be involved in membrane fusion and receptor binding. An evolutionary tree based on amino acid sequence homology of the E protein corresponds nicely to the established subdivision of flaviviruses into serocomplexes. Under natural ecological conditions the flavivirus E protein exhibits a low degree of variability, different isolates of the same virus generally sharing at least 98% of their amino acids.

Introduction

Members of the family *Flaviviridae* are characterized by a positive stranded RNA genome of almost 11000 nucleotides (for review, see [33]) and a common structural organization of virions. The nucleocapsid (composed of RNA and the capsid protein C) is surrounded by a lipid envelope containing two integrated membrane proteins. The major envelope protein E (MW 50–60000) of most but not all flaviviruses investigated so far is glycosylated and may contain one or two glycosylation sites. There is a second small membrane-associated protein (M; MW 7–8000) which is derived from a glycosylated precursor protein (prM) by proteolytic cleavage, causing the loss of its external carbohydrate containing part. The processing of prM—presumably by a Golgi protease—apparently represents a late event in the maturation of flaviviruses [40].

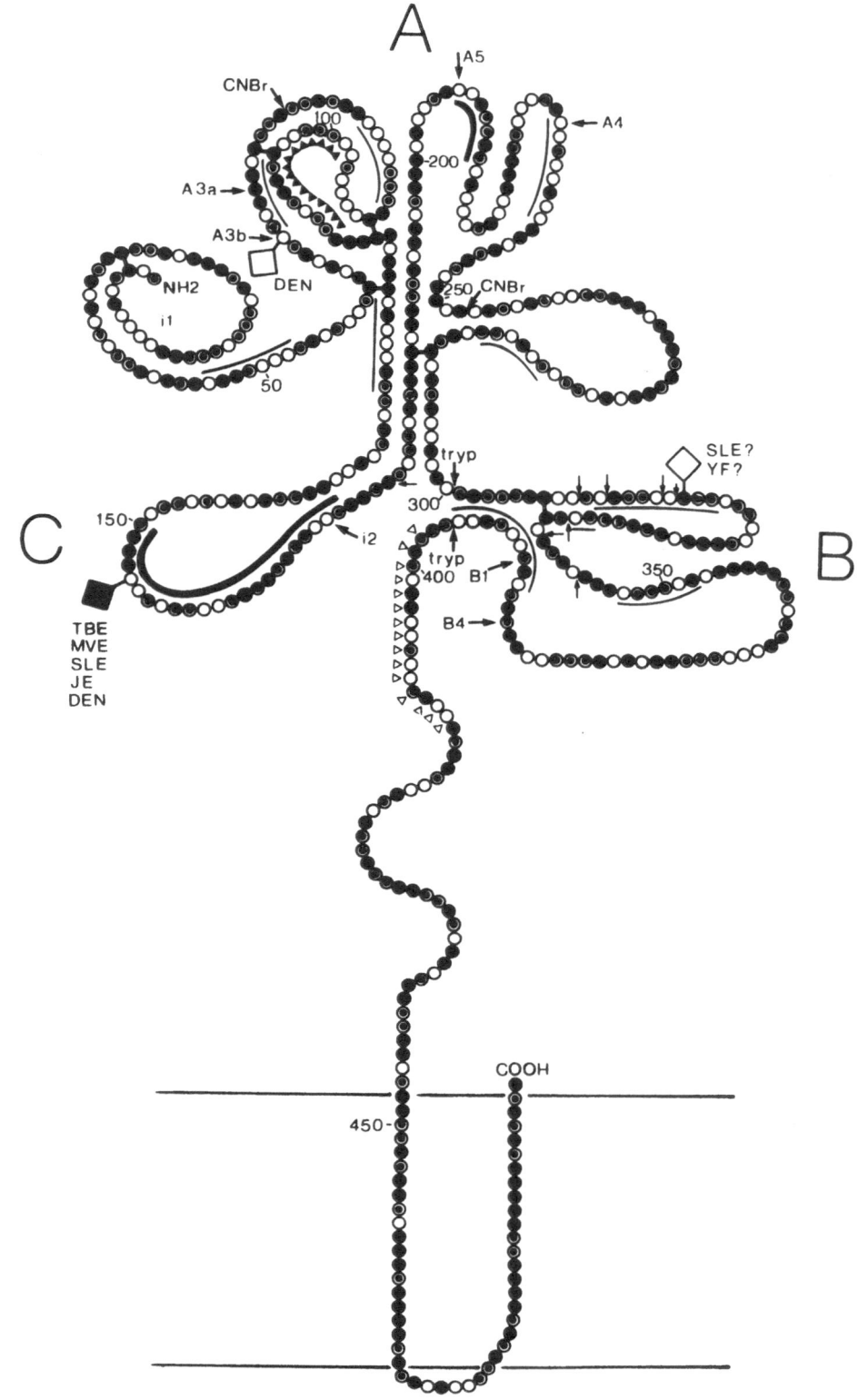

Fig. 1.

As deduced from the scheme of polyprotein processing both membrane proteins are believed to span the membrane twice at their carboxytermini, with the carboxy-terminus exposed outside the membrane. Detergent solubilized membrane proteins can be reassociated by virtue of their hydrophobic carboxytermini to form "rosette"-like structures which retain the native conformation of the E protein [15].

Several important biological activities are mediated by the E protein: it is the viral hemagglutinin, the receptor-binding protein and seems also to be involved in membrane fusion activity after acid-pH-induced conformational changes. It is of paramount importance for inducing a protective immunity and also represents the basis for the division of flaviviruses into several serocomplexes by crossneutralization with polyclonal immune sera [2, 6].

The antigenic structure and functions of protein E

Monoclonal antibodies (Mabs) have been used extensively in order to obtain a more precise description of the antigenic structure of protein E. These studies yielded a large amount of data including serological specificities, functional activities, and topological relationships of epitopes which were also used for the generation of epitope maps (for reviews, see [18, 34]). Extraordinary complex reactivity patterns of Mabs were obtained which revealed unexpected and previously unrecognized relationships between flaviviruses of different serocomplexes. These data are an apparent reflection of the heterogeneity of the antibody-accessible surface of protein E but did not, however, lead to a unifying picture of its antigenic structure.

Fig. 1. Model of the tick-borne encephalitis virus protein E. ○ Hydrophilic amino acid residues (Arg, Lys, Asn, Asp, Gln, Glu, His), ⊙ intermediate amino acid residues (Pro, Tyr, Ser, Trp, Thr, Gly), ● hydrophobic amino acid residues (Ile, Val, Leu, Phe, Cys, Met, Ala). Position numbers are shown every 50 amino acids. Cysteine residues forming disulfide bridges are connected by solid lines. Arrows depict cleavage sites that liberate immunoreactive fragments by the use of trypsin (tryp) and cyanogen bromide (CNBr). Small arrows indicate potential cleavage sites within these fragments that are not utilized. Two solid lines stand for the lipid membrane that is spanned by two transmembrane regions of protein E. The polypeptide chain is folded to indicate the antigenic domains A, B, and C (*A, B, C*). Arrows together with the names of neutralizing Mabs depict the locations of the mutations identified in the antigenic variants of tick-borne encephalitis virus selected by the respective Mab by sequence analysis. ▼ Almost perfectly conserved sequence within domain A. ▽ Region of a potential T-cell determinant. ◆ Carbohydrate side chain of tick-borne encephalitis virus. The Murray Valley encephalitis, St. Louis encephalitis, Japanese encephalitis, and dengue viruses have potential N-glycosylation sites at the homologous position. Yellow fever and St. Louis encephalitis viruses have such a site within domain B, dengue viruses an additional site within domain A. ◇ Homologous positions of tick-borne encephalitis virus. Thin solid lines indicate variable and thick solid lines hypervariable sequences as deduced from a sequence comparison of 15 flaviviral E proteins. Reproduced from [28] with permission from "Journal of Virology"

Using tick-borne encephalitis (TBE) virus as a model we have therefore
attempted to identify the location of individual epitopes in the sequence of
protein E and to correlate antigenic domains with distinct sequence elements
and protein domains [28]. The structural model shown in Fig. 1 is based on
an epitope map (Fig. 2) [11] that reveals the serological specificities, functional activities and topological relationships of 19 distinct epitopes. Based
on competitive binding studies most of those epitopes were shown to cluster
in three antigenic domains (A, B, C). The epitopes i1, i2, and i3 are defined by
Mabs for which no mutual blocking with other Mabs was observed.

Each of the domains contains epitopes of different serological specificities; broadly flavivirus-crossreactive epitopes, however, were only found
within domain A, whereas most of the domain C epitopes are subtype-
specific. Antibodies to certain epitopes within each of the domains were
shown to be involved in hemagglutination–inhibition and/or neutralization,
with those of domain A revealing the highest specific activities.

The binding of some Mabs may result in enhanced binding of other Mabs
to distant antigenic sites (Fig. 2, solid arrows). Such cooperative interactions
were described for other flaviviruses as well, both in binding assays and in
functional assays, such as neutralization and passive protection [19, 23].

The structural model shown in Fig. 1 includes the disulfide bridge
assignments determined by Nowak and Wengler [32] for the E protein of
West Nile virus and is based on data obtained by the following approaches:
selection and sequence analysis of Mab-escape mutants; generation of
defined protein fragments by the use of proteases, CNBr, or expression in
bacteria and assessment of their reactivity with each of the Mabs; characterization of the structural properties of each epitope with respect to conformation dependency, denaturation resistance, and involvement of disulfide
bridges.

Based on considerations described in more detail elsewhere [28] antigenic domain A apparently involves discontinuous sequence elements

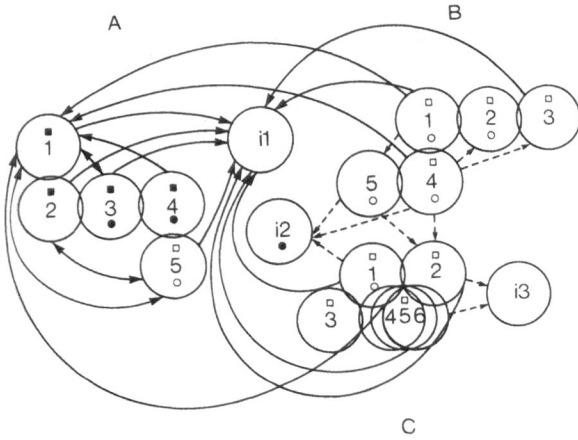

Fig. 2. Epitope map of the tick-
borne encephalitis virus glycoprotein
E showing three antigenic domains
(*A, B, C*) and three isolated epitopes
(*i1, i2, i3*). Overlapping circles indicate mutual blocking, dotted arrows
one-way blocking, and solid arrows
enhancement of binding as determined in competitive binding assays. Functional activities: ■ strong
HI activity; □ weak HI activity:
● strong NT activity; ○ weak NT
activity. Reproduced from [11] with
permission from "Virology"

separated by about 70 amino acids. This includes the highly denaturation-sensitive epitopes A3, A4, and A5 (mapped by use of Mab-escape mutants) and the epitopes A1 and A2 which are believed to involve the strongly disulfide bridge stabilized part of domain A. This structural element also contains the most highly conserved sequence of protein E (amino acids 98 to 111, located on both sides of a disulfide bridge). There are some indications that this sequence may be involved in acid pH-induced membrane fusion. It includes the sequence Gly–Leu–Phe–Gly which also forms the fusion-active amino terminus of the influenza virus HA2 (41). An inverse sequence (Phe–Leu–Gly) is also present at the amino terminus of the paramyxovirus F1 fusion protein and (as a tandem repeat) within the fusion-active gp41 of HIV-1, HIV-2 and SIV [8]. Low pH-induced conformational changes affecting the epitopes of domain A [11] indicate a rearrangement of the polypeptide chain which may be necessary for the exposure of the fusion-active site.

As revealed by bacterial expression, the isolated epitope i1 is located within the first 34 amino-terminal amino acids and is independent of the first disulfide bridge (Tuma W et al., unpubl. results). The epitopes of domain B are located on a characteristic structural element (amino acids 301 to 395) which contains a single disulfide bridge and can be isolated by trypsin digestion of the native protein [17, 42]. These epitopes are dependent on the disulfide bridge and also involve sequences near the carboxyterminus of this fragment. Antigenic domain C seems to be confined to a single hypervariable loop which also carries the carbohydrate side chain of TBE virus and several other flaviviruses.

Considering the conformational conservation of other evolutionary-related viral proteins [36] one can assume a common structural organization of flavivirus E proteins in general. This is also consistent with the absolute conservation of all 12 cysteine residues and the virtually superimposable hydrophilicity plots, despite considerable amino acid sequence divergence (discussed below).

Our model is in concordance with the results of experiments performed with other flaviviruses. The epitope recognized by the TBE virus-neutralizing Mab A3 corresponds exactly to the position of a neutralization site in the E protein of yellow fever (YF) virus [24]. A structural element homologous to antigenic domain B was identified in the E protein of Japanese encephalitis (JE) virus by Mason et al. [30].

Additional information was obtained by Roehrig et al. [35] who applied synthetic peptides to study the antigenic structure of the Murray Valley encephalitis (MVE) virus protein E. A neutralization site was identified which involves amino acids 35 to 50 and competitive binding studies using antipeptide sera and monoclonal antibodies provide evidence for a certain link between sequence elements containing residues 35–97, 305–319, and 356–376.

Each of seven TBE virus mutants selected in the presence of neutralizing Mabs [21] had only a single amino acid exchange in the E protein sequence as compared to the wild type (28). As revealed by studies using the mouse model, the amino acid exchange at position 384 (Tyr→His) in the mutant B4 resulted in a strong reduction of neurovirulence upon peripheral inoculation of adult mice [21a]. There is evidence that sequences within antigenic domain B may be of relevance for the attenuation of flaviviruses in general. Host range mutants which were selected by passaging MVE virus in human cells and had a single amino acid exchange at position 390 were also shown to be attenuated for 21 days old mice [26]. In addition, 3 of the 7 nonconservative amino acid exchanges in the E protein of the YF vaccine strain 17D are located within domain B [12]. Also, 3 of 5 nonconservative amino acid exchanges in an attenuated mutant of dengue virus type 2 map to this domain [13]. One may therefore speculate that sequences within antigenic domain B are involved in the putative receptor binding site and that changes within this site may alter receptor specificity and thus lead to attenuation.

Relationships among flaviviruses based on sequence comparisons

The recent elucidation of the genome structures of a considerable number of flaviviruses [4, 5, 7, 9, 14, 27, 29, 37, 38, 43, 44] allows us to further assess the relationships among flaviviruses by determining nucleic acid and amino acid sequence homologies. An evolutionary tree compiling the presently available amino acid sequence data of flavivirus E proteins is shown in Fig. 3. Considering the considerable confusion introduced by the application of monoclonal antibodies it is quite interesting to note that the homology data yield a picture that perfectly matches that of flavivirus serocomplexes as defined by cross-neutralization using polyclonal immune sera [2, 6].

YF virus and TBE virus are distantly related to the other flaviviruses sequenced so far, revealing an amino acid sequence homology of 41–46%, (depending on the pair of viruses compared). There is a somewhat closer relationship between members of the JE serocomplex and the dengue viruses (46–53%). The different dengue types share 62–69% of their amino acids, with den-1 and den-2 being more closely related to each other than to den-4. Members of the JE serocomplex reveal a higher degree of homology, with St. Louis encephalitis virus so far being the most distantly related virus, whereas West Nile and Kunjin virus share 93% of their amino acids.

The sequence data obtained for a European strain [27] and a far eastern strain [43] of TBE virus (sequence homology of 95.6%) are consistent with their classification as distinct subtypes.

Limited sequence information of Powassan virus E protein including about 10% of the whole protein (Mandl CW et al., unpubl. results) reveals less than 50% homology with mosquito borne flaviviruses and about 75%

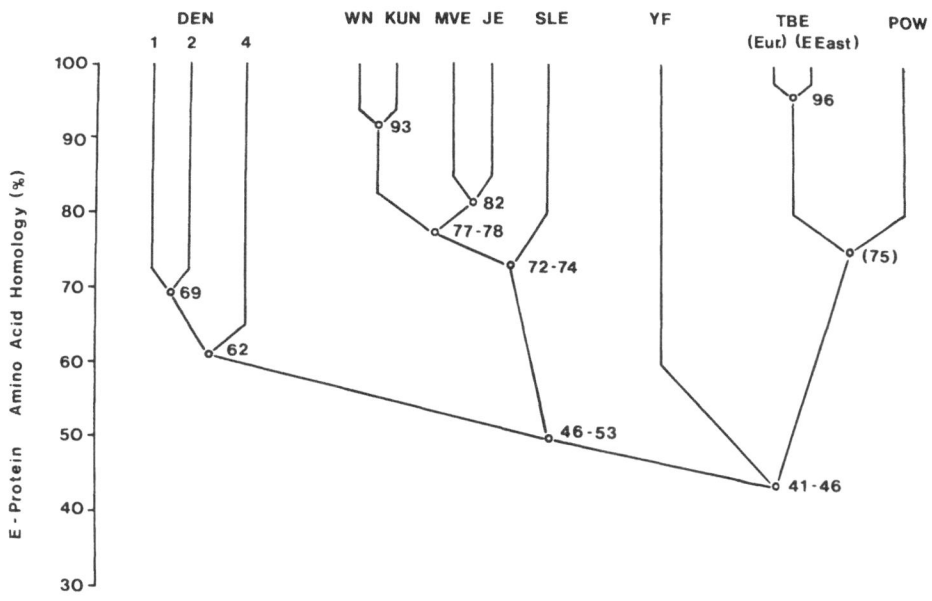

Fig. 3. Evolutionary tree of flaviviruses based on amino acid sequence homologies of their E proteins. Homology data were calculated by the Beckman Microgenie Software package (Version 4.0) as percentage identical residue after optimizing alignment of compared sequences counting gaps as mis-matches. Numbers at the branching points (○) relate to percentage homologies between those viruses shown to segregate at the respective point

homology with TBE virus. This is similar to homologies found within the JE serocomplex. Although this figure may change slightly when knowledge of the whole sequence becomes available, it is consistent with Powassan virus being a member of the TBE serocomplex, as proposed by Calisher et al. [2].

A discussion has recently been initiated concerning the term "TBE virus" and proposing its substitution by "Central European Encephalitis" virus for the western subtype and "Russian Spring Summer Encephalitis" virus for the far eastern subtype [1]. We are aware that the nomenclature of viruses bears intrinsic problems and may have its drawbacks (e.g., JE virus definitely not only occurs in Japan and is also not the only virus causing encephalitis in Japan). To avoid further confusion we suggest, however, to retain the term "TBE virus", which is generally accepted and used by the scientific community throughout the world. Recently published sequence data [27, 43] as well as serological data clearly allow the distinction of two closely related subtypes of the same virus (about 96% amino acid homology in the E protein). We have previously presented evidence for a high degree of homogeneity of strains isolated throughout Europe, including Finland. This is now corroborated by sequence comparisons and monoclonal antibody studies. The isolate A-52 from Kumlinge island (Finland) reveals only a single conservative amino acid exchange (I → V) in the E protein as compared to the Austrian isolate Neudoerfl (Mandl CW et al., unpubl. results). Also strain Absettarov (a human isolate from the Leningrad region) is absolutely indistinguishable from strain Neudoerfl by the panel of monoclonal antibodies available to date (Holzmann H et al., unpubl. obs.). Therefore the term "Central European" encephalitis virus certainly is not warranted and we suggest retention of the designations "European and far eastern subtype" of TBE virus.

Further analyses will be necessary to determine whether other isolates from the far eastern part of the U.S.S.R. are closely related to the prototype strain Sofyn or whether additional subtypes can be distinguished. We also propose that this nomenclature be adopted by the International Catalogue of Arboviruses [22]. Absettarov, Hanzalova, Hypr, and Kumlinge virus are clearly no more than different isolates of the same European subtype of TBE virus and should therefore not be listed as distinct viruses (compare Table I in [1]).

Variation of the E protein

By the use of peptide mapping and monoclonal antibody-reactivity profiles we have previously shown that the E proteins of TBE viruses isolated throughout Europe are quite homogeneous and apparently exhibit only a low degree of variation [10, 16]. These data are now corroborated by sequence analyses of two further Austrian TBE virus isolates (ZZ9 and Scharl) and an isolate from Kumlinge island, Finland (A-52) (Mandl CW et al., unpubl. results).

The E protein of the Finnish isolate differs from the completely sequenced Austrian strain Neudoerfl by a single conservative amino acid substitution (I→V). So far, a maximum of 6 substitutions have been found, corresponding to a sequence homology of 98.8%.

Similar to TBE virus mosquito-borne flaviviruses apparently also exhibit a low degree of variability in their E proteins. There is a maximum of 5 amino acid substitutions (99% homology) between any of the 3 sequenced strains of JE virus (Nakayama [31]; JaOAr S982 [37]; Beijing [14]). Australian isolates of MVE virus differ by no more than 3 amino acids (99.4% homology) and by 6 to 11 amino acid substitutions from Papua New Guinea isolates (at least 97.8% homology) [25]. A similar situation is also encountered with dengue viruses. Carribean isolates of den-1 were shown to differ by no more than 3 amino acids (99.4% homology) and by a maximum of 10 amino acids (98% homology) compared to isolates from the Philippines or Nauru island [3, 29]. Similarly, the New Guinea-C strain of den-2 differs from Carribean isolates by a maximum of 6 amino acid substitutions in their E proteins [9]. In summary, strain variations generally do not exceed a 2% amino acid sequence divergence in the E protein.

Given the potential for a high degree of variation intrinsic to RNA genomes [20] the observed stability of flavivirus E proteins is quite remarkable. Apparently the balanced ecological conditions necessary to maintain the virus in its natural cycle exert strong restraints that counteract extensive variations.

References

1. Calisher CH (1988) Antigenic classification and taxonomy of flaviviruses (family *Flaviviridae*) emphasizing a universal system for the taxonomy of viruses causing tick-borne encephalitis. Acta Virol 32: 469–478

2. Calisher CH, Karabatsos N, Dalrymple JM, Shope R, Porterfield JS, Westaway EG, Brandt WE (1989) Antigenic relationships between flaviviruses as determined by cross-neutralization tests with polyclonal antisera. J Gen Virol 70: 37–43

3. Chu MC, O'Rourke EJ, Trent DW (1989) Genetic relatedness among structural protein genes of dengue 1 virus strains. J Gen Virol 70: 1701–1712

4. Coia G, Parker MD, Speight G, Byrne ME, Westaway EG (1988) Nucleotide and complete amino acid sequences of Kunjin virus: definitive gene order and characteristics of the virus-specified proteins. J Gen Virol 69: 1–21

5. Dalgarno L, Trent D, Strauss JH, Rice CM (1986) Partial nucleotide sequence of Murray Valley encephalitis virus genome: comparison of the encoded polypeptides with yellow fever virus structural and nonstructural proteins. J Mol Biol 187: 309–323

6. deMadrid AT, Porterfield JS (1974) The flaviviruses (group B arboviruses): a cross-neutralization study. J Gen Virol 23: 91–96

7. Deubel V, Kinney RM, Trent DW (1988) Nucleotide sequence and deduced amino acid sequence of the non-structural proteins of dengue type 2 virus, Jamaica genotype: comparative analysis of the full-length genome. Virology 165: 234–244

8. Gallaher WR (1987) Detection of a fusion peptide sequence in the trans-membrane protein of human immunodeficiency virus. Cell 50: 327–328

9. Gruenberg A, Woo WS, Biedrcycka A, Wright PJ (1988) Partial nucleotide sequence and deduced amino acid sequence of the structural proteins of dengue virus type 2, New Guinea C and PUO-218 strains. J Gen Virol 69: 1391–1398

10. Guirakhoo F, Radda AC, Heinz FX, Kunz C (1987) Evidence for antigenic stability of tick-borne encephalitis virus in nature by the analysis of natural isolates. J Gen Virol 68: 859–864

11. Guirakhoo F, Heinz FX, Kunz C (1989) Epitope model of tick-borne encephalitis virus envelope glycoprotein E: analysis of structural properties, role of carbohydrate side chain, and conformational changes occurring at acidic pH. Virology 169: 90–99

12. Hahn CS, Dalrymple JM, Strauss JH, Rice CM (1987) Comparison of the virulent Asibi strain of yellow fever virus with the 17D vaccine strain derived from it. Proc Natl Acad Sci USA 84: 2019–2023

13. Hahn CS, Galler R, Hunkapiller T, Dalrymple JM, Strauss JH, Strauss EG (1988) Nucleotide sequence of dengue 2 RNA and comparison of the encoded proteins with those of other flaviviruses. Virology 162: 167–180

14. Hashimoto H, Nomoto A, Watanabe K, Mori T, Takezawa T, Aizawa C, Takegami T, Hiramatsu K (1988) Molecular cloning and complete nucleotide sequence of the genome of Japanese encephalitis virus Beijing-1 strain. Virus Genes 1: 305–317

15. Heinz FX, Kunz C (1980) Formation of polymeric glycoprotein complexes from a flavivirus: tick-borne encephalitis virus. J Gen Virol 49: 125–132

16. Heinz FX, Kunz C (1981) Homogeneity of the structural glycoprotein from European isolates of tick-borne encephalitis virus: comparison with other flaviviruses. J Gen Virol 57: 263–274

17. Heinz FX, Berger R, Tuma W, Kunz C (1983) Location of immunodominant antigenic

determinants on fragments of the tick-borne encephalitis virus glycoprotein: evidence for two different mechanisms by which antibodies mediate neutralization and hemagglutination inhibition. Virology 130: 485–501

18. Heinz FX (1986) Epitope model of flavivirus glycoproteins. Adv Virus Res 31: 103–106
19. Heinz FX, Mandl C, Winkler G, Tuma W, Kunz C (1986) Cooperative interactions between antibodies to structurally distinct antigenic sites. In: Vaccines 86. Cold Spring Harbor Laboratory, New York, pp 387–392
20. Holland J, Spindler K, Horodyski F, Graban E, Nichol S, Vanopol S (1982) Rapid evolution of RNA genomes. Science 215: 1577–1585
21. Holzmann H, Mandl C, Guirakhoo F, Heinz FX, Kunz C (1989) Characterization of antigenic variants of tick-borne encephalitis virus selected with neutralizing monoclonal antibodies. J Gen Virol 70: 219–222
21a. Holzmann H et al (1990) A single amino acid substitution in envelope protein E of tick-borne encephalitis virus leads to attenuation in the mouse model. J Virol (in press)
22. Karabatsos N (ed) (1985) International catalogue of arboviruses including certain other viruses of vertebrates, 3rd edn. American Society for Tropical Medicine and Hygiene, San Antonio, Texas
23. Kimura-Kuroda J, Yasui K (1986) Antigenic comparison of envelope protein E between Japanese encephalitis virus and some other flaviviruses using monoclonal antibodies. J Gen Virol 67: 2663–2672
24. Lobigs M, Dalgarno L, Schlesinger JJ, Weis RC (1987) Location of a neutralization determinant in the E protein of yellow fever virus (17D vaccine strain). Virology 161: 474–478
25. Lobigs M, Marshall ID, Weir RC, Dalgarno L (1988) Murray Valley encephalitis virus field strains from Australia and Papua New Guinea. Studies on the sequence of the major envelope protein gene and virulence for mice. Virology 165: 245–255
26. Lobigs M, Usha R, Nestorowicz A, Marshall ID, Weir RC, Dalgarno L (1990) Host cell selection of Murray Valley encephalitis virus variants altered at an RGD sequence in the envelope protein and in mouse virulence. Virology 176: 587–595
27. Mandl CW, Heinz FX, Kunz C (1988) Sequence of the structural proteins of tick-borne encephalitis virus (Western subtype) and comparative analysis with other flaviviruses. Virology 166: 197–205
28. Mandl CW, Guirakhoo F, Heinz FX, Kunz C (1989) Antigenic structure of the flavivirus envelope protein E at the molecular level, using tick-borne encephalitis virus as a model. J Virol 63: 564–571
29. Mason PW, McAda PC, Mason TL, Fournier MJ (1987) Sequence of the dengue-1 virus genome in the region encoding three structural proteins and the major nonstructural protein NS1. Virology 161: 262–267
30. Mason PW, Dalrymple JM, Gentry MK, McCown JM, Hoke CH, Burke DS, Fournier MJ, Mason TL (1989) Molecular characterization of a neutralizing domain of the Japanese encephalitis virus structural glycoprotein. J Gen Virol 70: 2037–2049
31. McAda PC, Mason PW, Schmaljohn CS, Dalrymple JM, Mason TL, Fournier MJ (1987) Partial nucleotide sequence of Japanese encephalitis virus genome. Virology 158: 348–360
32. Nowak T, Wengler G (1987) Analysis of disulfides present in the membrane proteins of the West Nile flavivirus. Virology 156: 127–137
33. Rice CM, Strauss EG, Strauss JH (1986) Structure of the flavivirus genome. In: Schlesinger S, Schlesinger MJ (eds) The Togaviridae and Flaviviridae. Plenum, New York, pp 279–326
34. Roehrig JT (1986) The use of monoclonal antibodies in studies of the structural proteins of togaviruses and flaviviruses. In: Schlesinger S, Schlesinger MJ (eds) The Togaviridae and Flaviviridae. Plenum, New York, pp 251–271

35. Roehrig JT, Hunt AR, Johnson AJ, Hawkes RA (1989) Synthetic peptides derived from the deduced amino acid sequence of the E-glycoprotein of Murray Valley encephalitis virus elicit antiviral antibody. Virology 171: 49–60
36. Rossmann MG, Rueckert RR (1987) What does the molecular structure of viruses tell us about viral functions? Microbiol Sci 4: 206–214
37. Sumiyoshi H, Morita K, Mori C, Fuke I, Shiba T, Sakaki Y, Igarashi A (1986) Sequence of 3000 nucleotides at the 5' end of Japanese encephalitis virus RNA. Gene 48: 195–201
38. Trent DW, Kinney RM, Johnson BJB, Vorndam AV, Deubel V, Rice CM, Hahn CS (1987) Partial nucleotide sequence of St. Louis encephalitis virus RNA: structural proteins, NS1, NS2a and NS2b. Virology 156: 293–304
39. Wengler G, Castle E, Leidner U, Nowak T, Wengler G (1985) Sequence analysis of the membrane protein V3 of the flavivirus West Nile virus and of its gene. Virology 147: 264–274
40. Wengler G, Wengler G (1989) Cell-associated West Nile flavivirus is covered with E + pre-M protein heterodimers which are destroyed and reorganized by proteolytic cleavage during virus release. J Virol 63: 2521–2526
41. Wharton SA, Martin SR, Ruigrok RWH, Skehel JJ, Wiley DC (1988) Membrane fusion by peptide analogs of influenza virus hemagglutinin. J Gen Virol 69: 1847–1857
42. Winkler G, Heinz FX, Kunz C (1987) Characterization of a disulfide bridge stabilized antigenic domain of tick-borne encephalitis virus structural glycoprotein. J Gen Virol 68: 2239–2244
43. Yamshchikov VF, Pletnev AG (1988) Nucleotide sequence of the genome region encoding the structural proteins and the NS1 protein of the tick-borne encephalitis virus. Nucleic Acids Res 16: 7750
44. Zhao B, Mackow E, Buckler-White A, Markoff L, Chanock RM, Lai CI, Makino Y (1986) Cloning full length dengue type 4 viral DNA sequences: analysis of genes coding for structural proteins. Virology 156: 77–88

Authors' address: Dr. F. X. Heinz, Institute of Virology, University of Vienna, Kinderspitalgasse 15, A-1095 Vienna, Austria.

Arch Virol (1990) [Suppl 1]: 137–152

Antigenicity of flaviviruses

E. A. Gould[1], A. Buckley[1], S. Higgs[1], and Sophia Gaidamovich[2]

[1] Institute of Virology and Environmental Microbiology, Oxford, U.K.
[2] WHO Collaborating Centre, D.I. Ivanovsky Institute of Virology, Moscow, U.S.S.R.

Accepted March 20, 1990

Summary. The importance of viral structural and non-structural antigens in flavivirus pathogenesis has not been satisfactorily assessed. This review examines the antigenic interrelationships of the flaviviruses, considers some of the implications of antigenic cross reactivity in virus virulence and highlights potential consequences of virus-antibody interactions.

Introduction

Viruses in the family *Flaviviridae*, of which yellow fever is the type species [94], are responsible for considerable human morbidity and mortality in particular areas of the world. There are currently 68 registered antigenically related, but distinct, flaviviruses, some of which are known to produce diseases causing uncomplicated febrile illness, perhaps accompanied by a rash, to hemorrhagic fever, shock syndrome, hepatitis or encephalitis (for reviews see [67, 57]). The morphological appearance of a single virion is illustrated schematically in Fig. 1. The particles are spherical with a diameter of 40–50 nm. There are three structural proteins: the capsid (C; M_r 13,000 to 16,000), which is surrounded by a host-derived membrane incorporating 2 viral proteins; the membrane protein (M; M_r 8,000), derived from a glycosylated precursor (prM; M_r 24,000 to 27,000), and the envelope (E; M_r 50,000 to 59,000), which is usually glycosylated. It is not yet clear if the C-terminal of the E protein consists of dual transmembrane elements, shown as a hooked structure [54, 91], or if there is a single transmembrane structure. This latter proposal derives from the observations of Aaskov et al. [2] and Innis et al. [48] who detected strong serological responses to the first hydrophobic transmembrane region, suggesting it could be exposed on the surface of the lipid envelope (both dual and single transmembrane elements are depicted in Fig. 1). The flaviviruses have a linear positive-stranded

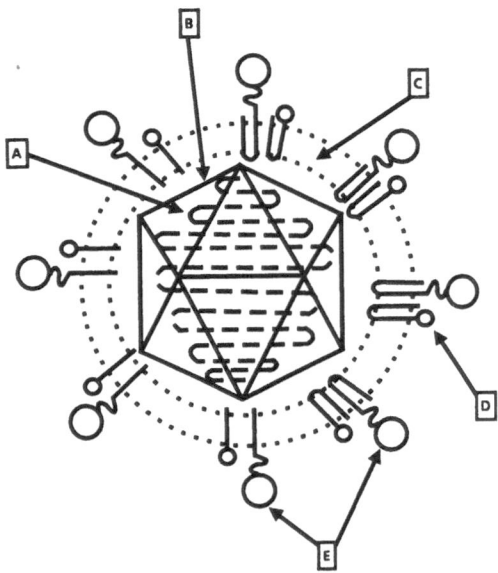

Fig. 1. Schematic representation of a flavivirus particle. Two forms of the virion are depicted; on the right, the virion has a small transmembranal span of appropriate polypeptide whereas on the left, the polypeptide is shown as a single transmembranal span; *A* viral RNA; *B* viral capsid; *C* lipid membrane; *D* membrane protein; *E* envelope protein

infectious RNA molecule between 10,000 and 11,000 nucleotides in length. The genomes of several flaviviruses have been sequenced and they have a similar organization [18, 24–26, 53, 54, 83, 85, 87]. The viral proteins are encoded in one open reading frame which represents about 90% of the genome. Seven non-structural virus-coded proteins are recognized and the entire gene order of both the structural and non-structural proteins is now known to be C-prM-E-NS1-NS2a-NS2b-NS3-NS4a-NS4b-NS5 [4, 5, 7, 19, 69, 70, 80].

The flaviviruses, originally defined as the group B arboviruses, were shown to cross-react in standard serological tests [17]. They were subsequently rearranged into 7 serogroups on the basis of their antigenic interrelationships in plaque reduction neutralization tests and also by taking into account the invertebrate vector, where recognized [65]. Additional viruses were recently examined in neutralization tests and the sub-groupings were updated accordingly [11].

With the advent of monoclonal antibody technology it became possible to examine these antigenic interrelationships in considerably more detail than was previously possible. Monoclonal antibodies (MAb) have therefore begun to be used in studies of flavivirus proteins. These studies contribute data on the possible significance of viral protein interactions with antibodies in flavivirus pathogenesis.

The viral envelope glycoprotein

Prior to the use of MAb it was recognized that the E glycoprotein of flaviviruses contained at least three classes of antigens, viz. type specific, complex reactive and flavivirus group reactive [16, 22, 86]. Subsequently, there have been several extensive analyses of antigenic cross reactivity between flaviviruses, particularly with E-specific MAb [28, 30, 33, 43, 44, 50, 51, 58, 63, 74, 82]. Antibody binding assays identify reactivity in nonfunctional tests and these have confirmed and extended the original observations. Some typical results obtained using indirect immunofluorescence tests with E-specific MAb and a large number of flaviviruses are presented in Table 1. Unique epitopes can distinguish between individual virus strains (e.g., 864), some epitopes are virus-specific (e.g., 825, 546 or 995), whilst others show intermediate reactivity with other flaviviruses (e.g., 126), particularly within individual serogroups. Finally, broadly cross reactive epitopes (e.g., 813) confirm the common ancestral origin of the flaviviruses. It is of

Table 1. Antigenic cross reactivity analyses between flaviviruses, using monoclonal antibodies (against serogroup 3 viruses) that identify the viral E protein

```
   MAb            56 Flaviviruses (group arranged)

          ----------  --------------   -------------------  --  ----  -----  ----
          GLLMNPRSCT  ACCDEJKKMMPRSS   AABIJJKKMNSSSUWWY    SZ  BINT  BBERU  DDDD
          GGIEEOFREY  PBRBNUAOOMPBAO   LRSLBUOUVJLETSSNF    PI  ATTM  AOHOG  EEEE
          YT AGW EEU  OC   TTDUDLB BK  FOQHEGKNENEPRUL      OK  G AU  NU CS  NNNN
          I                            A                    A                1234

   864                                                       +
   825                                                       +
   411                                                       +
   117                                                       +
   546                                                       +
   995                                            +
   541            +                                          +    +
   127                                        +              +    +
   538                                            +          +                 +
   549                            +                          +         ++
   427            +                       +      +           +
   623                                         +   +       + +              +
   140         +                     +         +            +         ++
  101/R3            +          +             +++ ++          +         ++  +
  F6/16A     +          +  +    +           +  ++ +          +         ++  +          +
   126            +          +     + +    + ++ ++  ++      + ++        ++ + ++   +
    38     ++ ++   +          +           +    + ++++++ ++ ++++        ++ + +  +
    39     +++ ++  +          +           ++   + ++++++ +  ++++        ++++ +  + ++
   110     +   ++ +++      ++ +  ++   ++++ +   +++ ++   + ++++ ++      ++ + ++ +
   810        ++++++   + ++ +++++++ ++      + ++++    +++++++ ++       +  +++ + ++++
   868     + +++ ++ +  ++ ++++++++++++ +   +++++   ++ ++++  + ++++     + ++ + +
   843     + +++ ++ +  ++++++++ +++++ +    +++++   ++ ++++ +  ++ + +++++  + ++
   528     +++++ +++   +++++ +++ ++++ +    ++++ +++++ ++++ ++ ++ + +++++     +++
   612     ++ ++ ++ +  +++++ +   ++++   ++++++++++++++++ ++ ++++  +++++  ++++
    T7     ++++++++++  ++++++++ +++++ +    ++++++++++ ++++ ++ ++++  +++++  ++++
   813     ++++++++++  ++++++++ +++++   +++++++++++++++++++ ++ ++++  +++++  ++++
   F7/3    +++++ ++++  ++++++++++++++   +++++++++++++++++ ++ ++++  +++++  ++++
   RH1     ++++++++++  +++++++++++++++  +++++++++++++++++ ++ ++++  +++++  ++++
   RH2     ++++++++++  +++++++++++++++  +++++++++++++++++ ++ ++++  +++++  ++++
```

interest that viruses within the tick-borne complex tend to exhibit relatively low levels of cross reactivity with viruses outside this complex. On the other hand, viruses in serogroup 3 tend to share E-specific epitopes with viruses in other serogroups. This is supported by the evidence already presented in Table 1 and further typified by results (Table 2) which show the relatively limited extent of cross reactivity with other flaviviruses when MAb prepared against the tick-borne virus louping ill, were analysed. Thus, MAb analyses of cross reactivities among flaviviruses have revealed both conserved and non-conserved epitopes in the envelope glycoprotein and exposed relationships that were not previously apparent. Such tests also demonstrate the usefulness of these antibodies to identify closely related viruses for diagnosis and in general support the serogroupings suggested by Porterfield [65].

Plaque reduction neutralization tests (functional analyses) with E-specific MAb reveal entirely different antigenic cross relationships from those seen in antibody binding tests. In general, MAb that are cross reactive in binding analyses show specificity for closely related flaviviruses in neutralization tests [9]. However, heterologous cross reactivity in neutralization tests has been seen. For example, a MAb derived using Japanese encephalitis virus neutralized several flaviviruses but failed to neutralize Japanese encephalitis virus. Similar heterogeneity in the neutralization test was also reported by Roehrig et al. [72] with St. Louis encephalitis, by Kimura-Kuroda and Yasui [50] with Japanese encephalitis, St. Louis encephalitis, West Nile and Murray Valley encephalitis, and by Yasui [95] with a dengue virus. It is also of interest that both de Madrid and Porterfield [27] and Calisher et al. [11] reported that yellow fever virus is not neutralized by polyclonal antisera

Table 2. Indirect immunofluorescence antigenic cross reactivity analyses between flaviviruses, using monoclonal antibodies against LI virus

```
MAb                    56 Flaviviruses (group arranged)

       ----------  ---------------  ------------------  --  ----  -----  ----
       GLLMNPRSCT  ACCDEJKKMMPRSS   ABIJJKKMNSSSUWWY    SZ  BINT  BBERU  DDDD
       GGIEEOFREY  PPRBNUAOOMPBAO   RSLBUOUVJLETSSNF    PI  ATTM  AOHOG  EEEE
       YT AGW EEU  OC  TTDUDLB BK   OQHEGKNELEPRUL      OK  G AU  NU CS  NNNN
                I                   A                   A                 1234

35.13a  + + + +                     +                      +
35.7      + +      +                +                      +     +      +
35.6a   + +        +  +                                    +     +      ++
35.10b  + + ++     +  +              +                      +
35.12a  + + +      +  + +            +                ++
35.1a   + + + +    + +  +            +               +            +
35.5b   + +        +      + + +      +                   +  +     +        +
35.3a    ++ ++     +      +  +  +        +                        +      +
35.4a   + + ++     +  + +                +                 +               + +
35.14b  + + +    +++ + +                                   +     +  +      +
35.11b  + +        +  + +  +     +       + +   ++        +  +     +
34.2    ++++++++++ +  +          ++ +       +  + +++   +                  ++ +
34.1    ++++++++++ +++ ++ + + + +       +++++ +++ +++                +    ++++
```

against other flaviviruses and that antiserum against yellow fever virus does not neutralize other flaviviruses. For this reason, yellow fever virus was not assigned to a serogroup in the most recent antigenic analysis of the flaviviruses [11]. The results presented by Buckley and Gould [9], with hyperimmune antisera, seem to contradict this observation since the hyper-immune antisera that were produced against either yellow fever or West Nile virus (RH1 and RH2 respectively) were positive in neutralization tests with both yellow fever and other flaviviruses. This difference in performance of antisera highlights one of the problems of using functional tests to classify viruses and almost certainly reflects the different methods used for pro-duction of the polyclonal hyperimmune antisera. In addition, neutralization of flavivirus infectivity depends not only upon the presence of the appropri-ate epitope on the envelope glycoprotein but also upon its precise pre-sentation [9]. Different yellow fever strains showed significant differences in their readiness to be neutralized by known monoclonal and polyclonal neutralizing antibodies. These observations demonstrate the importance of selecting the most suitable virus strain if the test is to be used as the basis for virus classification.

When administered passively, some antibodies against the viral E protein also have the capacity to protect animals against challenge with virulent virus. However, antibodies that show neutralization in vitro may or may not show protection in vivo against challenge virus. Furthermore, non-neutrali-zing E-specific antibodies may show protection [8, 9, 20, 34, 42, 45, 56]. In some cases, protection in vivo is probably mediated directly via neutraliza-tion of virus infectivity. In other cases, different mechanisms probably operate to protect the host. For example, virus aggregation would be expected to reduce virus infectivity. On the other hand complement-mediated cytotoxicity at the infected cell surface, could kill the infected cells, leading to protection of the infected host. In either situation the host would gain time to mount a successful immune response. Heterologous cross protection has also been observed in which MAb against either Japanese encephalitis or yellow fever were found to protect mice against challenge with other flaviviruses (unpubl. results). However, as was observed with neutral-ization, antibodies that protect against one strain of virus may fail to protect against another strain of the same virus, even though the antibody binds the different strains of virus with the same avidity [12]. The success or failure of E-specific MAb to protect mice against challenge with different strains of yellow fever virus was found to be partly dependent upon the inherent virulence of the virus strain, as measured by the time taken to kill mice. For example, the FNV strain of yellow fever virus killed weanling mice within 7 days [3] and we have consistently failed to identify MAb that can passively protect mice against this strain. However, if a strain of yellow fever that takes a longer time to kill mice was used, the passively administered E-specific MAb could protect mice. Protection in these cases is probably due to a combined effect of the passively administered antibody and the natural host immune response.

Monoclonal antibodies characterized as above, have been used in solid phase competition binding ELISA tests to map the topological arrangement of the antigenic determinants in the E glycoprotein of several flaviviruses [12, 20, 42, 46, 90]. Several facts emerge from these types of studies. The glycoprotein spike consists of at least 3 antigenic domains, each of which contains epitopes that haemagglutinate and/or induce antibodies responsible for neutralization of virus infectivity, protection of mice against virus challenge and enhancement of virus infectivity via Fc-receptor bearing cells. Epitopes that are conserved throughout the flaviviruses, map together and tend to exhibit poor neutralization activity, whereas non-conserved epitopes map in distinct regions and, in general, exhibit potent neutralizing activity. Some epitopes appear mobile in the sense that they may occupy different relative locations in different strains of the same virus. Finally, antibody binding to flavivirus E protein epitopes may induce conformational changes of the surface glycoprotein, revealed by enhanced antibody binding [43].

On the basis of nucleotide and amino acid sequence data, together with knowledge of the structural characteristics, assignment of disulphide bridges, epitope topology and studies with synthetic peptides, structural models of the E protein of West Nile, tick-borne encephalitis and Murray Valley encephalitis have been created [55, 60, 71]. These models predicted the existence of three antigenic domains in the primary structure, localised individual epitopes in the glycoprotein, identified 2 hydrophobic sequences that have properties typical of membrane anchor regions, revealed potential T-cell determinants conserved in all flaviviruses and revealed features of the folding of the polypeptide chain, including the generation of discontinuous protein domains. Future crystallographic analysis of this protein should elucidate the structure of the receptor site and the domain(s) for virus neutralization.

Non-structural flavivirus proteins

Extensive cross reactivity binding analyses also have been carried out using MAb specific for the NS1 protein. These MAb characteristically show only very limited cross reactivity with heterologous flaviviruses (Table 3). Studies with MAb against non-structural flavivirus-coded proteins, perhaps surprisingly, have revealed functional activity. Schlesinger et al. [75] and Gould et al. [34] reported protection of mice against virulent yellow fever virus, if the mice were given MAb against the NS1 protein of yellow fever virus immediately prior to virus challenge. Active immunization of mice and monkeys against yellow fever virus infection has been accomplished using either purified NS1 protein [76] or NS1 β-galactosidase fusion protein [13].

Table 3. Indirect immunofluorescence antigenic cross reactivity analyses between flaviviruses, using monoclonal antibodies against the NS1 protein of serogroup 3 viruses

MAb	GLLMNPRSCT GGIEEOFREY YT AGW EEU	ACCDEJKKMMPRSS PPRBNUAOOMPBAO OC TTDUDLB BK I	ABIJJKKMNSSSUWWY LSLBUOUVJLETSSNF FQHEGKNELEPRUL	SZ PI OK A	BINT ATTM G AU	BBERU AOHOG NU CS	DDDD EEEE NNNN 1234
863				+			
871				+			
979				+			
429				+			
423				+		+	
993			+	+			
999			+	+			
992		+	+	+			
917			+	+	+		
924			+	+			+
925			+	+		+	
615			+	+	+		
618			+	+	+		
492		+	+	+			
428		+	+	+			
109	++	+	+	+	+		

The NS1 protein is thought to be associated with the plasma membrane of infected cells [14, 34] and some antibodies to NS1 have shown cytolytic activity [75]. Protection might therefore result from lysis of infected cells bearing NS1 antigen.

The prM protein of most flaviviruses is considered to be a non-structural protein [77, 92]. However, both polyclonal antisera and MAb against dengue-1 virus prM bind to purified virions [1, 49] and at least one prM-specific MAb induces antibody dependent enhancement of infectivity in Fc-receptor bearing cells [46]. Therefore for dengue viruses at least, the prM protein is probably structural. In contrast with E-specific MAb, those that identify dengue prM protein do not appear to show cross reactivity in binding analyses with any other flaviviruses [44]. Nevertheless, functional activity with these MAb was recently observed [49]. These authors found that antibodies against dengue prM, administered passively, can protect mice against challenge with both homologous and heterologous dengue serotypes. Some prM-specific MAb also show low titer virus neutralization. The mechanisms of protection and neutralization are not clear but steric hindrance of E protein epitopes by added prM antibodies or antibody mediated cytolytic activity, probably in the presence of complement, seem to offer the most likely explanation for neutralization and protection by prM antibodies. Whether or not the prM protein has significance in the immuno-pathology of dengue haemorrhagic fever and dengue shock syndrome remains to be determined.

There are relatively few antigenic studies of flavivirus proteins other than those referred to above; however, a MAb which identified the non-structural p74 protein (NS3) of Japanese encephalitis virus showed antigenic cross reactivity with flaviviruses from several antigenic complexes [32]. This MAb was unusual since it also identified an epitope present in the nucleus of vertebrate cells. Such epitope mimicry might have significance in viral immunopathology.

Antibodies and pathogenesis

In addition, to neutralization of virus infectivity in vitro, antibodies specific for E protein may also enhance the susceptibility of Fc-receptor bearing cells to infection by viruses [36, 41] and this phenomenon is usually referred to as antibody dependent enhancement (ADE) of virus infectivity (for reviews, see [67]). With flaviviruses, ADE occurs in the presence of sub-neutralizing antibody concentrations when peripheral human blood monocytes or murine and human macrophage-like cell lines are used; these cells bear Fc-receptors [20, 21, 37, 47, 61, 64, 66, 73]. Enhanced virus infectivity is due to an increased number of infected cells in the presence of antibody complexed with virus [62]. Non-neutralizing antibodies may also elicit ADE [21]. Binding of the antibody-virus complexes to the Fc and/or complement receptors [15] leads to internalization and thus infection, by the non-neutralized virus, of the receptor bearing cells.

There is some evidence that an equivalent immune-mediated phenomenon exists in vivo with flaviviruses. Halstead [38] proposed that ADE was an important factor for the development of dengue haemorrhagic fever and dengue shock syndrome in the course of sequential infections of children possessing maternally acquired antibodies. There are no reports of a similar situation in nature with flaviviruses other than the dengue serotypes. However, an equivalent phenomenon with some other viruses has been seen. For example, "early death syndrome" after rabies virus infection in inadequately immunized humans or animals has been reported [6, 68, 79] and a similar early death effect was reported in kittens inoculated with feline infectious peritonitis virus following passive transfer of antiviral antibody [89].

Following the observations of Webb et al. [88], that increased viraemia occurred in mice challenged with Langat virus in the presence of antibody, attempts to develop an in vivo model with which to study virulence enhancement have met with some success using yellow fever and Japanese encephalitis viruses [31]. It was initially found that yellow fever E-specific MAb which do not protect mice against infection with yellow fever virus can shorten the average survival time of mice challenged with virulent yellow fever virus [3]. Subsequently, antibodies that could protect mice were shown to increase virus virulence if they were administered intraperitoneally after the virus had become established in the target tissue [35]. Detailed analysis of the growth characteristics of the viruses in mice immunocompromised with hydrocortisone, cobra venom or anti-thymocyte serum demonstrated that enhanced virulence probably results from the induction by virus-antibody complexes of an Arthus type 3 hypersensitivity reaction [35]. A similar observation has now been made with louping ill virus using louping ill virus-specific MAb (H. Reid, pers. comm., 1989). Whether or not antigenic cross-reactivity between flaviviruses has significance in the immunopathology of flavivirus disease characteristics remains to be determined. However, it may be significant that severe cases of human Japanese encephalitis infection often show high IgG antibody responses, implying an anamnestic response, whereas the less severe cases show low IgG but a high IgM response, implying a primary immune response (A. Mathur, pers. comm., 1986).

Studies of the replication cycle of flaviviruses have mostly shown that virus proteins occur only in the cytoplasm of infected cells. Unequivocal evidence that flavivirus-specific proteins can be found within the nuclei of flavivirus infected cells has only recently been forthcoming. Its significance in virus replication and/or pathogenesis has not yet been identified. Previous reports of dense particles in the nuclei of cells infected with flaviviruses [59, 85, 96] had not been confirmed and evidence of a virus-specific step in the nucleus was unconvincing [93]. Recently, Buckley and Gould [10] identified 2 MAb, prepared against yellow fever or West Nile virus, which produced fluorescent labelling of cytoplasmic antigen in cells infected with the virus used to derive the antibodies. However, when these antibodies were tested with Langat (for the West Nile derived virus) or Zika virus infected cells (for the yellow fever derived virus), they produced virus-specific nuclear as well as cytoplasmic fluorescence. Independently, Tadano et al. [84] also observed virus-specific nuclear fluorescence in dengue-4-infected cells tested with a core protein-specific MAb derived from dengue-4 virus. Further support for a role of the nucleus in the replication of flaviviruses comes from recent evidence obtained using MAb prepared against the French viscerotropic strain of yellow fever virus (yellow fever FVV). Many of these MAb produced very distinct nuclear fluorescent labelling of yellow fever-infected cells (Fig. 2). The presence of the same antigen at the peripheral part of the cytoplasm, on the plasma membrane, is also evident. Double fluorescence labelling of the infected cells using a nuclear and a cytoplasmic antibody confirmed that the nuclear fluorescence was present only in yellow fever-infected cells (data not shown).

In addition to viruses sharing common antigenic determinants with each other, they may also share antigenic determinants with the host cells that they infect [23, 29, 32, 39, 40, 52, 78]. This is often referred to as molecular mimicry. Currently, there is no direct evidence that flavivirus infections can be rendered more severe due to autoimmune responses. However, molecular

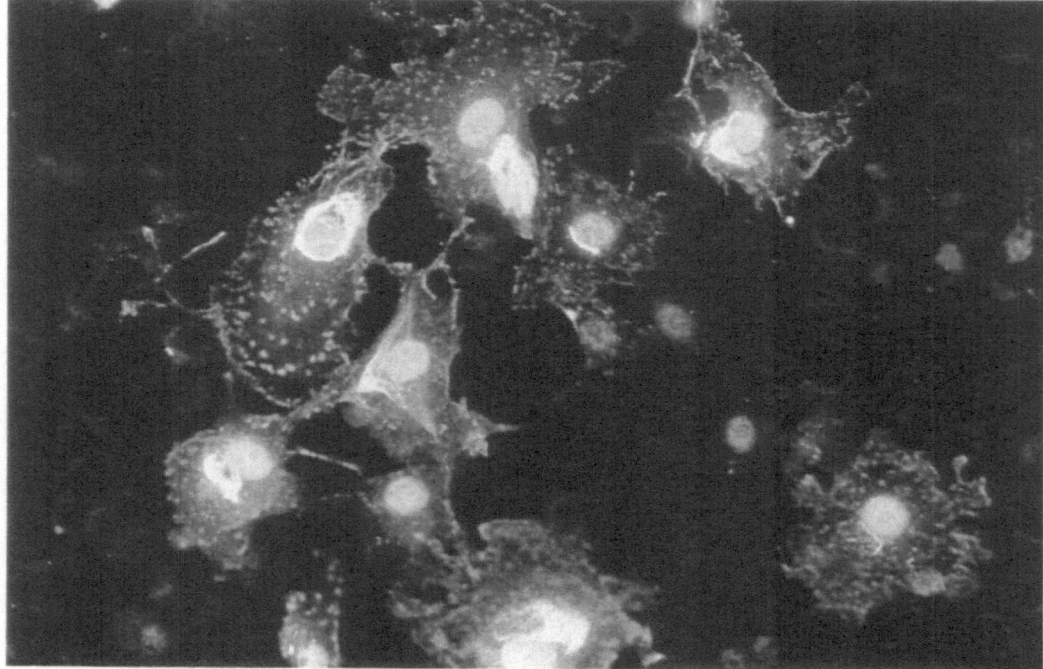

Fig. 2. Indirect immunofluorescence microscopy of cells infected with yellow fever virus and labelled with a monoclonal antibody that produces both nuclear and cytoplasmic fluorescence of infected cells. Note how the fluorescence appears in the plasma membrane as well as completely filling the nuclei

mimicry has been observed with Japanese encephalitis virus epitopes [32, 81] and human sera taken from Japanese encephalitis-infected patients with encephalitis contained anti-nuclear antibodies with the same fluorescence characteristics as the Japanese encephalitis MAb that showed anti-nuclear fluorescence [32]. In addition, a MAb against louping ill virus which identified both the viral envelope glycoprotein and a cellular nuclear epitope produced enhanced virulence of louping ill virus when given passively to mice prior to challenge with the virus (Gould, unpubl. results). Finally, anticellular antibodies, that do not react with the immunizing virus, are frequently detected in the course of deriving and identifying virus-specific MAb. On several occasions we have observed enhanced virus virulence when such anticellular antibodies are used in antibody mediated enhancement experiments with flaviviruses.

Passive transfer to mice of virus-specific anticellular antibodies mimics, to some extent, the development of an autoimmune response. We are beginning to accumulate evidence that these types of antibodies may alter disease severity at least under experimental conditions. Epitope mimicry amongst viral and cellular determinants is not uncommon, particularly with the flaviviruses [81] and such mimicry could lead to situations in which viral infections trigger an autoimmune response and thus aggravate the severity of the infection.

The antigenic cross reactivity of flaviviruses is evidently very complex and in many examples, the consequences of these antigenic interrelationships have not been assessed. It seems improbable that amongst such a large family of closely related viruses, the 4 dengue serotypes should be the only ones to exhibit enhanced immunopathology resulting from shared antigenic determinants. There is still considerable scope for studies of the antigenic characteristics of the flaviviruses particularly with respect to the immune response of the host.

Note added in proof. The entire nucleotide sequence of the envelope glycoprotein of louping ill virus has now been determined (Shiu, Ayres, and Gould, manuscript in prep.). The results show that louping ill virus shares a greater degree of homology with the western sub-type of tick-borne encephalitis virus than with the eastern sub-type and that both tick-borne sub-types are more closely related to louping ill virus than any of the known mosquito-transmitted flaviviruses.

References

1. Aaskov JG, Williams L, Fletcher J, Hay R (1988) Failure of a dengue 1 sub-unit vaccine to protect mice against a lethal dengue virus infection. Am J Trop Med Hyg 39: 511–518
2. Aaskov JG, Geysen HM, Mason TJ (1989) Serologically defined linear epitopes in the envelope protein of dengue 2 (Jamaica strain 1409). Arch Virol 105: 209–221
3. Barrett ADT, Gould EA (1986) Antibody-mediated early death in vivo after infection with yellow fever virus. J Gen Virol 67: 2539–2542
4. Bell JR, Kinney RM, Trent DW, Lenches EM, Dalgarno L, Strauss JM (1985) N-terminal amino acid sequences of structural proteins of three flaviviruses. Virology 143: 224–229
5. Biedrizycka A, Cauchi MR, Darthomoeusz A, Gorman JJ, Wright P (1987) Characterization of protease cleavage sites involved in the formation of the envelope glycoprotein and three nonstructural proteins of dengue virus type 2, New Guinea C strain. J Gen Virol 68: 1317–1326
6. Blancou J, Andral B, Andral L (1980) A model in mice for the study of the early death phenomenon after vaccination and challenge with rabies virus. J Gen Virol 50: 433–435
7. Boege U, Heinz PX, Wengler G, Kunz C (1983) Amino acid composition and amino terminal sequences of the structural proteins of a flavivirus, European tick-borne encephalitis virus. Virology 126: 651–657
8. Brandriss MW, Schlesinger JJ, Walsh EE, Briselli M (1985) Lethal 17D yellow fever encephalitis in mice. I. Passive protection by monoclonal antibodies to the envelope proteins of 17D yellow fever and dengue 2 viruses. J Gen Virol 67: 229–234
9. Buckley A, Gould EA (1985) Neutralization of yellow fever virus studied using monoclonal and polyclonal antibodies. J Gen Virol 66: 2523–2531
10. Buckley A, Gould EA (1988) Detection of virus-specific antigen in the nuclei or nucleoli of cells infected with Zika or Langat virus. J Gen Virol 69: 1913–1920
11. Calisher CH, Karabatsos N, Dalrymple JM, Shope RE, Porterfield J, Westaway EG, Brandt WE (1989) Antigenic relationships between flaviviruses as determined by cross-neutralization tests with polyclonal antisera. J Gen Virol 70: 37–43

12. Cammack N, Gould EA (1986) Topographical analysis of epitope relationships on the envelope glycoprotein of yellow fever 17D vaccine and the wild type Asibi parent virus. Virology 150: 333–341

13. Cane PA, Gould EA (1988) Reduction of yellow fever virus mouse neurovirulence by immunization with a bacterially synthesized non-structural protein (NS1) fragment. J Gen Virol 69: 1241–1246

14. Cardiff RD, Lund J (1976) Distribution of dengue 2 antigens by electron immunocytochemistry. Infect Immun 13: 1699–1709

15. Cardosa MJ, Porterfield JS, Gordon S (1983) Complement receptor mediates enhanced flavivirus replication in macrophages. J Exp Med 158: 258–263

16. Casals J (1963) Relationships among arthropod-borne viruses determined by cross-challenge tests. Am J Trop Med Hyg 12: 587–596

17. Casals J, Brown LV (1954) Haemagglutination with arthropod borne viruses. J Exp Med 99: 429–449

18. Castle E, Leidner U, Nowak T, Wengler G, Wengler G (1986) Primary structure of the flavivirus West Nile genome region coding for all nonstructural proteins. Virology 149: 10–26

19. Castle E, Nowak T, Leidner U, Wengler G, Wengler G (1985) Sequence analysis of the viral core protein and the membrane-associated proteins V1 and NV2 of the flavivirus West Nile virus and of the genome sequence for these proteins. Virology 145: 227–236

20. Cecilia D, Gadkari DA, Kedarnath N, Ghosh SN (1988) Epitope mapping of Japanese encephalitis virus envelope protein using monoclonal antibodies against an Indian strain. J Gen Virol 69: 2741–2747

21. Chanas AC, Gould EA, Clegg JCS, Varma MGR (1982) Monoclonal antibodies to Sindbis virus glycoprotein E1 can neutralize, enhance infectivity and independently inhibit haemagglutination or haemolysis. J Gen Virol 58: 37–46

22. Clarke DH (1964) Further studies on antigenic relationship among the viruses of the group B tick-borne complex. Bull WHO 31: 45–65

23. Dales S, Fujinami RS, Oldstone MBA (1983) Infection with vaccinia favors the selection of hybridomas synthesising autoantibodies against intermediate filaments, one of them cross-reacting with the virus haemagglutinin. J Immunol 131: 1546–1553

24. Dalgarno L, Trent DW, Strauss JH, Rice CM (1986) Partial nucleotide sequence of the Murray Valley encephalitis virus genome. Comparison of the encoded polypeptides with Yellow Fever virus structural and non-structural proteins. J Mol Biol 187: 309–323

25. Deubel V, Kinney RM, Trent DW (1986) Nucleotide sequence and deduced amino acid sequence of the structural proteins of dengue type 2 virus, Jamaican genotype. Virology 155: 365–377

26. Deubel V, Kinney RM, Trent DW (1988) Nucleotide sequence and deduced amino acid sequence of the nonstructural proteins of Dengue type 2 virus, Jamaica genotype: comparative analysis of the full length genome. Virology 165: 234–244

27. de Madrid AT, Porterfield JS (1974) The Flaviviruses (group B arboviruses) a cross-neutralization study. J Gen Virol 23: 91–96

28. Dittmar D, Haines HG, Castro A (1980) Monoclonal antibodies specific for dengue virus type 3. J Clin Microbiol 12: 74–78

29. Fujinami RS, Oldstone MBA, Wroblewska Z, Frankel ME, Koprowski H (1983) Molecular mimicry in virus infection: cross reaction of measles virus phosphoprotein or of herpes simplex virus protein with human intermediate filaments. Proc Natl Acad Sci USA 80: 2346–2350

30. Gentry MK, Henchal EA, McCown JM, Brandt WE, Dalrymple JM (1982) Identification of distinct antigenic determinants on dengue 2 virus using monoclonal antibodies. Am J Trop Med Hyg 31: 548–555

31. Gould EA, Buckley A (1989) Antibody dependent enhancement of yellow fever and Japanese encephalitis virus neurovirulence. J Gen Virol 70: 1605–1608

32. Gould EA, Chanas AC, Buckley A, Clegg CS (1983) Monoclonal immunoglobulin M antibody to Japanese encephalitis virus that can react with a nuclear antigen in mammalian cells. Infect Immun 41: 774–779

33. Gould EA, Buckley A, Cammack N, Barrett ADT, Clegg JCS, Ishak R, Varma MGR (1985) Examination of the immunological relationships between flaviviruses using yellow fever virus monoclonal antibodies. J Gen Virol 66: 1369–1382

34. Gould EA, Buckley A, Barrett ADT, Cammack N (1986) Neutralizing (54K) and non-neutralizing (54K and 48K) monoclonal antibodies against structural and non-structural yellow fever virus proteins confer immunity in mice. J Gen Virol 67: 591–595

35. Gould EA, Buckley A, Groeger BK, Cane PA, Doenhoff M (1987) Immune enhancement of yellow fever virus neurovirulence for mice: Studies of mechanism involved. J Gen Virol 68: 3105–3112

36. Halstead SB, Chow JS, Marchette MJ (1973) Immunological enhancement of dengue virus replication. Nature 243: 24–25

37. Halstead SB, O'Rourke EJ (1977) Dengue viruses and mononuclear phagocytes. I. Infection enhancement by non-neutralizing antibody. J Exp Med 146: 201–217

38. Halstead SB (1981) The pathogenesis of dengue: molecular epidemiology in infectious disease. Am J Epidemiol 114: 632–648

39. Haspel MV, Onodera T, Prabhakar BS, Horita M, Suzuki H, Notkins AL (1983) Virus induced autoimmunity: monoclonal antibodies that react with endocrine tissues. Science 220: 304–306

40. Haynes BF, Robert-Gurhoff M, Metzgar RS, Franchini G, Kalyanamaran VS, Palker TJ, Gallo RC (1983) Monoclonal antibody against human T cell leukaemia virus p19 defines a human thymic epithelial antigen acquired during ontogeny. J Exp Med 157: 907–920

41. Hawkes RA, Lafferty KJ (1967) The enhancement of virus infectivity by antibody. Virology 33: 250–261

42. Heinz FX, Berger R, Tuma W, Kunz C (1983) Location of immunodominant antigenic determinants on fragments of the tick-borne encephalitis virus glycoprotein: evidence for two different mechanisms by which antibodies mediate neutralization and haem-agglutination–inhibition. Virology 126: 525–537

43. Heinz FX, Mandl C, Berger R, Tuma W, Kunz C (1984) Antibody-induced conformational changes result in enhanced avidity of antibodies to different antigenic sites on the tick-borne encephalitis virus glycoprotein. Virology 133: 25–34

44. Henchal EA, Gentry MK, McCown JM, Brandt WE (1982) Dengue virus-specific and flavivirus group determinants identified with monoclonal antibodies by indirect immunofluorescence. Am J Trop Med Hyg 31: 830–836

45. Henchal EA, McCown JM, Seguin MC, Gentry MK, Brandt WE (1983) Rapid identification of dengue virus isolates by using monoclonal antibodies in an indirect immunofluorescence assay. Am J Trop Med Hyg 32: 164

46. Henchal EA, McCown JM, Burke DS, Seguin MC, Brandt WE (1985) Epitope analysis of antigenic determinants on the surface of dengue 2 virions using monoclonal antibodies. Am J Trop Med Hyg 34: 162–169

47. Hotta H, Wiharta AS, Hotta S (1984) Antibody-mediated enhancement of dengue virus infection in mouse macrophage cell lines Mk1 and Mm1. Proc Soc Exp Biol Med 175: 320–327

48. Innis BL, Thirawuth V, Hemachuda C (1989) Identification of continuous epitopes of the envelope glycoprotein of dengue type 2 virus. Am J Trop Med Hyg 40: 676–687

49. Kaufman BM, Summers PL, Dubois DR, Cohen WH, Gentry MK, Timchak RL, Burke

150   E. A. Gould et al.

DS, Eckels KH (1989) Monoclonal antibodies for dengue virus prM glycoprotein protect mice against lethal dengue infection. Am J Trop Med Hyg 41: 576–580

50. Kimura-Kuroda J, Yasui K (1983) Topographical analysis of antigenic determinants on envelope glycoprotein V3 (E) of Japanese encephalitis virus, using monoclonal antibodies. J Virol 45: 124–132

51. Kobayashi Y, Hasegawa H, Oyama T, Tamai T, Kusaba T (1984) Antigenic analysis of Japanese encephalitis virus by using monoclonal antibodies. Infect Immun 44: 117–123

52. Lane DP, Hoeffler WK (1980) SV40 large T shares an antigenic determinant with a cellular protein of molecular weight 68,000. Nature 288: 167–170

53. McAda PC, Mason PW, Schmaljohn CS, Dalrymple JM, Mason TL, Fournier MJ (1987) Partial nucleotide sequence of the Japanese encephalitis virus genome. Virology 158: 348–360

54. Mandl CW, Heinz FX, Kunz C (1988) Sequence of the structural proteins of tick-borne encephalitis virus (Western subtype) and comparative analysis with other flaviviruses. Virology 166: 197–205

55. Mandl CW, Guirhakoo F, Holzman H, Heinz FX, Kunz C (1989) Antigenic structure of the flavivirus envelope protein E at molecular level, using tick-borne encephalitis virus as a model. J Virol 63: 564–571

56. Mathews JM, Roehrig JT (1984) Elucidation of the topography and determination of the protective epitopes on the E glycoprotein of St Louis encephalitis virus by passive transfer with monoclonal antibodies. J Immun 132: 1533–1537

57. Monath TP (1986) Pathobiology of the flaviviruses. In: Schlesinger S, Schlesinger MJ (eds) The Togaviridae and Flaviviridae. Academic Press, New York, pp 375–440

58. Monath TP, Schlesinger JJ, Brandriss MW, Cropp CB, Prange WB (1984) Yellow fever monoclonal antibodies: type-specific and cross-reactive determinants identified by immunofluorescence. Am J Trop Med Hyg 33: 695–698

59. Murphy FA (1980) Morphology and morphogenesis. In: Monath TP (ed) St. Louis encephalitis. American Public Health Association, Washington, DC, pp 65–103

60. Nowak T, Wengler G (1987) Analysis of disulphides present in the membrane proteins of the West Nile flavivirus. Virology 156: 127–137

61. Peiris JSM, Porterfield JS (1979) Antibody-mediated enhancement of flavivirus replication in macrophage-like cell lines. Nature 282: 509–511

62. Peiris JSM, Gordon S, Unkeless JC, Porterfield JS (1981) Monoclonal anti Fc receptor IgG blocks enhancement of viral replication in macrophages by antiviral antibody. Nature 289: 189–191

63. Peiris JSM, Porterfield JS, Roehrig JT (1982) Monoclonal antibodies against the flavivirus West Nile. J Gen Virol 58: 283–289

64. Phillpotts RJ, Stephenson JR, Porterfield JS (1985) Antibody-dependent enhancement of tick-borne encephalitis virus infectivity. J Gen Virol 66: 1831–1837

65. Porterfield JS (1980) Antigenic characteristics and classification of Togaviridae. In: Schlesinger RW (ed) The Togaviruses. Academic Press, New York, pp 13–46

66. Porterfield JS (1982) Immunological enhancement and the pathogenesis of dengue haemorrhagic fever. J Hyg 89: 355–364

67. Porterfield JS (1986) Antibody-dependent enhancement of viral infectivity. Adv Virus Res 31: 335–352

68. Prabhakar BS, Nathanson N (1981) Acute rabies death mediated by antibody. Nature 290: 590–591

69. Rice CM, Lenches EM, Eddy SR, Shin SJ, Sheets RL, Strauss JH (1985) Nucleotide sequence of yellow fever virus: Implications for flavivirus gene expression and evolution. Science 229: 726–733

70. Rice CM, Strauss EG, Strauss JH (1986) Structure of the flavivirus genome. In:

Schlesinger S, Schlesinger MJ (eds) The Togaviridae and Flaviviridae. Plenum, New York, pp 279–326

71. Roehrig JT, Hunt AR, Johnson AJ, Hawkes RA (1989) Synthetic peptides derived from the deduced amino acid sequences of the E glycoprotein of Murray Valley encephalitis virus elicit antiviral antibody. Virology 171: 49–60

72. Roehrig JT, Mathews JH, Trent DW (1983) Identification of epitopes on the E glycoprotein of Saint Louis encephalitis virus using monoclonal antibodies. Virology 128: 118–126

73. Schlesinger JJ, Brandriss MW (1981) Growth of 17D yellow fever virus in a macrophage-like cell line, U937: role of Fc and viral receptors in antibody-mediated infection. J Immunol 127: 659–665

74. Schlesinger JJ, Brandriss MW, Monath TP (1983) Monoclonal antibodies distinguish between wild and vaccine strains of yellow fever virus by neutralization, haemagglutination–inhibition and immunoprecipitation of virus envelope protein. Virology 125: 8–17

75. Schlesinger JJ, Brandriss MW, Walsh EE (1985) Protection against 17D yellow fever encephalitis in mice by passive transfer of monoclonal antibodies to the non-structural glycoprotein gp48 and by active immunization with gp48. J Immunol 135: 2805–2809

75. Schlesinger JJ, Brandriss MW, Cropp CB, Monath TP (1986) Protection against yellow fever in monkeys by immunization with yellow fever virus nonstructural NS1. J Virol 60: 1153–1155

77. Shapiro D, Brandt WE, Russell PK (1972) Changes involving a viral membrane glycoprotein during morphogenesis of group B arboviruses. Virology 50: 906–911

78. Sheshberadaran H, Norrby E (1984) Three monoclonal antibodies against measles virus F protein cross-react with cellular stress proteins. J Virol 52: 995–999

79. Sikes RK, Cleary WF, Koprowski H, Wiktor TJ, Kaplan MM (1971) Effective protection of monkeys against death from street rabies by post-exposure administration of tissue culture rabies vaccine. Bull WHO 45: 1–11

80. Speight S, Coia G, Parker MD, Westaway EG (1988) Gene mapping and positive identification of the nonstructural proteins NS2A, NS2B, NS3 and NS5 of the flavivirus Kunjin and their cleavage sites. J Gen Virol 69: 23–34

81. Srinivasappa J, Saegusa J, Prabhakar BS, Gentry MK, Buchmeier MJ, Wiktor TJ, Koprowski H, Oldstone MBA, Notkins AL (1986) Molecular mimicry: frequency of reactivity of monoclonal antiviral antibodies with normal tissues. J Virol 57: 397–401

82. Stephenson JR, Lee JM, Wilton-Smith PD (1984) Antigenic variation among members of the tick-borne encephalitis complex. J Gen Virol 65: 81–89

83. Sumiyoshi H, Mori C, Morita K, Kuhara S, Kondou J, Kukushi Y, Nagamatu H, Igarashi A (1987) Complete nucleotide sequence of the Japanese encephalitis virus genome RNA. Virology 161: 497–510

84. Tadano M, Makino Y, Fukunaga T, Okuno Y, Fukai K (1989) Detection of Dengue 4 virus core protein in the nucleus. I. A monoclonal antibody to dengue 4 virus reacts with the antigen in the nucleus and cytoplasm. J Gen Virol 70: 1409–1415

85. Tikhomirova TI, Karpovich LG, Reingold VN, Levkovich EN, Shestopalova NM (1968) Comparative study of Langat and tick-borne encephalitis viruses in cultured cells. Acta Virol 12: 529–534

86. Trent DW (1977) Antigenic characterization of flavivirus structural proteins separated by isoelectric focusing. J Virol 22: 608–618

87. Trent DW, Kinney RM, Johnson BJB, Vordnam AV, Grant JA, Deubel V, Rice CM, Hahn C (1987) Partial nucleotide sequence of St Louis encephalitis virus RNA: structural proteins, NS1, ns2a and ns2b. Virology 156: 293–304

88. Webb HE, Wight DGD, Platt GS, Smith CEG (1968) Langat virus encephalitis in mice.

I. The effect of the administration of specific antiserum. J Hyg 66: 343–354

89. Weiss RC, Scott FW (1981) Antibody mediated enhancement of disease in feline infectious peritonitis: comparisons with dengue haemorrhagic fever. Comp Immunol Microbiol Infect Dis 4: 175–289

90. Wengler G, Castle E, Leidner U, Nowak T, Wengler G (1985) Sequence analysis of the membrane protein V3 of the flavivirus West Nile virus and of its gene. Virology 147: 264–274

91. Wengler G, Wengler G, Nowak T, Wahn K (1987) Analysis of the influence of proteolytic cleavage on the structural organization of the surface of the West Nile flavivirus leads to the isolation of a protease-resistant E protein oligomer from the viral surface. Virology 160: 210–219

92. Westaway EG (1973) Proteins specified by group B Togaviruses in mammalian cells during productive infection. Virology 51: 454–465

93. Westaway EG (1980) Replication of flaviviruses. In: Schlesinger RW (ed) The Togaviruses. Academic Press, New York, pp 531–577

94. Westaway EG, Gaidamovich SYa, Horzinek MS, Igarashi A, Kääriäinen L, Lvov DK, Porterfield JS, Russell PK, Trent DW (1985) Flaviviridae. Intervirology 24: 183–192

95. Yasui K (1984) The antigenic structure of Japanese encephalitis virus glycoprotein VB(E). In: Proceedings of the International Workshop on the Molecular Biology of Flaviviruses, Fort Detrick, pp 16–17

96. Yasuzumi G, Tsubo I (1965) Analysis of the development of Japanese B encephalitis (JBE) virus. III. Electron microscope studies on inclusion bodies appearing in neurons and microglial cells infected with JBE virus. J Ultrastruc Res 12: 317–327

Authors' address: E. A. Gould, Institute of Virology and Environmental Microbiology, Mansfield Road, Oxford OX1 3SR, England.

Arch Virol (1990) [Suppl 1]: 153–159

Laboratory diagnosis of tick-borne encephalitis

H. Hofmann, C. Kunz, and **F. X. Heinz**

Institute of Virology, University of Vienna, Austria

Accepted February 5, 1990

Summary. As isolation of tick-borne encephalitis (TBE) virus is successful only from blood in the first, nonspecific phase of disease or from autopsy material, laboratory diagnosis is done usually by serological means. Classical serologic tests have been replaced by ELISA for detecting IgM antibodies to TBE virus. We tested ELISA formats that demonstrated differences in both sensitivity and susceptibility to interfering factors, e.g., rheumatoid factors or heterophile antibodies. In our experience the anti-μ test using enzyme-labelled antigen proved to be the method best suited for routine laboratory diagnosis of TBE virus infection.

In Austria between 200,000 and 300,000 primary TBE vaccinations are performed each year. This may cause diagnostic problems because IgM antibodies persist in the vaccinees for as long as 8 months. In such cases confirmatory diagnosis may require demonstrating locally formed antibodies in the brain. For that purpose a special anti-μ ELISA was developed.

Introduction

Tick-borne encephalitis (TBE) is the most important human virus infection of the central nervous system in Central Europe [1, 12, 15], although the prevalence of TBE is decreasing in Austria because of an aggressive vaccination campaign (Table 1) [13]. The disease usually takes a biphasic course [10, 14], shown schematically in Fig. 1. In the first phase of disease the patient develops viremia and suffers from a mild febrile illness which usually lasts only a few days. However, after a symptomless interval of a few days, fever rises again and in this second phase of disease the patient develops meningitis or encephalitis. By the beginning of the second phase, antibodies are detectable in the serum and therefore virus can no longer be detected in blood.

Table 1. Numbers of laboratory confirmed infections with tick-borne encephalitis virus in Austria 1981–1989

Year	No. of cases	Incidence[a]	
		A	B
1981	294	5.9	3.9
1982	612	12.2	8.2
1983	240	4.8	3.2
1984	337	6.7	4.5
1985	300	6.0	4.0
1986	258	5.2	3.4
1987	215	4.3	2.9
1988	201	4.0	2.7
1989	132	2.6	1.8

[a] Per 100,000 capita. Austria has a population of about 7.5 million inhabitants; however, not all live in infected areas. Population at risk therefore is estimated about 5 million. Incidence A is for population at risk, incidence B for total population.

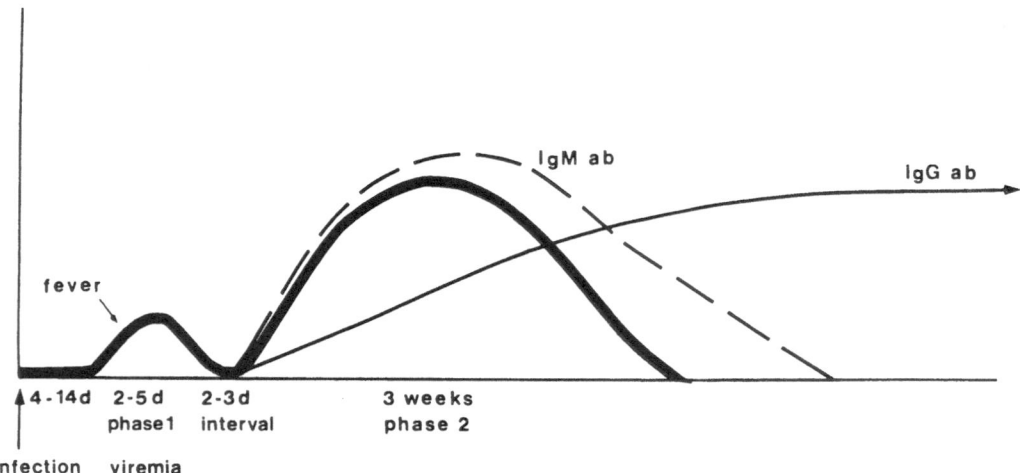

Fig. 1. Schematic course of tick-borne encephalitis infection in humans

Virus isolation

Theoretically, in the first phase of disease virus can be isolated from the blood. However, in practice this does not occur because clinical signs at this time are so mild that there is no demand for diagnostic confirmation.

Alternatively, in the rare fatal case, virus may be isolated from brain tissue. As has been the case for several decades, intracerebral inoculation of suckling mice is the most sensitive method of isolating TBE virus.

Serology

TBE therefore usually is diagnosed by serological means. The classical tests [9, 11], neutralization, complement fixation, and hemagglutination–inhibition, have been replaced by enzyme-linked immunosorbent assays (ELISA), because this test can be configured to distinguish between IgM and IgG antibodies. Figure 2 shows four formats of IgM antibody ELISA which we have used. The simplest method (Fig. 2, A) is to bind the antigen to the solid phase [4]. However, this test gives false positive results in patients with IgG antibodies to TBE and with rheumatoid factor (RF). This became increasingly important in Austria as the number of vaccinees (with IgG antibodies against TBE virus) increased. We therefore changed formats [3]. In this test (Fig. 2, B), an anti-μ serum is bound to the solid phase, then the patient's serum containing IgM antibodies is added. Thereafter the antigen is bound, and an enzyme-labelled antibody against the virus is added. However, in this test heterophile antibodies of the IgM type may cause false positive results because they bind to the enzyme-labelled antibody, which is usually produced in rabbits. The RF may also cause false positive results when it has activity against rabbit IgG.

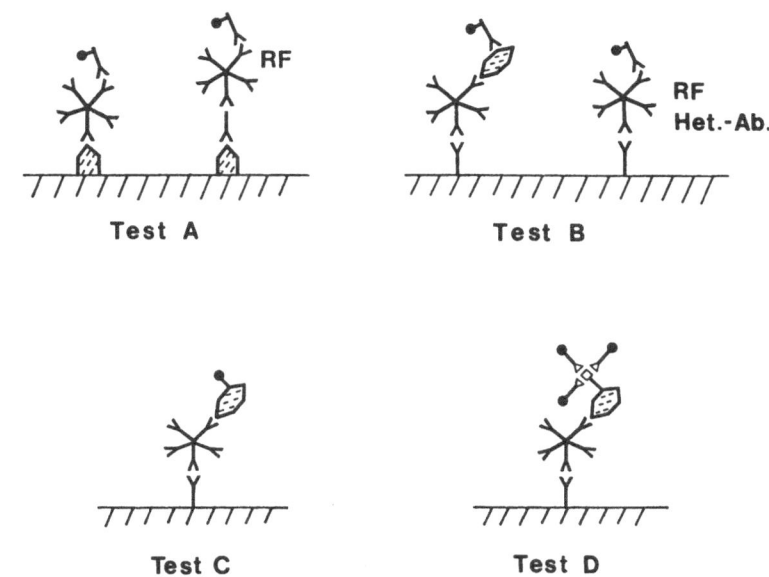

Fig. 2. IgM ELISA formats for detecting IgM antibody to TBE virus: Test A: Antigen on the solid phase, rheumatoid factor may interfere. Tests B, C, D: Anti-μ on the solid phase. Test B: Rheumatoid factor and heterophile antibodies may interfere. Test C: Enzyme-labelled antigen. Test D: Biotine labelled antigen, avidin-labelled with enzyme

Finally, we developed assays using TBE antigen labelled either with enzyme (Fig. 2, C) or biotin (Fig. 2, D).

Influence of rheumatoid factor and heterophile antibodies on different ELISA test formats

As shown in Table 2, we tested 11 sera containing RF from persons with IgG but not IgM antibody to TBE virus (patients with past TBE virus infections and vaccinees) in the 4 ELISA tests designated A–D (Fig. 2). Nine sera gave a positive result in tests A and B but not in tests C and D. Two of these 9 sera were also tested by configuration B with omission of the antigen and again the test showed a positive result. However, 2 other sera gave positive results in test A but not in tests B, C, or D. For controls we tested 7 sera with RF from patients with recent TBE (TBE IgM and IgG antibodies positive). These 7 were positive in the IgM tests of all 4 formats.

The influence of heterophile antibodies on TBE IgM ELISA is shown in Table 3. Twelve serum samples from patients with infectious mono-

Table 2. Serum samples from 18 patients with rheumatoid factor tested by ELISA for IgM antibody to tick-borne encephalitis virus

No. of sera	IgM ELISA test			
	A	B	C	D
Post TBE, Vaccinees (IgG +, IgM −)				
7	pos	pos	neg	neg
2	pos	pos	neg	neg
2	pos	neg	neg	neg
Recent TBE (IgM +, IgG +)	pos	pos	pos	pos

For test description, see legend of Fig. 2

Table 3. Serum samples from 12 people with heterophile antibody (infectious mononucleosis), tested by ELISA for IgM antibody to tick-borne encephalitis virus

No.	IgM ELISA test positive			
	A	B	C	D
12	0	10	0	0

nucleosis, and without IgM or IgG antibodies to TBE virus were tested by the 4 different methods. Tests A, C, and D showed no reactivity: however, 10 sera were reactive in test B, even when the antigen was omitted.

Due to the fact that the tests applying labelled antigen are not influenced by RF or heterophile antibody and because the antigen is more easily labelled with enzyme than with biotin, we now employ test C in our laboratory for routine diagnosis of TBE.

Persistence of IgM antibodies to TBE virus

Previously we reported that after natural infection IgM antibodies (determined by test B) may be detected for as long as 10 months after onset of acute illness [7]. These may combine to cause diagnostic problems in patients who contract diseases caused by other neurotropic viruses some months after a diagnosed or undiagnosed TBE virus infection.

However, IgM antibodies to TBE virus are not only found after natural infection but also after the initial two vaccination doses [7]. In Austria between 200,000 and 300,000 primary TBE vaccinations are performed each year. Similar to those found in natural infections, these IgM antibodies may also persist for as long as 8 months. This is of great importance because a non-TBE neurotropic virus infection in a TBE vaccinee may be misdiagnosed and because serologic diagnosis of TBE in a vaccinee may cast doubt on the efficacy of the TBE vaccine. TBE infections in vaccinees, while extremely rare, do occur [13]. Whereas the causes of such apparent vaccine failures may, in fact, be vaccination failures or may be due to other failures, they undoubtedly pose difficult diagnostic problems.

Diagnosis of TBE in vaccinees

In patients presenting with meningitis or encephalitis and who had received the initial two TBE vaccinations and therefore have IgM antibodies to TBE virus detectable in the serum, one must decide whether the patient contracted TBE despite vaccination or has been infected by another neurotropic virus and has IgM antibodies as a result of vaccination. This is of significance in Austria [13].

Patients suffering from TBE produce antibodies not only in the serum but also locally in the brain, such antibodies being detectable in the cerebrospinal fluid (CSF) [5, 6]. The presence of IgM antibodies to TBE virus in CSF is, however, not always conclusive evidence of an infection with this virus, as these antibodies also may be found if the blood–brain barrier is damaged, such as by another neurotropic virus. Burke et al. [2] have shown for Japanese encephalitis that one can distinguish between the two possibilities.

In instances of an intact blood–brain barrier with local production of antibody to TBE virus, the relative amount of specific IgM antibody to TBE virus in the CSF is (very) high, whereas in patients with compromised blood–brain barriers we find antibodies similar in concentration to that in serum, suggesting that the relative amount of specific IgM is low. Therefore, as reported previously [8], we constructed an ELISA with anti-µ antibody bound to the solid phase but in low concentration. IgM antibodies to TBE virus in the CSF, which are actually serum antibodies that pass through a damaged blood–brain barrier, compete with non-TBE IgM, resulting in a low absorbance value. However, there is no such competition by non-TBE IgM in the case of locally produced IgM in the brain and an intact blood–brain barrier. We have detected high absorbance values, although the

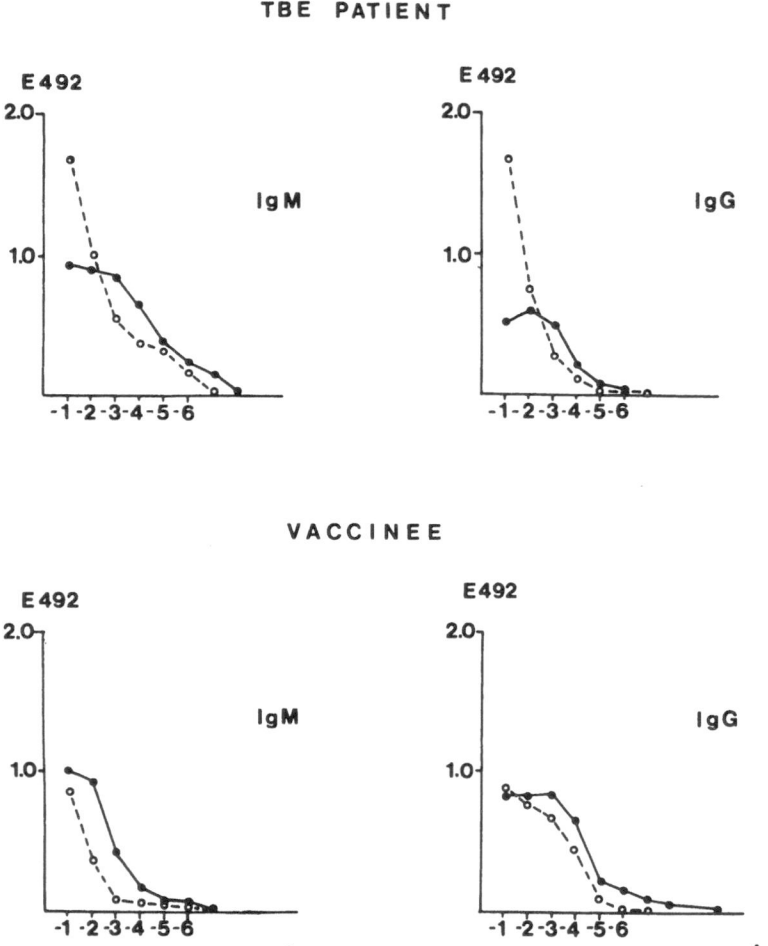

Fig. 3. Capture ELISA (low concentration on solid phase). Determination of TBE IgM (**a**) and IgG (**b**) antibodies in serum (●———●———●) and CSF (○– – – –○– – –○) of a TBE patient and a TBE vaccinee suffering from an enterovirus meningitis. Absorbance measured at 492 nm. Serum dilution (log 10) from −1 to −6

concentration of the anti-μ antibody on the plate is low. An example is shown in Fig. 3.

Applying the methods described, we are sure that we can diagnose nearly every case of TBE infection, even those rare cases, which are complicated by previous vaccination. Occasionally a second serum sample is necessary, sometimes we employ an additional ELISA format, but in the end we can confirm or exclude the diagnosis of an infection with TBE virus.

References

1. Ackermann R, Krüger K, Roggendorf M, Rehse-Küpper B, Mortter M, Schneider M, Vukadinovic I (1986) Die Verbreitung der Frühsommer-Meningoenzephalitis in der Bundesrepublik Deutschland. Deutsch Med Wochenschr 24: 927–933
2. Burke DS, Nisalak A, Ussery NM (1982) Antibody capture immunoassay detection of Japanese encephalitis virus immunoglobulin M and G antibodies in cerebrospinal fluid. J Clin Microbiol 16: 1034–1042
3. Heinz FX, Roggendorf M, Hofmann H, Kunz C, Deinhardt F (1981) Comparison of two different enzyme immunoassays for detection of immunoglobulin M antibodies against tick-borne encephalitis virus in serum and cerebrospinal fluid. J Clin Microbiol 14: 141–146
4. Hofmann H, Frisch-Niggemeyer W, Heinz FX (1979) Rapid diagnosis of tick-borne encephalitis by means of enzyme-linked immunosorbent assay. J Gen Virol 42: 505–511
5. Hofmann H, Frisch-Niggemeyer W, Heinz FX, Kunz C (1979) Immunoglobulin to tick-borne encephalitis in the cerebrospinal fluid of man. J Med Virol 4: 241–245
6. Hofmann H, Popow-Kraupp T (1982) Tick-borne encephalitis (TBE) infection: antiviral antibodies in cerebrospinal fluid. Zentralbl Bakteriol Mikrobiol Hyg [A] 253: 305–311
7. Hofmann H, Kunz C, Heinz FX, Dippe H (1983) Detectability of IgM antibodies against TBE virus after natural infection and after vaccination. Infection 11: 164–166
8. Hofmann H, Heinz FX, Dippe H, Kunz C (1985) Tick-borne encephalitis: a simple method for detection of antibody production in the brain. Zentralbl Bakteriol Mikrobiol Hyg [A] 260: 132–138
9. Hubinger MG, Hofmann H, Krausler J (1970) Das Überdauern von Antikörpern gegen das Frühsommer-Meningoenzephalitis (FSME)-Virus beim Menschen. Zentralbl Bakteriol Mikrobiol Hyg [A] 213: 145–151
10. Krausler J (1981) 23 years of TBE in the district of Neunkirchen (Austria). In: Kunz C (ed) Tick-borne encephalitis. Facultas, pp 6–12
11. Kunz C, Moritsch H (1961) Zur serologischen Diagnostik der Frühsommer-Meningoenzephalitis (FSME). Arch Ges Virusforsch 11: 568–582
12. Kunz C, Hofmann H (1972) Arboviren in Europa. Geogr Z 1972: 47–62
13. Kunz C, Heinz FX, Hofmann H (1990) Vaccination against tick-borne encephalitis in Austria. (In preparation)
14. Reisner H (1981) Clinic and treatment of tick-borne encephalitis. In: Kunz C (ed) Tick-borne encephalitis. Facultas, pp 1–5
15. WHO (1986) Tick-borne encephalitis and haemorrhagic fever with renal syndrome in Europe. WHO-EURO Reports and Studies 104

Authors' address: H. Hofmann, Institute of Virology, University of Vienna, Kinderspitalgasse 15, A-1095 Vienna, Austria.

Arch Virol (1990) [Suppl 1]: 161–168

New perspective vaccines from tick-borne encephalitis virus propagated in green monkey kidney cell cultures

M. P. Chumakov, S. G. Rubin, I. V. Semashko, M. N. Matrosovich, L. L. Mironova, L. I. Martyanova, Yu. A. Kniaginskaya, Ya. A. Salnikov, A. S. Gambaryan, A. S. Karavanov, and O. V. Kurennaya

Institute of Poliomyelitis and Viral Encephalitides, U.S.S.R. Academy of Medical Science, Moscow, U.S.S.R.

Accepted May 4, 1990

Summary. A new inactivated absorbed tick-borne encephalitis (TBE) vaccine from TBE virus propagated in primary green monkey kidney cells (MKC) is offered. The utilization of this cellular substrate provides a stable high accumulation of the virus in consecutive harvests of culture fluid made at 2–3-day intervals and results in the following advantages of the new TBE vaccine:

— abundance of a cheap virus source for vaccine production;
— an absence of necessity to concentrate the virus;
— considerable lowering of content of total protein and host-derived impurities;
— the high specific seroconversion and non-reactogenicity of the vaccine as judged by observation of volunteers.

The technology of TBE vaccine production in MKC-derived continuous cell line 4647 using simple cellular DNA-decontamination procedure was also developed.

The advantages of the technology used for preparation of this TBE vaccine could be employed for production of vaccines against other viral infections.

*

Tick-borne encephalitis (TBE) is a severe disease occurring in the USSR and many European countries. It is well proved that inactivated tissue culture vaccines produced in the U.S.S.R. and Austria from TBE virus propagated in primary chick embryo cells (CEC) enable safe protection against the illness [3, 5].

At the present time two types of inactivated, absorbed chick embryo cell TBE vaccines are in use in the U.S.S.R. The first vaccine is manufactured from non-concentrated, unpurified virus-containing culture fluid [5]. The second vaccine, which is produced in scanty quantities, is manufactured by twenty five to thirty-fold concentration of the initial culture fluid with simultaneous purification of the virus [2]. This type of TBE vaccine corresponds approximately to the preparation produced in Austria [4].

The shortcomings of the purified CEC TBE vaccine are a rather complex production technology and a relatively high cost. As to the non-concentrated CEC vaccine, it contains residual CEC-derived impurities and requires a longer vaccination schedule.

To overcome the above mentioned shortcomings of existing TBE vaccines we offer an alternative substrate for the propagation of TBE virus: primary African green monkey kidney cell (MKC) culture [1].

The use of this new substrate provides uniformly higher TBE virus titers in culture fluid, about ten times more than in CEC cell culture. Another essential advantage of MKC cultures is the absence of cell monolayer destruction during TBE virus propagation. This enables the repeated harvesting of virus-containing fluid from culture vessels and increases 4- to 6-fold the yield of starting material for vaccine production. As a result, up to 600,000 doses of TBE vaccine can be produced from every pair of green monkey kidneys.

Table 1 illustrates the results of several typical experiments on TBE virus (Sofjin strain) propagation in MKC cultures. The first three experiments were performed with flask cultures, experiments no. 4 and 5, with roller bottle cultures. The cultures were infected at MOI of 1–10 PFU/cell and maintained in medium 199 with 0.02% human serum albumin and 5% aminopeptide at 37 °C. Every 48–72 h the culture fluid was harvested and replaced by fresh nutrient medium. The results presented evidence of a stable high

Table 1. TBE virus titres in consecutive harvests from MKC cultures ($\log_{10}$ PFU/ml)

Experiment no.	Harvest no.					
	1	2	3	4	5	6
1	8.4	8.8	8.6	9.2	8.3	–
2	8.6	8.4	9.0	9.3	8.4	8.5
3	8.1	8.3	8.3	8.9	8.3	–
4	10.0	9.7	9.5	9.1	8.6	8.6
5	9.9	9.8	10.3	9.8	9.8	–

accumulation of the virus in consecutive harvests of culture fluid, making it unnecessary to concentrate it for preparation of the vaccine.

As a result the technology of the vaccine production includes the following steps:

1. Preparation of seed virus.
2. Preparation of MKC cultures.
3. Inoculation of cultures with seed virus, their incubation, and consecutive harvesting of virus suspension at 2–3 day intervals (3–6 harvests).
4. Inactivation of virus with formalin.
5. Clarification of virus suspension by microfiltration.
6. Adsorption of virus on aluminium hydroxide.
7. Filling, sealing, and labelling of final containers.

Using this technology we produced more than 30 experimental lots of MKC TBE vaccine. The control tests demonstrated regular reproducibility of their quality, complete inactivation of TBE virus, sterility and innocuity of preparations. Total protein content was within the range from 75 to 170 μg per immunizing dose (0.5 ml). It must be noted that most of this total protein is actually non-reactogenic human serum albumin, a constituent of the nutrient medium used for virus propagation. To evaluate the real content of host-derived potentially reactogenic impurities in the vaccine we developed an ELISA test system for the quantitative analysis of total green monkey kidney cell antigens (MN Matrosovich and AS Gambaryan, unpubl.). According to ELISA the content of substrate-derived antigens in the vaccine did not exceed 2 μg per dose. Such unexpectedly low content of substrate antigens in a vaccine which had not undergone any special purification procedures seems to be associated with an absence of a cytopathic effect of TBE virus in green monkey cell culture. Whichever the case, it should be stated, that by the above mentioned properties this vaccine meets requirements for purified preparations produced according to a more complex technology.

Table 2 shows comparative data on potency testing in mice of three MKC vaccine experimental lots and two types of the commercial TBE vaccines which are in use in the U.S.S.R. presently. The mouse challenge potency test was performed by giving 3 subcutaneous applications (1st, 3rd, and 7th days) of serial threefold dilutions of vaccine to groups of 10 mice each (0.5 ml per mouse). Fourteen days after the last immunization mice were challenged by intraperitoneal inoculation of 100 to 1,000 LD_{50} of TBE virus (Absettarov strain). PD_{50} (protective dilution of the vaccine which protects 50% of mice) was calculated after observation for 14 days according to Reed and Muench. Then a volume of minimal immunizing dose (MID) was calculated by division of 1 ml by PD_{50}. The number of MIDs in a human immunizing dose of the vaccine is shown in the last column of Table 2. As can be seen, all three MKC vaccine lots were approximately 2-fold more

Table 2. Potency of TBE vaccines in mouse challenge test represented in values of protective dilution (PD$_{50}$) and mouse minimal immunizing doses (MID)

Vaccine	Lot no.	Human dose (ml)	PD$_{50}$	MID (ml)	MID per human dose
MKCV	7755-f	0.5	1:144.9	0.0068	73
	7855-f		1:151.5	0.0066	76
	7855-r		1:141.3	0.0070	71
CECV (Tomsk)	158-5	1.0	1:30.0	0.033	30
CECV-concentr.	4	0.5	1:77.1	0.013	38
	5		1:54.5	0.018	28
(Moscow)	43		1:76.6	0.013	38

immunogenic than four control commercial vaccines prepared from the virus propagated in CEC-cultures. It should be noted that the first two lots of MKC vaccine were produced in flask cultures, and only lot 7855-r was prepared in roller bottle cultures, whereas all commercial vaccines mentioned were produced in roller bottle cultures.

For the controlled trials on volunteers we used 3 lots of MKC vaccine mentioned in Table 2. As a control, the commercial unconcentrated chick embryo cell vaccine lot No 158-5 produced in Tomsk was employed.

Trials were carried out on 180 volunteers aged 18–20 years, 50 persons for each lot of MKC vaccine and 30 persons for the control preparation. The volunteers were recruited according to standard and strict informed consent procedures from those who were to work in the near future in the epidemiological areas, i.e., with a potential risk of attack by infected ticks. MKC vaccine was administered in a 0.5 ml dose by the intramuscular route, control vaccine was applied in a 1.0 ml dose subcutaneously, according to recommendations of the manufacturer. The second dose was given three months after the first one, using the same preparations.

Blood was taken from vaccinees at the time of the first and the second vaccinations and one month thereafter. Sera were analyzed for TBE antibodies by hemagglutination inhibition (HI), neutralization test by plaque reduction (NT), and ELISA—separately for IgG and IgM antibodies.

The serum was considered positive in HI if its titer was not less than 10, in ELISA, if its titer was not less than 160, and in NT, if at a dilution of 1:8 its neutralization index was not less than $2 \log_{10}$ PFU/ml. Prevaccination sera were negative in all three tests used.

The trials showed that by the incidence and the rate of side reactions the MKC vaccine lots did not differ from the control commercial vaccine. Some subjects had small erythemas up to 10 mm around the injection site but without infiltrations. Swelling of the regional lymphoid nodules was not observed. A transient increase in body temperature following the first immunization was registered in 2% to 6% of vaccinated persons. After the second immunization, febrile reactions were absent.

Serological investigations evidenced high immunogenic activity of TBE vaccine prepared in the green monkey kidney cells (Table 3). Neutralizing anti-TBE antibodies appeared in sera of 93–100% of vaccinated persons after the first inoculation with MKC vaccine. The response in volunteers to the first inoculation with the control CEC vaccine was only 54%. Following the second immunization the sera from recipients of MKC vaccine were 100% positive, irrespective of the test used.

As can be seen from Tables 4 and 5, during the period between the first and the second vaccinations with MKC vaccine there was no considerable drop in neutralizing antibody levels or in the levels of antibodies detected by ELISA. After the second vaccination with MKC vaccine the high neutralization index levels of serum antibodies were achieved, from 4.5 to 5.6 $\log_{10}$ PFU/ml. The same value for the control vaccine was 3.2 $\log_{10}$ PFU/ml. The titers of antibodies detected by ELISA were also higher in groups vaccinated by MKC vaccine than in the control group. The neutralization indices of sera after one injection of MKC vaccine were equal to those obtained after two immunizations with the commercial vaccine.

The high percent of seroconversions, the high levels of virus-neutralizing antibodies and their good maintenance following the first immunization with

Table 3. Percentage of volunteers with anti-TBE serum antibodies after immunization with TBE vaccines

Test	Vaccine lot	Time of response		
		after 1st injection	before 2nd injection	after 2nd injection
HAI	7755-f*	32	7	100
	7855-f*	46	39	100
	7855-r*	75	55	100
	158-5**	9	0	75
NT	7755-f	93	95	100
	7855-f	97	97	100
	7855-r	100	100	100
	158-5	54	56	92
ELISA	7755-f	95	90	100
	7855-f	90	81	100
(IgG)	7855-r	100	89	100
	158-5	56	42	100

  * MKC vaccine
 ** CEC vaccine (Tomsk)

Table 4. Serological response in volunteers given TBE vaccines. Mean geometric titre of serum antibodies

Test	Vaccine lot	Time of response		
		after 1st injection	before 2nd injection	after 2nd injection
ELISA	7755-f*	785	< 160	< 160
	7855-f*	1450	< 160	< 160
(IgM)	7855-r*	995	< 160	< 160
	158-5**	330	< 160	< 160
ELISA	7755-f	550	604	4100
	7855-f	860	770	4970
(IgG)	7855-r	1280	861	5485
	158-5	195	180	1920

   * MKC vaccine
  ** CEC vaccine (Tomsk)

Table 5. Serological response in volunteers given TBE vaccines. Mean neutralization index of serum antibodies $\log_{10}$ PFU/ml

Test	Vaccine lot	Time of response		
		after 1st injection	before 2nd injection	after 2nd injection
NT	7755-f	2.9	3.1	4.5
	7855-f	3.1	3.0	4.9
	7855-r	3.6	3.8	5.6
	158-5	2.1	1.7	3.2

MKC vaccine permit in urgent situations (a lack of time) the use of only one inoculation of that vaccine immediately before the epidemiological season.

Currently, industrial lots of MKC vaccine have been prepared which are 3–5 times more potent than commercial CEC vaccines. They will be investigated in field trials in the coming epidemiological season.

In conclusion, the following advantages of the MKC TBE vaccine can be noted:

— cheap source of virus for vaccine production;
— simple production technology without concentration of the initial material;
— low content of host-derived impurities;
— good stability of the vaccine during storage—not less than 2 years;
— high immunogenic activity, low reactogenicity

In our opinion the advantages of the technology used for preparation of this TBE vaccine could be also employed for production of vaccines against other viral infections.

Simultaneously with primary cell cultures, continuous cell lines are used for the vaccine production on an increasing scale. In our institute the continuous cell line 4647 derived from green monkey kidney cell was developed [6]. We studied the possibility of TBE vaccine production in this cell line and found it to be suitable for the preparation of TBE vaccine by using the above described technology.

According to WHO requirements parenteral preparations produced in continuous cell lines must be purified from cellular DNA, the concentration of residual DNA in such preparations must not exceed 100 pg per single dose [7]. Taking this into account we developed a new simple and effective method for the purification of virus from cellular DNA (MN Matrosovich et al., in prep.). The antigen loss during the DNA-decontamination step does not exceed 10–20 percent as measured by ELISA.

The characteristics of three experimental lots of TBE vaccine produced in 4647 continuous cell line are presented at Table 6. As can be seen, these lots are analogous to the TBE vaccine produced in primary green monkey kidney cells in respect to virus accumulation titers, concentrations of total protein and host-derived impurities, and potency. The content of residual cellular DNA is far lower than that allowed by WHO requirements. The main future benefits of possible utilization of 4647 continuous cell line for TBE vaccine production as compared to the primary monkey kidney cells are its better standardization, lower cost and the possibility of eliminating the need for monkeys.

Table 6. Control tests on TBE vaccines produced in 4647 continuous cell line

Control test	Vaccine lot no		
	1	2	3
Virus titre before inactivation ($\log_{10}$ PFU/ml)	9.8	10.1	10.1
Test for inactivation[a]	—	—	—
Protein (µg/dose)	110	95	110
Host-derived impurities (µg/dose)	1	1.5	1
Cellular DNA (pg/dose)	15	10	10
Innocuity test[b]	—	—	—
Potency (MID/dose)	90	135	137

[a] No live virus present

[b] Innocuous for mice and guinea pigs

References

1. Chumakov MP, Rubin SG, Avdeeva LI, Semashko IV, Salnikov YaA, Martyanova LI, Gagarina AV (1986) Vaccine against tick-borne encephalitis. USSR Invention Certificate No 1,349,757
2. Elbert LB, Krasilnikov IV, Drozdov SG, Grachev VP, Pervikov YuV, Krutyanskaya GL, Khanina MK, Mchedlishvily BM, Kolikov VM, Bresler SE, Vorobyeva MS (1985) Concentrated purified vaccine against tick-borne encephalitis produced by ultrafiltration and chromatography. Vopr Virusol 1: 90–93 (in Russian)
3. Kunz C, Hofmann H, Heinz FX, Dippe, H (1980) The efficacy of vaccination against tick-borne encephalitis. Wien Med Wochenschr 92: 809–813
4. Heinz FX, Kunz C, Fauma H (1980) Preparation of a highly purified vaccine against tick-borne encephalitis by continuous flow zonal ultracentrifugation. J Med Virol 6: 213–221.
5. Lvov DK, Gagarina AV (1965) Immunoprophylaxis of tick-borne encephalitis. In: Chumakov MP (ed) Tick-borne encephalitis and other arboviral diseases. VNIIMI, Moscow, pp 97–127 (Viruses and viral diseases, vol 1) (in Russian)
6. Mironova LL, Shalunova NV, Lomanova GA, Nikolaeva MA, Khapchaev YuH, Stobeckij VI, Karmysheva VYa, Popova VD, Alpatova GA, Sobolev SG, Grachev VP (1987) The bank of green monkey kidney cell continuous line 4647. Vopr Virusol 6: 740–743 (in Russian)
7. WHO (1987) Requirements of continuous cell lines used for biological production. WHO Tech Rep Ser 747, Annex 3

Authors' address: M. P. Chumakov, Institute of Poliomyelitis and Viral Encephalitides of the U.S.S.R. Academy of Medical Sciences, 142 782 Moscow Region, U.S.S.R.

Arch Virol (1990) [Suppl 1]: 169–179

Characterization of Dugbe virus by biochemical and immunochemical procedures using monoclonal antibodies

A. A. El-Ghorr, A. C. Marriott, V. K. Ward, T. F. Booth, S. Higgs, E. A. Gould,
and **Patricia A. Nuttall**

NERC Institute of Virology and Environmental Microbiology Oxford, U.K.

Accepted February 6, 1990

Summary. Dugbe virus is assigned to the family *Bunyaviridae*, genus *Nairovirus* and is related to the tick-borne Crimean-Congo hemorrhagic fever and Nairobi sheep disease viruses. The proteins of Dugbe virus were studied biochemically and immunochemically using monoclonal and polyclonal antibodies. The G1, N and G2 proteins were detected by PAGE at 73, 49, and 35 kDa, respectively. On a Western blot, polyclonal antisera to Dugbe virus reacted with the G1, N and G2 proteins and with several additional proteins (210, 45, 40, 33, and 30 kDa). The G1 and G2 proteins were shown to be located on the surface of virus particles and to be glycosylated while the N protein was internal and non-glycosylated. The 40 kDa protein also was found to be glycosylated. This 40 kDa glycoprotein may represent an additional glycoprotein, as found in Hazara virus, or may be a breakdown product of G1. A monoclonal antibody (McAb H28.89) against the N protein specifically labelled purified nucleocapsids in an immunogold EM reaction. This McAb did not neutralise viral infectivity in an in vitro assay and did not label whole virus particles with colloidal gold. Another McAb (H28.17) against the G1 protein neutralised virus infectivity and labelled whole virus particles by immunogold EM.

This work, combined with ongoing genome sequencing and expression research, should lead to a better understanding of the molecular biology of nairoviruses and to the production of useful diagnostic tools.

Introduction

Dugbe (DUG) virus is a member of the genus *Nairovirus* in the family *Bunyaviridae*. The genus *Nairovirus* is further subdivided on the basis of antigenic reactivity into seven serogroups: Crimean-Congo hemorrhagic fever (CCHF) group, Dera Ghazi Khan group, Hughes group, Nairobi sheep disease (NSD) group, Qalyub group, Sakhalin group [7], and Thiafora group [17]. Several nairoviruses are of concern to humans, e.g., CCHF virus causes severe human illness in parts of Africa, Europe, and Asia [3] while NSD virus causes considerable economic losses of cattle and sheep in Africa [9]. These two nairoviruses require high levels of containment if they are to be safely studied in the laboratory. DUG virus is a member of the NSD serogroup but causes only mild disease in humans and other animals and is an ideal model for the study of nairoviruses [4].

Nairoviruses are transmitted by certain species of ticks. Comparatively little effort has been spent on their study and little is known of their molecular biology and of the evolutionary and molecular significance of the different serogroups.

Members of the family *Bunyaviridae* possess a tripartite single stranded RNA genome (L, M, and S segments) and are enveloped viruses ranging between 90–120 nm in diameter [1]. Some nairoviruses have been shown to possess three structural proteins by radio-immune precipitation: a nucleoprotein (N: 48–54 kDa) and two external glycoproteins (G1: 72–84 kDa and G2: 30–40 kDa). For some members of the genus *Nairovirus*, however, a G1 protein has not been detected, while the G2 has only been detected in members of the Qalyub and CCHF serogroups [7]. Additionally, Qalyub has been shown to possess two large non-structural glycopeptides (115 kDa and 85 kDa) which are considered to be precursors of G1 and G2 [6]. Hazara virus, a member of the CCHF serogroup, has been shown to possess three structural glycoproteins and a nucleocapsid protein [10]. Until 1981 only one structural protein had been detected in DUG virus: the nucleocapsid protein at 49 kDa [7, 8]. More detailed work in 1985 detected six DUG virus-induced polypeptides [4]. Radiolabelled infected cell lysates contained 92 kDa, 83 kDa, 52 kDa, and 48 kDa viral-induced intracellular polypeptides. The 52 kDa and 48 kDa polypeptides were also detected in purified virion preparations. Purified virus also contained two other structural proteins of 77 kDa and 34 kDa. Using ^{3}H-glucosamine only one of these proteins (77 kDa) was shown to be glycosylated. Neither Western blotting experiments nor investigations utilising monoclonal antibodies have been reported with members of the genus *Nairovirus*. The structural polypeptides of a model nairovirus, DUG, were investigated using monoclonal and polyclonal antibodies in polyacrylamide gel electrophoresis (PAGE), Western blotting, immunosorbent electron microscopy (ISEM), and immunogold electron microscopy (IGEM) experiments. Effort was concentrated on detecting the G1 and G2 glycoproteins and locating them on the virus particle. The possible presence of a third glycoprotein was also investigated.

Materials and methods

Virus and cells

The DUG ArD44313 strain was isolated from a pool of 20 male *Amblyomma variegatum* ticks collected in November 1985 at Bouroufaye, Senegal. This virus was identified using the complement fixation test and confirmed to be DUG virus by cross-immunofluorescence and cross-neutralisation with the prototype strain, DUG lbAr 1792. The ArD44313 strain was then grown in the pig kidney epithelial (PS) cell line at 35 °C for 3 days using an input multiplicity of infection of 0.05 plaque forming units per cell. The cell culture supernatant fluid was then separated from the infected monolayer and was used as a source of antigen for immune electron microscopy.

Cell culture supernatant was clarified by centrifugation at 2,000 × **g** for 15 min to remove gross debris. Virus in the supernate was pelleted through a 10% (w/w) sucrose cushion by centrifugation at 100,000 × **g** for 45 min at 4 °C. The virus pellet was resuspended in 1/100th the original volume and was used as a source of DUG viral structural proteins.

For the purification of viral nucleocapsids, the DUG virus infected PS monolayer was lysed with 1% Triton N101, clarified at 6,000 × **g** for 30 min at 4 °C and centrifuged on a CsCl gradient at 200,000 × **g** overnight at 20 °C. The nucleocapsid band, seen approximately half way down the tube, was harvested and pelleted at 400,000 × **g** for 2 h at 20 °C.

Virus surface proteins were investigated by protease digestion. Virus pellets were incubated with 0.1 mg/ml α-chymotrypsin (Sigma, Dorset, U.K.) for 30 min at room temperature (RT), boiled for 2 min in sample buffer (see below) and then loaded onto a 10% sodium dodecyl sulphate (SDS) polyacrylamide gel.

The enzyme N-Glycanase (Genzyme, Suffolk, U.K.) was used to hydrolyse asparagine-linked oligosaccharides from glycoproteins. The glycoprotein sample was boiled for 3 min in the presence of 0.5% SDS – 0.1 M 2-mercaptoethanol and was incubated with the enzyme (10 units/ml final concentration) overnight at 37 °C before analysis by SDS-PAGE.

For the detection of sugars in glycoproteins, a glycan detection system (Boehringer Mannheim, Philadelphia, U.S.A.) was used. Adjacent hydroxyl groups in saccharides were oxidized to aldehyde groups by mild periodate treatment. The spacer-linked steroid hapten digoxigenin was then covalently linked to these aldehydes via a hydrazide group. The samples were resolved by SDS-PAGE and electro-blotted onto nitrocellulose as described below. Subsequently, digoxigenin-labelled glycoproteins were detected in an enzyme immunoassay using an antibody to digoxigenin conjugated to alkaline phosphatase.

Antisera

The DUG strain KT281/75 [4] was used to raise polyclonal hyperimmune antiserum and monoclonal antibodies to DUG virus. This KT281/75 strain was found to be identical to the ArD44313 and the prototype strain in binding assays (Dr. EA Gould, pers. comm.). A 40% suckling mouse brain suspension of DUG was centrifuged at 90,000 × **g** for 3 h at 4 °C on a 30–40% glycerol/potassium tartrate gradient. The visible virus band (lowest band) was collected and stored at −70 °C. Hyperimmune antiserum was prepared by intramuscular inoculation of rabbit with 0.5 ml of purified virus mixed with an equal volume of complete Freund's adjuvant. A second intramuscular inoculation was given four weeks later. At three and six weeks after this, 1.0 ml of the virus without adjuvant was administered intradermally. After a further two weeks the serum was collected and stored at −20 °C.

In order to raise monoclonal antibodies, female Balb/c mice were inoculated intraperitoneally with 0.1 ml of suckling mouse brain suspension containing DUG virus. After one week, the spleen was removed and a fusion with NS0 cells was performed [5].

172   A. A. El-Ghorr et al.

SDS-PAGE and Western blotting

The polyacrylamide gel electrophoresis technique utilising sodium dodecyl sulphate [13] was used for this investigation. A 10 μl aliquot of each sample was diluted with 10 μl sample buffer (2% SDS, 4% 2-mercaptoethanol, 0.04% bromophenol blue), boiled for 2 min, and loaded onto a 10% gel. The Western blotting technique of electrophoretic transfer of proteins from polyacrylamide gels onto nitrocellulose sheets was used [2, 14] utilising the Sartorius semi-dry blotting apparatus. The nitrocellulose was blocked with 3% skimmed milk powder containing 0.05% Tween 20 in phosphate buffered saline (PBS/T) for a minimum of 30 min at ambient temperature and then reacted with rabbit polyclonal antiserum to DUG virus or mouse monoclonal antibodies to DUG viral proteins. The PBS/T was used as a washing solution and diluent in order to reduce non-specific binding. After the primary antibody reaction and a washing step, we added anti-rabbit or anti-mouse immunoglobulins conjugated to alkaline phosphatase (Sigma, Dorset, U.K.). The substrates Nitroblue tetrazolium chloride (NBT) and 5-bromo-4-chloro-3-indolyl-phosphate p-toluidine salt (BCIP) were then prepared according to the manufacturer's instructions (Bethesda Research Laboratories, Uxbridge, U.K.) and used simultaneously to visualise immunoreactive protein bands.

Direct electron microscopy

Formvar carbon coated 400 mesh copper EM grids were used throughout. A 5 μl drop of sample to be examined was placed on each grid and excess fluid was blotted with filter paper.

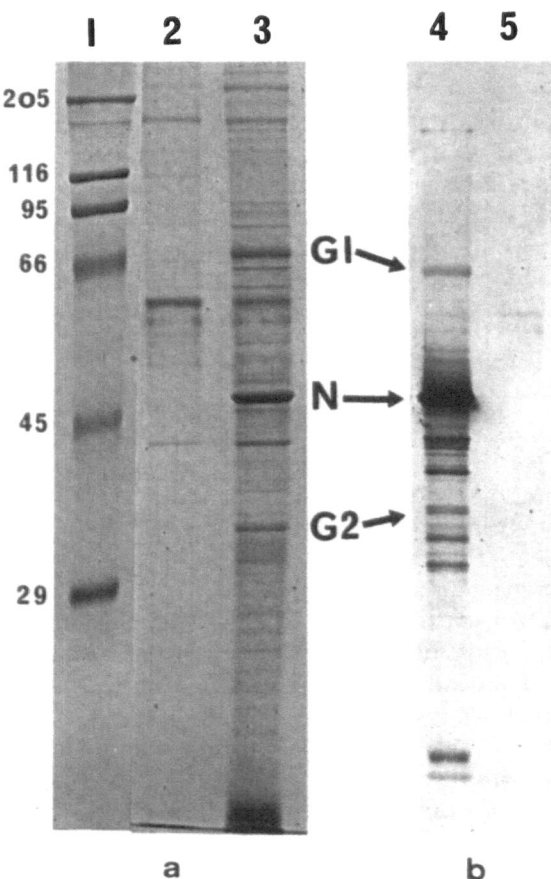

Fig. 1. a A Coomassie-stained 10% poly-acrylamide gel of molecular weight markers (*1*), control mock-infected PS cells pellet (*2*) and DUG virus pellet (*3*). The G1, N, and G2 proteins were readily detected in the virus pellet. **b** On Western blotting and probing with a polyclonal rabbit antiserum to DUG virus, the G1, N, and G2 proteins were still detected in the virus pellet. In addition, 5 other viral proteins appeared at 210, 45, 40, 33, and 30 kDa (*4*). This rabbit antiserum did not react with any protein bands in the mock-infected control (*5*)

The sample was fixed with 2% glutaraldehyde, negatively stained using phosphotungstic acid and ammonium molybdate (2:1), and examined in a Jeol (Tokyo, Japan) JEM-100C electron microscope.

Immunosorbent electron microscopy

The IgG fraction was purified from all sera and monoclonal antibodies using a protein A column prior to ISEM and IGEM. Copper EM grids were floated on a 10 µl drop of purified IgG diluted 1/50 in 0.1 M phosphate buffer (PB) pH 7.2 for 1 h at 36 °C in a moist chamber. The phosphate buffer did not contain any sodium salt since this salt would interfere with the negative stain. The grids were then washed twice with PB and incubated on a drop of antigen. The grids were finally fixed and negatively stained as described.

Immunogold electron microscopy

Copper EM grids were coated with antigen by floating each grid on cell culture supernatant fluid containing DUG virus (ArD44313). After incubation the grids were washed and reacted with mouse polyclonal or monoclonal gamma immunoglobulins to DUG virus. After a second washing step the grids were floated on a drop of goat anti-mouse IgG conjugated to 10 nm colloidal gold (BioCell, Cardiff, U.K.) or protein A conjugated to 20 nm gold particles. Colloidal gold particles were prepared by citric acid reduction of chloroauric acid [11] and complexed with protein A [12]. The grids were finally fixed and negatively stained as described. All incubations were performed in a moist chamber at 36 °C for 1 h.

Results

Viral structural proteins

SDS-PAGE of pelleted virus showed 3 main protein bands, at 73, 49, and 35 kDa, that were not present in the mock-infected cell culture control lane (Fig. 1a). The molecular weights of these proteins agreed with the reported molecular weights of G1, N, and G2 respectively [4]. When this gel was electro-blotted onto a nitrocellulose sheet and probed with rabbit polyclonal antiserum to DUG virus, only virus specific proteins were detected. No protein bands were seen in the mock-infected cell culture control lane while in the lane containing pelleted virus several polypeptide bands were consistently detected in addition to the G1, N, and G2 bands. The main extra bands appeared at 210, 45, 40, 33, and 30 kDa (Fig. 1b).

The intensity of the Coomassie blue staining of the SDS-polyacrylamide gel showed that all three viral proteins were present in similar quantities. However, the strong color reaction that was observed for the 49 kDa nucleocapsid protein in the Western blotting experiment indicated that this N protein was probably the most immunogenic of the viral proteins because most of the immunoglobulins in the polyclonal antisera were directed against it. The N protein also was seen occasionally to migrate as a double band.

A summary of the results of the biochemical investigations is presented in Fig. 2. When 2-mercaptoethanol was omitted from the dissociation buffer,

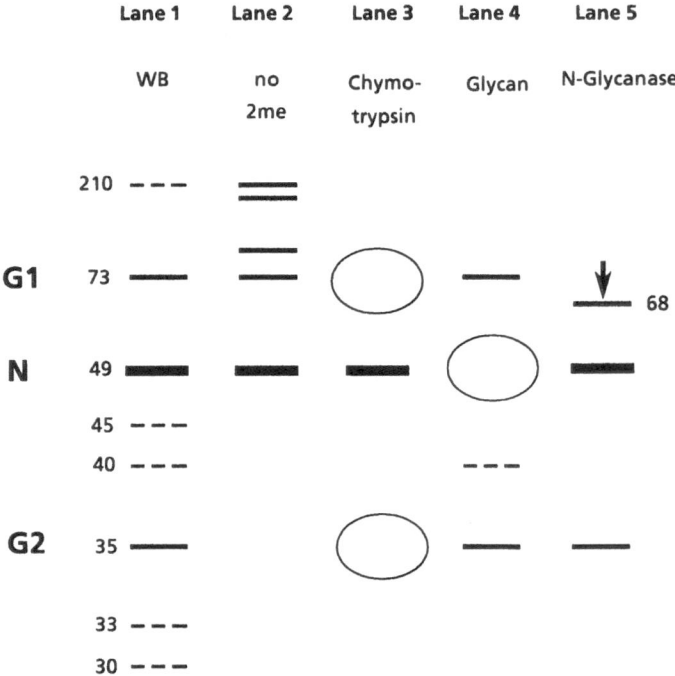

Fig. 2. *1* Western blotting result of DUG virus pellet. In the absence of 2-mercaptoethanol, the low molecular weight proteins were replaced by other proteins at 99 and 170 kDa (*2*) on Western blots. After protease treatment (*3*) the G1 and G2 proteins could not be detected in Coomassie-stained gels. The glycan detection technique gave a positive result with the G1, G2 and 40 kDa proteins (*4*). N-Glycanase digestion of DUG virus pellet leads to a detectable shift only in the mobility of the G1 protein by 5 kDa (*5*)

leaving disulphide bridges intact, none of the viral bands below the N protein appeared on Western blots. In their places, other protein bands were detected at 99 and 170 kDa. The low molecular weight proteins seemed to form complex tertiary structures or were breakdown products of other proteins. The mobility of the G1 and N proteins was not altered.

Protease treatment of DUG virus pellet with α-chymotrypsin resulted in the removal of the G1 and G2 proteins from the virus while the N protein was unaffected (Fig. 2, lane 3). This indicated that the G1 and G2 proteins were present on the outside of the virus while the N protein was internal to the viral envelope.

Using the glycan detection technique to identify glycoproteins, three glycoprotein bands were observed in DUG virus infected cells over mock infected controls. In addition to the 73 kDa G1 and the 35 kDa G2 proteins, a 40 kDa glycoprotein was also observed (Fig. 2, lane 4).

When the enzyme N-glycanase was used to digest asparagine-linked oligosaccharide side chains, the G1 glycoprotein migrated differently on SDS-PAGE. A shift of 5 kDa could be calculated for the G1 protein but no detectable difference was observed in the migration of N or G2 (Fig. 2, lane 5). Additionally, no shift was detected in the 40 kDa protein on Western blots after treatment with this enzyme. The G1 protein was therefore heavily glycosylated while the G2 and 40 kDa proteins were only lightly glycosylated or not asparagine linked. The N protein was not glycosylated because it did not react in the glycan detection system.

Electron microscopic studies

Under the EM using negative staining, DUG virus particles could readily be detected in supernatant fluid from infected cell cultures (Fig. 3a). A mixture of phosphotungstic acid and ammonium molybdate in a 2:1 ratio was found to give the best resolution in negative staining. The enveloped DUG virus particles appeared spherical in shape, 90–100 nm in diameter and surrounded by 7–9 nm long structures projecting from the viral envelope. There seemed to be a distinctive texture to the virion, indicating that the surface structures may be cylindrical in shape. When a monoclonal antibody (McAb H28.17) was used in an IGEM test, it specifically labelled the surface projections on intact virus particles (Fig. 3b). This McAb also neutralised virus infectivity in an in vitro test and protected mice from DUG virus following intracranial inoculation. On Western blots McAb H28.17 reacted mainly with the 73 kDa protein (data not shown). The G1 protein was therefore located on the outside of the virus, had a molecular weight of 73 kDa, and bore some neutralising epitopes.

Another monoclonal antibody (H28.89), which reacted only with the 49 kDa N protein on Western blots, did not neutralise virus infectivity in vitro or protect mice. This monoclonal antibody did not gold label whole virus particles in the IGEM test (Fig. 3c). Disrupted particles, however, were labelled.

Further confirmation of the specificity of McAb H28.89 was obtained when this antibody specifically gold-labelled purified nucleocapsids (Fig. 4), whereas McAb H28.17 resulted in only minor labelling of the nucleocapsid preparation.

Using the ISEM technique the anti-G1 McAb concentrated the number of virus particles seen on EM grids 35-fold (Table 1), again indicating that

Table 1. The concentration of virus particles on 4 grid squares by immunosorbent electron microscopy using purified immunoglobulins

Antibody	No. of virus particles	Virus concentration
Buffer control	17	
anti-N McAb	91	5×
anti-G1 McAb	590	35×
poly anti-DUG	69	4×

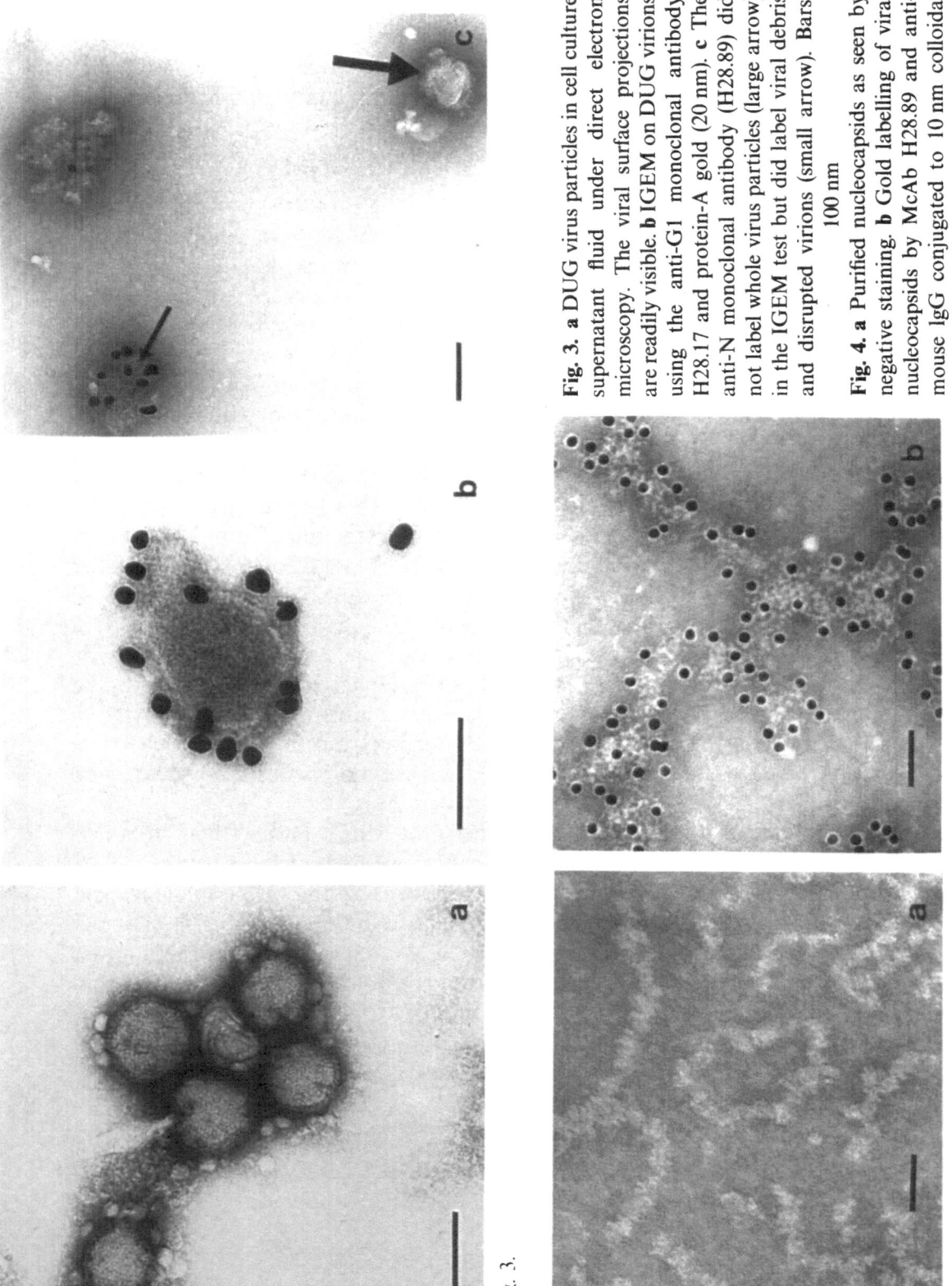

Fig. 3. a DUG virus particles in cell culture supernatant fluid under direct electron microscopy. The viral surface projections are readily visible. **b** IGEM on DUG virions using the anti-G1 monoclonal antibody H28.17 and protein-A gold (20 nm). **c** The anti-N monoclonal antibody (H28.89) did not label whole virus particles (large arrow) in the IGEM test but did label viral debris and disrupted virions (small arrow). Bars: 100 nm

Fig. 4. a Purified nucleocapsids as seen by negative staining. **b** Gold labelling of viral nucleocapsids by McAb H28.89 and anti-mouse IgG conjugated to 10 nm colloidal gold. Bars: 50 nm

Fig. 3.

Fig. 4.

G1 was a surface protein. With the anti-N McAb only a 5-fold concentration effect was seen, probably due to the presence of disrupted particles and partly damaged virions. The polyclonal antiserum produced fourfold concentration. This result confirmed our earlier observations that polyclonal antisera contained mainly anti-N immunoglobulins and relatively little anti-glycoprotein activity.

Discussion

Nairoviruses are among the least well studied members of the family *Bunyaviridae*. Dugbe virus was used as a model of more pathogenic nairoviruses, such as those that cause CCHF and NSD, with the initial aim being to obtain diagnostic probes for detecting these diseases. Current diagnostic techniques rely on lengthy cell culture procedures which make it difficult to rapidly obtain an accurate assessment of the epidemiology of these diseases, the lack of which being a hinderance to control measures.

These initial studies concentrated on investigating the structural proteins of the virus we consider a model for the genus. DUG virus was shown to possess 2 glycoproteins present on the surface of the virion, a 73 kDa G1 glycoprotein which is heavily glycosylated and a 35 kDa G2 protein which is only slightly glycosylated. The nucleocapsid protein N, at 49 kDa is not glycosylated and was found to be contained entirely within intact virus particles. Many virions, however, were found to be fragile and easily disrupted, revealing the N protein. Several other low molecular weight bands were detected by Western blotting. These probably reflect breakdown products of the structural proteins because such proteins have not been detected with other nairoviruses [7]. One of these proteins, at 40 kDa, was found to be glycosylated. This may have been a breakdown product of G1 or a true third glycoprotein, as has been detected for Hazara virus [9]. It is not yet clear which is the case but this protein could only be detected on Western blots or in the glycan detection system and was only lightly glycosylated.

Polyclonal antisera to DUG virus were found to contain mainly anti-N activity. The N protein was the most immunogenic of the virus structural proteins but did not bear any neutralising epitopes. The G1 protein did, however, elicit neutralising antibodies. No monoclonal antibodies have been raised to other proteins of DUG virus.

The N protein may provide a useful tool for developing diagnostic tests for nairoviruses because most of the antibody response is directed towards this protein. The N protein of DUG virus has been shown to be coded for by the S-RNA segment, now completely sequenced [15]. Preliminary experiments have utilised Dugbe N protein expressed in a baculovirus vector to detect antibody to CCHF virus in an enzyme-linked immunosorbent assay [16].

By analogy with other *Bunyaviridae* the M-RNA segment of DUG virus is considered to code for the surface glycoproteins. This segment currently is being sequenced. When this is completed there should be sufficient information to determine the exact number of surface glycoproteins and in conjunction with more monoclonal antibodies, to map neutralising epitopes on these proteins. Sophisticated diagnostic techniques could then be based on nucleic acid hybridisation probes, expressed proteins, group specific antibodies or a combination of these. M segment sequence data may also provide crucial information about methods of controlling and limiting the spread of pathogenic nairoviruses by providing data about the molecular basis of pathogenicity or virulence.

References

1. Bishop DHL, Shope RE, Beaty BJ (1987) Molecular mechanisms in arbovirus disease. In: Russel WC, Almond JW (eds). Molecular basis of virus disease. Cambridge, Cambridge University Press, pp 135–166
2. Burnette W (1981) "Western blotting": electrophoretic transfer of proteins from sodium dodecyl sulfate polyacrylamide gels to unmodified nitrocellulose and radiographic detection with antibody and radioiodinated protein A. Anal Biochem 112: 195–203
3. Casals J, Tignor GH (1974) Neutralization and haemagglutination–inhibition tests with Crimean haemorrhagic fever-Congo virus. Proc Soc Exp Biol Med 145: 960–966
4. Cash P (1985) Polypeptide synthesis of Dugbe virus, a member of the *Nairovirus* genus of the *Bunyaviridae*. J Gen Virol 66: 141–148
5. Chanas AC, Gould EA, Clegg JCS, Varma MGR (1982) Monoclonal antibodies to Sindbis virus glycoprotein E1 can neutralize, enhance infectivity, and independently inhibit haemagglutination or haemolysis. J Gen Virol 58: 37–46
6. Clerx JPM, Bishop DHL (1981) Qalyub virus, a member of the newly proposed *Nairovirus* genus (*Bunyaviridae*). Virology 108: 361–372
7. Clerx JPM, Casals J, Bishop DHL (1981) Structural characteristics of nairoviruses (genus *Nairovirus*; *Bunyaviridae*). J Gen Virol 55: 165–178
8. David-West TS (1974) Polyacrylamide gel electrophoresis of Dugbe virus infected cells. Microbios 11: 21–23
9. Davies FG, Casals J, Jesset DM, Ochieng P (1978) The serological relationships of Nairobi sheep disease virus. J Comp Pathol 88: 519–523
10. Foulke RS, Rosato RR, French GR (1981) Structural polypeptides of Hazara virus. J Gen Virol 53: 169–172
11. Frens G (1973) Controlled nucleation for the regulation of particle size in monodisperse gold suspension. Nature 241: 20–22
12. Horisberger M, Clerc MF (1985) Labelling of colloidal gold with protein A. A quantitative study. Histochemistry 82: 219–223
13. Laemmli UK (1970) Cleavage of structural proteins during assembly of the head of bacteriophage T4. Nature 227: 680–685
14. Towbin H, Staehelin T, Gordon J (1979) Electrophoretic transfer of proteins from polyacrylamide gels to nitrocellulose sheets: procedure and some applications. Proc Natl Acad Sci USA 76: 4350–4354
15. Ward VK, Marriott AC, El-Ghorr AA, Nuttall PA (1990) Coding strategy of the S RNA segment of Dugbe virus (*Nairovirus*; *Bunyaviridae*). Virology 175: 518–524
16. Ward VK, Marriott AC, Polizoui T, El-Ghorr AA, Antoniadis A, Nuttall PA (1990) Expression of the nucleocapsid protein of Dugbe virus (*Nairovirus*; *Bunyaviridae*) and employment as a diagnostic reagent for Crimean-Congo hemorrhagic fever. (In prep)

17. Zeller HG, Karabatsos N, Calisher CH, Digoutte JP, Cropp CB, Murphy FA, Shope RE (1989) Electron microscopic and antigenic studies of uncharacterized viruses. II. Evidence suggesting the placement of viruses in the family *Bunyaviridae*. Arch Virol 108: 211–227

Authors' address: Patricia A. Nuttall, NERC Institute of Virology and Environmental Microbiology, Mansfield Road, Oxford, OX1 3SR, England.

Arch Virol (1990) [Suppl 1]: 181–195

Studies of the pathogenesis of Dugbe virus in normal and in immunosuppressed mice

C. Sweet and **D. M. Coates***

Microbial Molecular Genetics and Cell Biology Group, School of Biological Sciences,
University of Birmingham, Birmingham, U.K.

Accepted January 18, 1990

Summary. Susceptibility to lethal infection with the KT281/75 strain of the tick-borne nairovirus, Dugbe (DUG) virus, was similar in an outbred strain and several inbred strains of mice. For the outbred strain, both neural and extraneural routes of inoculation of virus resulted in lethal infection, but susceptibility decreased with age and only intracranial inoculation produced a lethal infection in adults. In newborn mice, subcutaneous (s.c.) inoculation of virus (analogous to a tick-bite) produced a slowly developing disseminated infection with virus not reaching maximum titres in the brain until 8 days after inoculation. In contrast, in intranasally (i.n.) inoculated animals virus spread rapidly from the upper respiratory tract to the brain by 2 days after inoculation, in the absence of a detectable viremia. No viremia was detected in s.c. or i.n. inoculated adults: in the former, virus replicated only at the site of inoculation; in the latter, virus replicated in the respiratory tract, later spreading to the brain. Immunosuppression of i.n. inoculated adult mice with cyclophosphamide produced some mortality, indicating that host defenses are important in protecting the adult. Differences in virulence were exhibited between two tick isolates of Dugbe virus (strains ArD16095 and ArD44313), a human isolate (IbH11480) and the laboratory strain KT281/75. Virulence differences varied with the route of inoculation. The human isolate, unlike the other isolates, was lethal for adult mice by the intranasal route without immunosuppression. The similarity between the pattern of DUG virus infection in the mouse with that of other, more pathogenic, nairoviruses suggests that, although hemorrhagic disease was not observed, ours may be a

---

* Present address: PHLS Centre for Applied Microbiology and Research, Division of Pathology, Porton Down, Salisbury, Wiltshire, U.K.

useful model for studying the genetic basis of nairovirus virulence and for testing vaccines and anti-viral drugs.

Introduction

Viruses from several families cause severe and often fatal hemorrhagic fevers [10]. However, because laboratory work with these agents presents particular problems of safety, knowledge of their pathogenetic mechanisms is rudimentary.

Crimean-Congo hemorrhagic fever virus (CCHF), a tick-borne arbovirus belonging to the genus *Nairovirus* of the family *Bunyaviridae*, is found throughout most of sub-Saharan Africa, eastern Europe, the Middle East, and Asia [15], and causes sporadic, but sometimes epidemic, outbreaks of severe human disease with mortality rates between 15 and 50%. In nosocomial outbreaks, the case fatality rate is often high, perhaps related to respiratory spread. Patients present with fever, chills and prostration, sometimes followed by severe hemorrhagic signs. The primary pathophysiological event appears to be erythrocyte and plasma leakage through vascular endothelium. Edema, focal necrosis, hemorrhage and vascular congestion are found in brain, liver, heart, and other organs, but inflammation is minimal. Death may result from severe circulatory disturbances, extensive loss of blood and functional conduction defects in the myocardium due to necrosis.

An animal model for CCHF would allow a more thorough investigation of the pathogenesis of the etiologic agent. Recently, a suckling mouse model infected with CCHF virus has been described [9], but the hazards of working with this pathogen presents a problem in most laboratories. Consequently, we have investigated the pathogenesis of the related nairovirus, Dugbe (DUG) virus, in the mouse as an alternative. This virus, a member of the Nairobi sheep disease (NSD) serogroup [2], produces only a moderately severe febrile illness in man [15]. It has been used previously as a representative nairovirus in studies of structure and replication [3, 4].

In the present study, we compared the susceptibilities of outbred and inbred laboratory strains of mice to infection with a laboratory strain (KT281/75) of DUG virus by various routes of inoculation. Also, we examined the spread of virus in subcutaneously and intranasally inoculated adult and newborn outbred mice, the former route being taken as analogous to the route of infection after feeding by infected ticks, and the latter to aerosol spread, which may be important in nosocomial outbreaks of CCHF [10, 16]. In addition, the relative importance of host defences in determining DUG virus pathogenesis was examined in immunosuppressed animals. Finally, the virulence of three other isolates of DUG virus for newborn and adult CD1 mice was compared with that of strain KT281/75. The potential for the use of this model in future studies is discussed.

Materials and methods

Cells and viruses

BS-C-1 cells were grown at 37 °C in Glasgow modification of Eagle's medium (GMEM) containing 10% heat-inactivated newborn bovine serum (NBS), single-strength non-essential amino acids supplement (NEAA), 4 mM L-glutamine, 100 IU/ml penicillin, 100 μg/ml streptomycin and 1.125 mg/ml sodium bicarbonate as buffer (Flow Laboratories, Irvine, Scotland). Xenopus XTC cells (kindly supplied by Dr. P. A. Nuttall, NERC Institute of Virology and Environmental Microbiology, Oxford, England) were grown at 28 °C in Leibovitz L-15 medium (Gibco, Paisley, Scotland) containing 10% heat-inactivated foetal bovine serum (FBS), 10% tryptose phosphate broth (TPB), 100 IU/ml penicillin and 100 μg/ml streptomycin (Flow Laboratories). Strain KT281/75, was supplied by Prof. N. Dimmock, University of Warwick, as a stock which had been plaque-purified twice in BS-C-1 cells. We passaged this plaque-purified virus twice in BS-C-1 cells and twice in suckling mouse brain to produce the final working stock, which was stored at -70 °C as 0.5 ml aliquots of a 20% suckling mouse brain (SMB) preparation with a titre of 10^7 plaque forming units (pfu)/ml. Stocks of DUG virus isolates ArD16095, ArD44313 (both tick isolates from Senegal), and IbH11480 (a human isolate) were prepared by single passage in SMB; these isolates were supplied by Dr. P. A. Nuttall as infected SMB stocks at passage numbers 7, 6, and 7, respectively.

Assay for infectious virus

Infectivity of the laboratory strain KT281/75 was assayed for most experiments by plaque formation in BS-C-1 cell monolayers. Samples were diluted in maintenance medium consisting of GMEM containing 2% NBS, NEAA, L-glutamine, penicillin, streptomycin, sodium bicarbonate buffer (concentrations as above for growth medium) and 200 μg/ml DEAE-dextran (Sigma Chemical Company, Poole, England), and 0.25 ml was inoculated per 35 mm diameter well (containing approximately 7×10^5 cells) of Nunclon (Gibco, Paisley, Scotland) 6 well multidishes. After adsorption at room temperature for 30 min, excess inoculum was removed and the cells overlaid with maintenance medium (containing DEAE-dextran) solidified with 0.9% (w/v) Noble agar (Difco, East Molesey, England). After incubation at 37 °C for 4–5 days, plaques were visualised by staining with neutral red. As the human isolate of Dugbe virus did not plaque in BS-C-1 cells, for virulence comparisons the four isolates were assayed by plaque formation in XTC cell monolayers. Samples were diluted in maintenance medium (2% FBS) and 0.04 ml of each dilution placed in wells of a Nunclon 96-well flat-based microtitre plate; 0.04 ml of a suspension of XTC cells (4×10^6/ml) was then added to each well and the plate incubated at 28 °C for 3 h. An overlay of 0.08 ml of maintenance medium containing 1.5% carboxymethylcellulose (CMC) was then added to each well and plaques were visualised after 4 days incubation at 28 °C by removing the overlay medium and staining the cells with crystal violet (20% (v/v) methanol, 10% (v/v) formalin and 1% (v/v) of a 10% (w/v) preparation of crystal violet in methanol, in water).

Animals

Outbred adult mice and litters were taken from a colony of Crl:CD-1 (ICR) BR mice (CD-1). Inbred strains were purchased as adults and pregnant females from Harlan Olac Ltd., Bicester, England (NIH/Ola, C57BL/6/Ola, A2G/Ola) and Charles Rivers U.K. Ltd., Margate, England (BALB/cAnCrlBR, C3H/HeNCrlBR). After inoculation mice were maintained in an Isotec 12134 isolator (Bicester, England) under negative pressure, with HEPA-filtered incoming air.

Inoculation of mice and assay of susceptibility

Groups of 10 newborn (12 to 24 h old), 1 week-old, and adult mice were inoculated intracranially (i.c.), intranasally (i.n.), intraperitoneally (i.p.) or subcutaneously (s.c.) with 10-fold dilutions of virus stock, and deaths were recorded to 14 days post-inoculation (p.i.); the 50% lethal dose (LD_{50}) was then calculated by the method of Spearman-Karber [18]. Volumes of inocula used with newborn, week-old and adult animals were, respectively, as follows: i.c., 10, 10, and 30 μl; i.n., 10, 10, and 50 μl; i.p., 10, 50, and 100 μl; and s.c., 10, 10, and 50 μl. Adult animals were anesthetised by inhalation of ether vapour before i.c., i.n., and s.c. inoculations. Subcutaneous inoculation was at a site at the back of the neck. Intranasal inoculation consisted of instillation of portions of the inoculum into alternate nostrils [5].

Measurement of virus replication in infected mice

Newborn CD-1 mice were inoculated either s.c. or i.n. with $4.3 \log_{10}$ pfu (measured in BS-C-1 cells) of strain KT281/75 (approximately 250 LD_{50}) and killed by i.p. administration of sodium pentobarbitone (Sagatal, May & Baker Ltd., Dagenham, England) on days 1 to 9 p.i. Tissues were dissected and stored at $-70\,°C$ prior to homogenisation for 30 s, on ice, in a final volume of 3 ml medium/tissue, using a Sorvall Omni-Mixer (Camlab Ltd., Cambridge, England). Homogenate supernatant fluids were stored at $-70\,°C$ before assay for infectious virus. Adult CD-1 mice were treated similarly except that they were inoculated with $5.7 \log_{10}$ pfu of strain KT281/75 and killed daily on days 1 to 8 p.i.

Immunosuppression of adult mice

Young adult CD-1 mice (15 to 20 g body weight) were immunosuppressed by i.p. inoculation of a single dose of 0.3 ml Endoxana (WB Pharmaceuticals Ltd., Bracknell, England), reconstituted with water, containing 6 mg cyclophosphamide [14], either as controls or at various times pre- or post-inoculation, i.n. or s.c., with $5.7 \log_{10}$ pfu (measured in BS-C-1 cells) of DUG KT281/75 virus. Deaths were recorded up to 4 weeks p.i., and mice killed to determine the level of infectious virus in their tissues at various times p.i.

Histopathology

Organs and tissues from DUG virus-infected and control mice were fixed in neutral formal-saline for at least 48 h at room temperature before sectioning and staining with hematoxylin and eosin.

Results

Comparative susceptibility of different strains and ages of mice to Dugbe virus

The susceptibilities of different mouse strains to infection with strain KT281/75 was assessed in terms of pfu/LD_{50} in newborn mice inoculated i.p. The inbred mouse strains tested were representative of five different H-2 haplotypes but they were not much less susceptible than outbred CD-1 mice (Table 1). Consequently, subsequent studies concentrated on CD-1 mice.

In general, newborn mice were more susceptible than week-old mice, and much more susceptible than adults (Table 2). At all ages, animals were more

Table 1. Susceptibility of different strains of newborn mice to lethal infection with Dugbe virus (strain KT281/75) following intraperitoneal inoculation

Mouse strain	H-2 haplotype	pfu/LD$_{50}$[a]	
CD-1	outbred	71	(34-153)[b]
NIH	q	17	(6-44)
C57BL	b	54	(13-222)
BALB/c	d	83	(16-417)
C3H/He	k	77	(5-1179)
A2G	a	111	(52-240)

[a] Amount of infectious virus (in terms of pfu measured in BS-C-1 cells) required to produce 50% mortality
[b] 95% confidence limits

Table 2. Susceptibility of different ages of outbred CD-1 mice to lethal infection with Dugbe virus (strain KT281/75) inoculated via different routes

Age	Route of inoculation	pfu/LD$_{50}$[a]
Newborn	i.c.	0.3
	i.p.	71
	i.n.	43
	s.c.	1522
Week-old	i.c.	0.6
	i.p.	4870
	i.n.	570
	s.c.	$> 3.4 \times 10^4$
Adult	i.c.	3340
	i.p.	$> 8.4 \times 10^4$
	i.n.	$> 1.0 \times 10^6$
	s.c.	$> 1.0 \times 10^6$

[a] Amount of infectious virus (in terms of pfu measured in BS-C-1 cells) required to produce 50% mortality
$>$ No deaths occurred at the lowest dilution of virus tested

susceptible by the intracranial route of inoculation, although i.p., i.n., and s.c. routes with newborns, and i.p. and i.n. routes with week-olds, also produced lethal infection.

Infection in newborn CD-1 mice inoculated with DUG strain KT281/75 virus by subcutaneous and intranasal routes

Most newborn CD-1 mice, inoculated either s.c. or i.n. with 4.3 log$_{10}$ pfu of strain KT281/75 (approximately 250 LD$_{50}$), died between 7 and 9 days p.i.

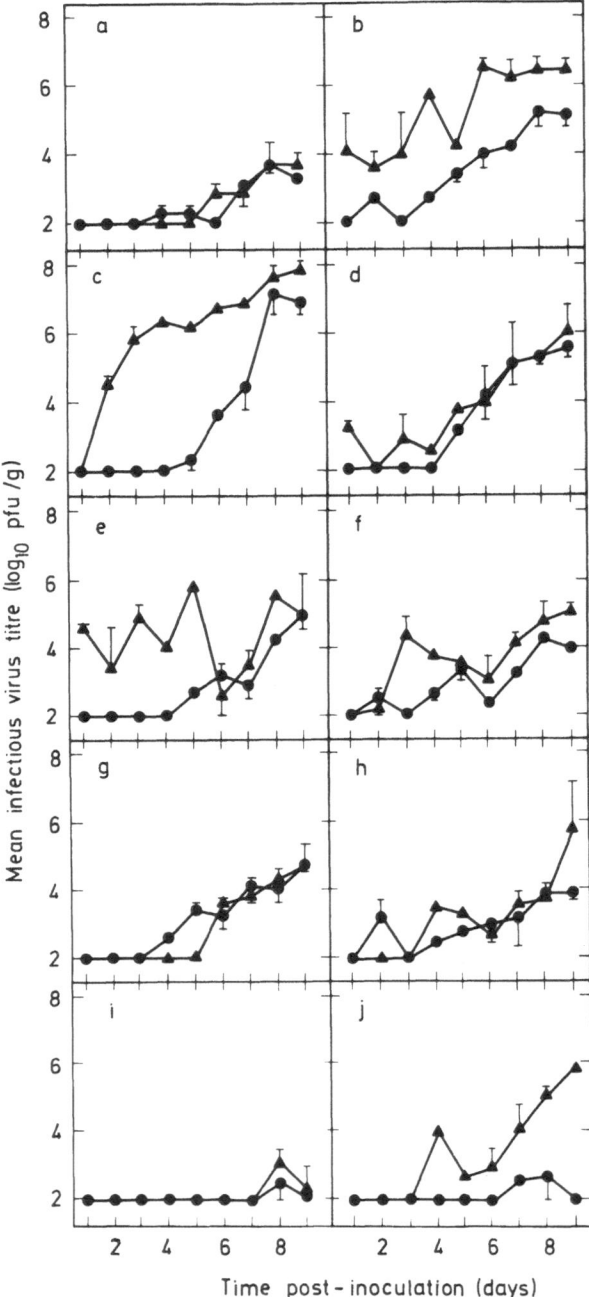

Fig. 1. Mean infectious virus titres, in $\log_{10}$ pfu per g (or ml, for blood) of wet tissue, in **a** blood, **b** upper respiratory tract, **c** brain, **d** heart, **e** lower respiratory tract, **f** liver, **g** spleen, **h** kidneys, **i** gut, and **j** carcass of newborn CD-1 mice 1 to 9 days after subcutaneous (●) or intranasal (▲) inoculation with 4.3 $\log_{10}$ pfu of suckling mouse brain-grown DUG strain KT 281/75. Titres below 2 $\log_{10}$ pfu/g were undetectable in BS-C-1 cells. Bars represent SEM

Infection patterns (Fig. 1), however, differed markedly between the two routes of inoculation. In s.c. inoculated animals, virus was not isolated consistently from the upper respiratory tract (URT), liver, kidney, blood or spleen until 4 days p.i. Virus appeared in brain, lower respiratory tract (LRT) and heart on day 5 p.i., the carcass on day 7 and gut on day 8. While virus titres were greater in the URT, spleen and liver on day 5 p.i. (3.4, 3.4 and

$3.3 \log_{10}$ pfu/g wet tissue, respectively), and by 7 days p.i., in the heart ($5.0 \log_{10}$ pfu/g), subsequently, however, virus titres increased more rapidly in the brain to reach $7.1 \log_{10}$ pfu/g by day 8.

In contrast, with neonates inoculated i.n., virus was present in moderately high titres in URT and LRT on day 1 p.i. (4.1 and $4.6 \log_{10}$ pfu/g, respectively), reaching similar titres, by day 2, in brain ($4.5 \log_{10}$ pfu/g), in which it continued to replicate, reaching $6.3 \log_{10}$ pfu/g by day 4 and $7.8 \log_{10}$ pfu/g by day 9 (Fig. 1). Virus was first detected in liver on day 2 ($2.2 \log_{10}$ pfu/g), heart on day 3 ($2.9 \log_{10}$ pfu/g), and kidney on day 4 ($3.5 \log_{10}$ pfu/g), but viraemia was not detected until 6 days p.i. In general, the course of infection in i.n. inoculated animals was more rapid than in those inoculated s.c., and virus titres were higher.

Histopathological examination of tissue sections from sick s.c. inoculated mice at 10 days p.i. revealed areas of neuronal necrosis in the hippocampus and cerebral cortex of the brain (Fig. 2a and b), mild abnormalities (consistent with virus infection) in the bronchiolar epithelium of the lung (Fig. 2c), and mild myocarditis in the heart (Fig. 2d). The liver, spleen, kidney, and gut did not show pathological changes.

Infection in adult CD-1 mice inoculated by subcutaneous and intranasal routes

In contrast to newborn animals, no deaths occurred in adult CD-1 mice inoculated s.c. or i.n. with $5.7 \log_{10}$ pfu of strain KT281/75. No virus was detectable in blood or other tissues of s.c.-inoculated adults up to 8 days p.i., except in subcutaneous tissue at the site of inoculation, where virus was detected ($2.1 \log_{10}$ pfu/g) on day 4 p.i. but not on days 1 to 3 or 5 to 8 p.i.

In i.n. inoculated adult mice, virus was detected in the LRT on day 1 p.i., the URT on day 2 and the brain on day 3, with titres in all three tissues peaking at 2.7, 2.6 and $2.4 \log_{10}$ pfu/g wet tissue, respectively, on day 4 p.i. (Fig. 3). Virus was cleared from these tissues by 7 days p.i., and remained undetected in any other tissue, again suggesting that spread of virus from the URT to the brain, in i.n.-inoculated animals, is via the olfactory nerve.

Infection in immunosuppressed adult CD-1 mice

To determine whether immunosuppression would increase susceptibility of adult mice to infection with DUG virus, s.c. and i.n. inoculated mice were inoculated i.p. with a single dose of 6 mg cyclophosphamide 3–4 h pre-inoculation or 2–3 h, 24 h, 48 h, and 72 h after inoculation with strain KT281/75. Deaths were monitored up to 40 days p.i. No deaths occurred in adult CD-1 mice given cyclophosphamide alone, or in those inoculated s.c. with virus, but mortality was as high as 80% in i.n. inoculated animals when cyclophosphamide was administered at 24 h p.i. and 60% when cyclophosphamide was given 2–3 h, 48 h, or 72 h p.i. No deaths occurred when cyclophosphamide was given 3–4 h prior to intranasal inoculation of virus.

Virus isolations on days 2, 5, 8, 12, and 16 p.i. from blood and organs of i.n.-inoculated mice, administered with cyclophosphamide at 24 h p.i., showed some variation in individual mice (Table 3) probably because not all mice died or became sick. In some mice, cyclophosphamide treatment lead to

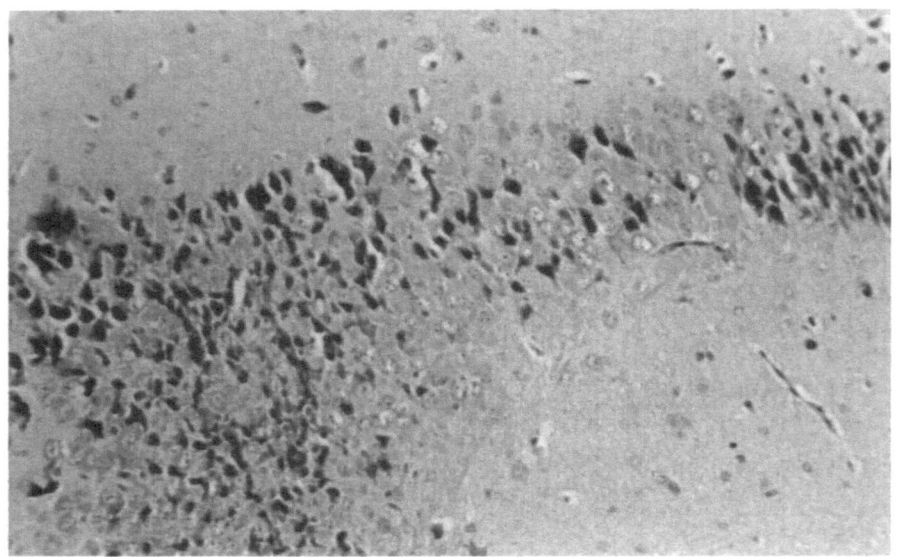

a

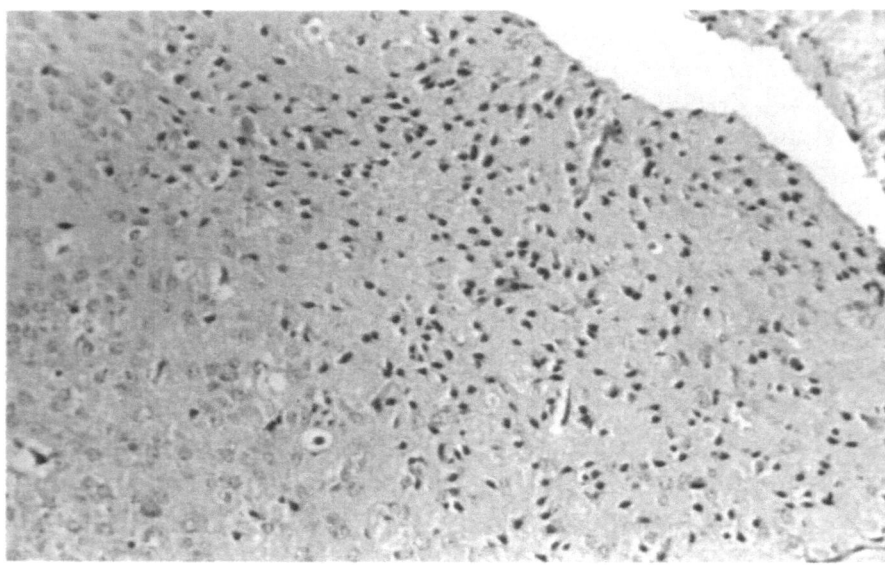

b

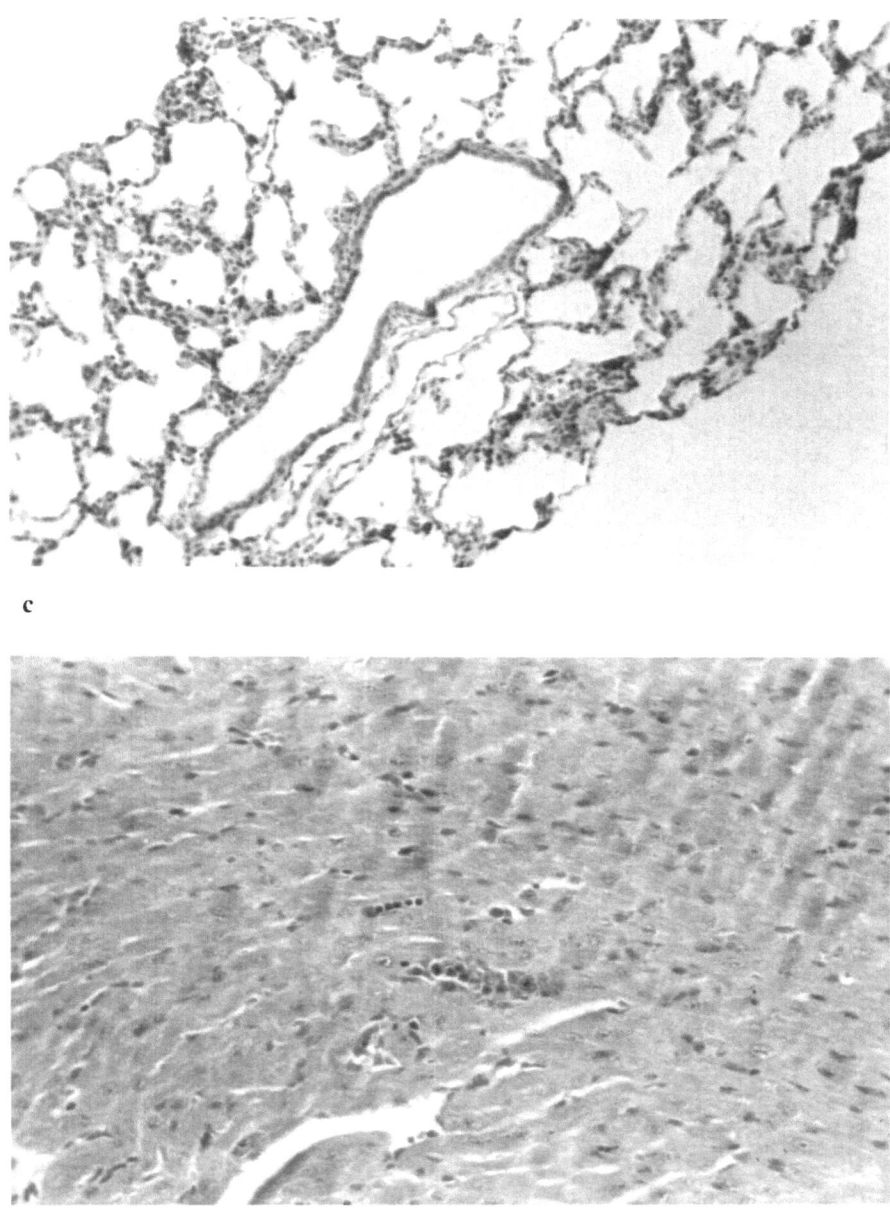

Fig. 2. Sections of tissue from 11 day old CD-1 mice s.c. inoculated when 1 day old with DUG strain KT281/75, showing neuronal necrosis in **a** the hippocampus and **b** cerebral cortex of the brain, **c** mild abnormalities in the bronchiolar epithelium of the lung, and **d** mild myocarditis in the heart. × 156

a more disseminated infection compared with normal mice (see Fig. 3), with virus persisting in the brain for at least 16 days. Histopathological examination of brain and lung tissue from a mouse showing moderate levels of infectious virus in these organs at 8 days p.i. (as well as lower levels in the URT, liver and spleen) revealed areas of neuronal damage in the hippocam-

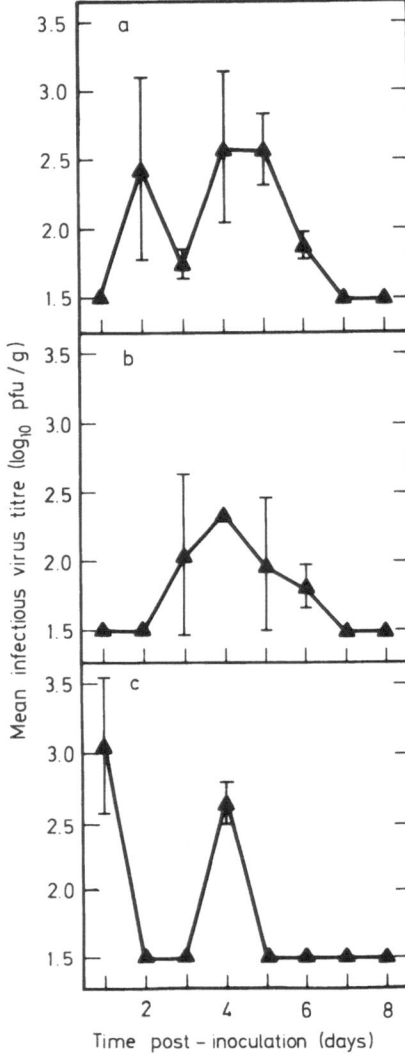

Fig. 3. Mean infectious virus titres, in $\log_{10}$ pfu per g of wet tissue, in **a** upper respiratory tract, **b** brain, and **c** lower respiratory tract of adult CD-1 mice on days 1 to 8 following intranasal (▲) inoculation with 5.7 $\log_{10}$ pfu of suckling mouse brain-grown DUG strain KT281/75. Titres below 1.5 $\log_{10}$ pfu/g were undetectable in BS-C-1 cells. Infectious virus was not detected in the blood or other organs tested. Bars represent SEM

pus of the brain and pulmonary edema and atypical bronchiolar epithelium in the lung (Fig. 4).

Comparative virulence of Dugbe virus isolates in newborn and adult mice

The virulence of different isolates of DUG virus in mice was assessed in terms of amount of infectious virus, assayed in XTC cells, required to produce 50% mortality (pfu/LD_{50}) in newborn mice inoculated i.c., i.n., or s.c., and adult mice inoculated i.n. and s.c. (Table 4). In newborn mice, virulence was greatest for all four isolates when inoculated via the i.c. route, although there were significant differences in their comparative neurovirulence: strain ArD16095 was eightfold more neurovirulent than strain ArD44313, 63-fold more neurovirulent than the human isolate strain IbH11480 and 197-fold

Table 3. Virus isolation from organs of adult CD-1 mice immunosuppressed with cyclophosphamide and inoculated intranasally with Dugbe virus strain KT281/75

Tissue	Time p.i. (days)											
	2			5			8		12		16	
	a[a]	b	c	a	b	c	a	b	a	b	a	b
Blood	–	–	–	–	–	–	–	–	–	–	–	–
URT	–	–	2.6	–	–	3.4	2.4	3.8	–	–	–	–
Brain	–	–	–	–	–	3.1	4.7	1.4	6.5	–	3.8	1.4
Heart	–	–	–	–	–	–	–	–	–	–	–	–
LRT	2.9[b]	–	3.6	–	2.9	5.1	3.7	1.5	–	–	–	–
Liver	1.3	–	–	–	–	4.8	2.7	–	–	–	–	–
Spleen	–	–	–	–	–	6.0	2.8	–	–	–	–	–
Kidneys	–	–	–	–	–	2.8	–	–	–	–	–	–

[a] Results for two or three mice (a, b, or c) tested at each time point. Individual results, rather than means, are given due to variation in the pattern of infection between individual mice

[b] Infectious virus titres expressed in terms of $\log_{10}$ pfu (assayed in BS-C-1 cells) per g of wet tissue. Titres less than $1.2 \log_{10}$ pfu/g were below the level of detectability of the assay

more neurovirulent than strain KT281/75. Strain ArD16095 was also more virulent via extraneural routes of inoculation than the other isolates (Table 4) but the relative differences in virulence of the latter varied with route of inoculation. Thus, isolate strain KT281/75, the least neurovirulent isolate, was as virulent by the intranasal route as isolate strain IbH11480 and fivefold more virulent than strain ArD44313. However, strain KT281/75 was of low virulence when inoculated subcutaneously. Similarly, while strain ARD44313 was only eightfold less neurovirulent than strain ArD16095, it was avirulent via the subcutaneous route (Table 4).

Adult mice did not suffer a lethal infection when inoculated subcutaneously with any isolate. However, strain IbH11480, the human isolate, was virulent for adult mice by the i.n. route, the level of virulence being comparable with newborn mice inoculated i.n. with strain ArD44313 (Table 4).

No evidence was obtained either from histological or macroscopic examination of tissues that any of these isolates inoculated by any route produced hemorrhagic manifestations in any of the mice.

Discussion

These results show that the laboratory mouse is susceptible to infection with the nairovirus, DUG virus. This susceptibility is age-related, young animals being more susceptible than adults. Although mice in each age group were

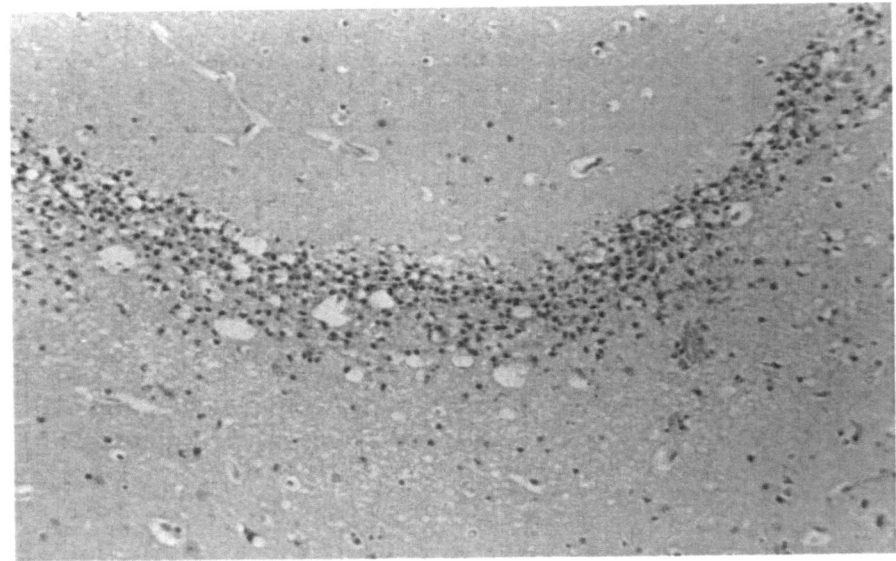

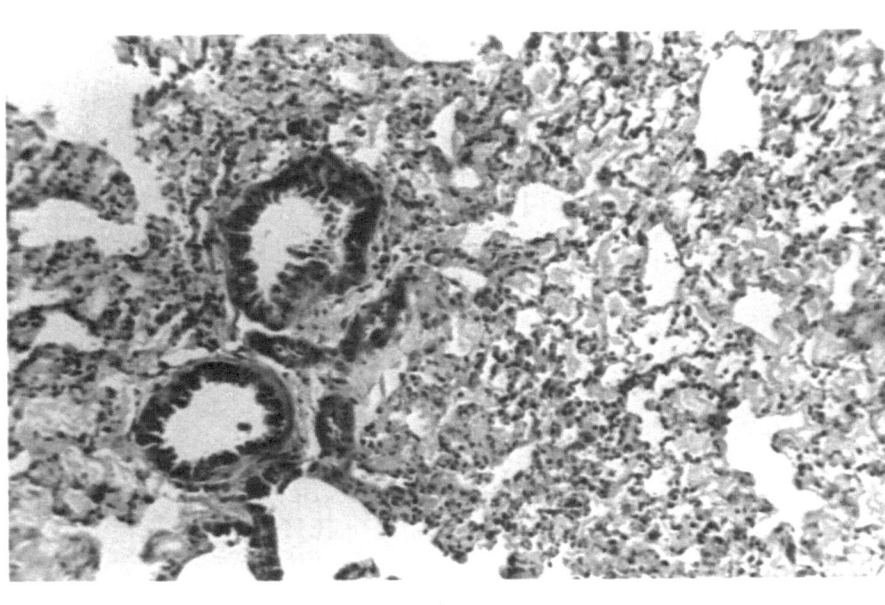

Fig. 4. Sections of tissue from an adult CD-1 mouse i.n.-inoculated with DUG strain KT281/75 and immunosuppressed by i.p. inoculation of a single dose of cyclophosphamide at 24 h p.i., showing **a** neuronal damage in the hippocampus of the brain and **b** pulmonary edema and atypical bronchiolar epithelium in the lung. × 156

most susceptible following intracranial inoculation, infection also occurred via peripheral routes, deaths occurring in young mice at higher doses. In young mice inoculated subcutaneously (a route of inoculation taken as being analogous to infection after feeding by infected ticks) with strain KT281/75, virus was detectable in most organs, including the blood, by 4 to 5 days p.i.,

Table 4. Virulence of four strains of Dugbe virus for newborn and adult mice

Strain	pfu/LD_{50} (range[a])				
	Newborn			Adult	
	i.c.	i.n.	s.c.	i.n.	s.c.
KT281/75	1.97 (0.65–6)	248 (93–665)	8798 (3694–2×10^4)	$>6 \times 10^6$ –	$>6 \times 10^6$ –
ArD16095	0.01 (0.01–0.03)	16 (8–32)	65 (24–178)	$>4 \times 10^4$ –	$>4 \times 10^4$ –
ArD44313	0.08 (0.04–0.2)	1320 (622–2801)	$>3 \times 10^4$ –	$>1 \times 10^5$ –	$>1 \times 10^5$ –
IbH11480	0.63 (0.34–1.1)	125 (59–264)	44 (12–165)	2488 (510–1×10^4)	$>5 \times 10^7$ –

[a] 95% confidence limits

and by 8 to 9 days p.i. virus titres were greatest in the brain, with moderate titres also present in upper respiratory tract, lower respiratory tract, heart, liver, spleen, and kidney. In contrast, in intranasally-inoculated newborn mice, moderately high titres were present in the upper and lower respiratory tracts from 1 day p.i., with virus spreading to the brain by 2 days p.i.; a viremia was not detected until 6 days p.i., suggesting that virus may spread directly from the upper respiratory tract to the brain, probably via the olfactory nerve as occurs with rabies [6], Mount Elgon bat [12], herpes simplex type 1 [1], Semliki Forest [8], mouse hepatitis strain JHM [13], and St. Louis encephalitis viruses [11]. Intranasal infection of adult mice with strain KT281/75 and strain IbH11480 also supports this conclusion, as viremia was again not detected and yet virus spread from the upper respiratory tract to the brain. The susceptibility of mice to infection by the intranasal route supports the suggestion that other nairoviruses, such as CCHF virus, may infect by this route in hospital and laboratory situations [10, 16], and has implications for the management of CCHF (and DUG) patients and the safety procedures employed when working with nairo-viruses in the laboratory.

Lack of viremia in adult mice may not represent an unnatural situation, because another tick-borne arbovirus, Thogoto virus, is transmitted to co-feeding ticks in the absence of detectable viremia in the host [7].

The pattern of DUG virus infection in mice bears similarities to the few descriptions of infection of mice with other nairoviruses. Nairobi sheep disease virus (NSD) produces a lethal infection of newborn mice following inoculation via the i.p. and i.c. routes (the latter being the most effective), but only produces deaths in adult mice when inoculated i.c. (i.n. route not tested) [17]. In adults infected with NSD, highest virus titres were in the brain (with little or none detectable in liver or blood) producing an encephalitis with marked neuronal necrosis in the hippocampus. CCHF virus also produces a lethal infection of young mice, following i.p. inoculation, with virus predominantly in the liver, lung and, later, brain [9]. The only adult strain of mouse found to suffer a lethal infection with CCHF virus following i.p. inoculation was the severe combined immune deficient (SCID) derivative of BALB/c.17 mice, but, as with the cyclophosphamide-treated CD-1 mice inoculated with DUG virus, mortality was less than 100%. The DUG virus-mouse model may thus provide an alternative or complimentary system for the study of factors important in virulence of nairoviruses and the use of various techniques of immunosuppression may aid elucidation of factors important in resistance in the adult.

Although the pulmonary edema seen in immunosuppressed adult mice infected with DUG virus is a feature of both NSD in sheep and CCHF in man, hemorrhage (an important feature of these latter two diseases) was not observed in mice infected with the KT281/75 strain of Dugbe virus. Neither was it observed with three other isolates of DUG virus, although the human isolate (strain IbH11480) was lethal for nonimmunosuppressed adult mice inoculated i.n. It may be possible to genetically analyze the virulence difference observed between the isolates by using genetic reassortment between virulent and attenuated strains to indicate important virulence properties of DUG virus and perhaps of other members of the nairovirus genus. This model may also be useful for identifying immunogens of DUG virus and for examining putative antiviral compounds. Whereas none of the isolates we used produced a hemorrhagic disease, it is possible that other nairoviruses may differ in mouse virulence and may produce a more suitable model for the hemorrhagic diseases.

Acknowledgments

We gratefully acknowledge the technical assistance of Miss G. Beckett, Mrs. J. Lichfield, Mr. J. Atkinson, and Mr. J. Martin, and would like to thank Dr. A. J. Howie of the Department of Pathology, University of Birmingham, for assistance with the interpretation of the histopathology.

References

1. Anderson JR, Field HJ (1983) The distribution of herpes simplex type 1 antigen in mouse central nervous system after different routes of inoculation. J Neu Sci 60: 181–195
2. Bishop DHL, Shope RE (1979) Bunyaviridae. In: Fraenkel-Conrat H, Wagner RR (eds) Comprehensive virology, vol 14. Plenum, New York, pp 1–156
3. Cash P (1985) Polypeptide synthesis of Dugbe virus, a member of the *Nairovirus* genus of the *Bunyaviridae*. J Gen Virol 66: 141–148
4. Clerx JPM, Casals J, Bishop DHL (1981) Structural characteristics of nairoviruses (genus *Nairovirus*; *Bunyaviridae*). J Gen Virol 55: 165–178
5. Collie MH, Sweet C, Husseini RH, Smith H (1980) Influenza of pregnant mice: infection with non-passaged viruses. FEMS Microbiol Lett 9: 49–51

6. Constantine DG (1971) Bat rabies: current knowledge and future research. In: Nagan Y, Davenport FM (eds) Rabies. Tokyo, University of Tokyo Press, p 253
7. Jones LD, Davies Cr, Steele GM, Nuttall PA (1987) A novel mode of arbovirus transmission involving a nonviraemic host. Science 237: 775–777
8. Kaluza G, Lell G, Reinacher M, Stitz L, Willems WR (1987) Neurogenic spread of Semliki Forest virus in mice. Arch Virol 93: 97–110
9. Kenyon RH, Thurman JD, Peters CJ (1988) Is there a suitable animal model for CCHF virus? In: Proc First Int Symp on Hantaviruses and Crimean-Congo Hemorrhagic Fever Virus, Porto Carras, Halkidiki, Greece. Aristotolian University of Thessaloniki Greece, p 38
10. McCormick JB, Johnson KM (1984) Viral hemorrhagic fevers. In: Warren KS, Mahmoud AAF (eds) Tropical and geographic medicine. McGraw-Hill, New York, pp 676–697
11. Monath TP, Cropp B, Harrison AK (1983) Mode of entry of a neurotropic arbovirus into the central nervous system. Reinvestigation of an old controversy. Lab Invest 48: 399–410
12. Patel JR (1979) Invasion of mouse brain by Mount Elgon bat virus. J Gen Virol 44: 587–597
13. Perlman S, Jacobsen G, Afifi A (1989) Spread of a neurotropic murine coronavirus into the CNS via the trigeminal and olfactory nerves. Virology 170: 556–560
14. Rager-Zisman B, Allison AC (1973) Effects of immunosuppression of Coxsackie B-3 virus infection in mice and passive protection by circulatory antibody. J Gen Virol 19: 339–351
15. Shope RE (1985) Bunyaviruses. In: Fields BN, Knipe DM, Chanock RM, Melnick JL, Roizman B, Shope RE (eds) Virology. Raven, New York, pp 1055–1082
16. Van Eeden PJ, Van Eaden SF, Joubert JR, King JB, Van De Wal BW, Michell WL (1985) A nosocomial outbreak of Crimean-Congo haemorrhagic fever at Tygerberg Hospital. S Afr Med J 68: 718–721
17. Weinbren MP, Gourlay RN, Lumsden WHR, Wienbren BM (1958) An epizootic of Nairobi sheep disease in Uganda. J Comp Pathol 68: 174–187
18. Zar JH (1984) Biostatistical analysis. Prentice-Hall, Englwood Cliffs, NJ

Authors' address: C. Sweet, School of Biological Sciences, Biology West Building, University of Birmingham, P.O. Box 363, Birmingham B15 2TT, England.

Arch Virol (1990) [Suppl 1]: 197–205

Dugbe virus susceptibility to neutralization by monoclonal antibodies as a marker of virulence in mice

A. Buckley, S. Higgs, and **E. A. Gould**

NERC Institute of Virology and Environmental Microbiology, Oxford, U.K.

Accepted February 12, 1990

Summary. A panel of monoclonal antibodies (MAb), reactive with the nairovirus Dugbe, was prepared and characterized. When used to compare seven strains of Dugbe virus by indirect immunofluorescence, they were antigenically indistinguishable. On the basis of in vitro plaque reduction neutralization and in vivo mouse protection tests by passive immunization, these strains could be divided into two groups. Five of the seven strains were neutralized by most or all of the neutralizing MAb and mice were protected from lethal infection by these viruses in the presence of passively administered MAb. The other two strains were not neutralized by most of the MAb and most of the MAb failed to protect mice from lethal infection by them. The binding capacity of the antibodies with these viruses in ELISA tests showed no correlation with their neutralization or protective capacity, therefore differences in avidity of antibody binding did not account for the differences in neutralization or protection. On the other hand, there was a direct relationship between neurovirulence of these strains for mice and their readiness to be neutralized or protected.

Introduction

Dugbe virus is a member of the family *Bunyaviridae*, genus *Nairovirus* [2, 5]. Nairoviruses are tick-borne, most being associated with sea bird colonies. Notable exceptions are Nairobi sheep disease, Crimean-Congo hemorrhagic fever and Dugbe, which infect humans and animals. They are enveloped viruses, about 90–120 nm in diameter with a well defined arrangement of surface glycoproteins [1, 6, 7] protruding from the viral surface, and a tripartite, single stranded, RNA genome consisting of large (L), medium (M), and small (S) RNA molecules.

Although it is assumed, by analogy with other viruses of the family *Bunyaviridae*, that at least one of the viral surface glycoproteins elicits a neutralizing antibody response, the exact identity of this glycoprotein has not yet been ascertained. Moreover, the antigenic interrelationships of nairoviruses have not been investigated in detail and the precise mechanisms of neutralization of these viruses have not been defined.

There are as yet few investigations into the factors that determine nairovirus virulence. We prepared monoclonal antibodies (MAb) against Dugbe virus and here we describe their use in studying virus neutralization, mouse protection, and virus virulence. Our results indicate a correlation between the ability of the virus to be neutralized and its capacity to produce a neurovirulent infection in mice.

Materials and methods

Viruses

The Dugbe virus strains were kindly provided by Professor N. Dimmock (University of Warwick), Dr. N. Moore (Ministry of Defence, Porton), and Dr. P. Nuttall (Institute of Virology, Oxford). They were designated IbAr 1792, ArD 16095; ArD 16769; ArD 44313; KT 281/75; IbH 11480; and 547, which was plaque purified from Dugbe KT 281/75 and given an additional three subcultures. All viruses were passed once by intracranial inoculation of outbred Swiss white TO suckling mice; stock virus was stored at $-70\,^{\circ}$C as 20% suckling mouse brain suspensions.

Monoclonal antibodies

Monoclonal antibodies were prepared by inoculating Balb/c mice with suckling mouse brain preparations of Dugbe virus (547) using the procedures described previously [11].

Indirect immunofluorescence tests

Virus at an input multiplicity of 0.01 was used to infect SW13 cells in suspension. The cells were incubated in medium on glass coverslips at $35\,^{\circ}$C for 3 days, washed in phosphate buffered saline (PBS) and fixed in cold acetone for 5 min. The dried coverslips were stored at $-20\,^{\circ}$C in sealed boxes until required. The method of indirect immunofluorescence with biotin/streptavidin was described previously [10].

Plaque reduction neutralization test

Aliquots of tenfold dilutions of MAb in PBS were placed in each well of 24 well plastic plates (Sterilin, U.K.) and an equal 0.25 ml volume of appropriately diluted virus (100 plaques per well) was added. After incubation at $35\,^{\circ}$C for 1 h, 0.25 ml of SW13 cells (1.6×10^6 cells/ml) were added and the plates were re-incubated at $35\,^{\circ}$C for a further 4 h. Carboxymethylcellulose (1.5%) in L15 medium containing 3% fetal bovine serum was then added and the plates incubated for seven days at $35\,^{\circ}$C. They were then fixed with 10% formol saline and stained with 0.1% naphthalene black. The largest dilution of MAb that reduced plaque numbers by 50% was the antibody neutralization titre.

If complement (Wellcome reagents Ltd, U.K.) at 1/40 dilution or rabbit anti-mouse immunoglobulin (Amersham, U.K.) at 1/100 dilution was used to mediate the neutralization, it was added 30 min after virus and antibody were mixed together and the mixtures were incubated for a further 30 min.

Passive protection of mice

Groups of ten newborn, 3-week or 6-week old mice were given virus either intracranially (0.02 ml) or intraperitoneally (0.1 ml) and they were monitored daily. Average survival times (AST) and virus infectivities (LD_{50}/ml) were obtained for each strain in each age group of mice. Three week old mice were given 0.1 ml undiluted MAb as ascitic fluid 2 h prior to challenge with 100 LD_{50} of virus given intracranially and then monitored twice daily. Treatment of mice with hydrocortisone, which suppresses immunocompetence, followed the method described previously [12].

Assessment of virus-antibody binding

Virus concentrated and purified by centrifugation at $100,000 \times g$ through 10% sucrose in Tris/EDTA was titrated in checkerboard fashion against the appropriate antibody dilutions using the ELISA streptavidin/peroxidase system described previously [10].

Treatment of mice with hydrocortisone

Mice were given 1.25 mg of hydrocortisone acetate 24 h before virus challenge and subsequently 0.5 mg/mouse on alternate days.

Results

Comparison of the antigenic relationships of seven strains of Dugbe virus by indirect immunofluorescence

Thirty MAb prepared against Dugbe virus (547) were tested by indirect immunofluorescence for their reactivity with each of the seven different Dugbe viruses defined in Materials and methods. There were no antigenic differences between the viruses, all MAb reacted with all viruses displaying the same patterns of fluorescence throughout.

Plaque reduction neutralization tests (PRNT)

Each MAb was tested initially for neutralization activity with Dugbe virus strain 547. Of 30 MAb tested, ten showed neutralizing activity; only these neutralizing MAb are shown. They were then tested for the capacity to neutralize each of the other Dugbe virus strains. The results are shown in Table 1. Two categories of virus were identified, those that were neutralized by most or all MAb and, those that were not neutralized by any MAb or were only neutralized very poorly. These results were reproducible with different passages of the same virus and antibodies.

A. Buckley et al.

Table 1. Comparison of Dugbe strains in neutralization tests with monoclonal antibodies

Antibody	Dugbe virus strain						
	547	ArD 16095	ArD 16769	ArD 44313	KT 281/75	IbH 11480	Ar 1792
28.5	3.0	2.0	1.3	2.0	0.0	0.0	0.0
28.7	4.0	1.4	1.8	2.0	1.8	1.0	0.0
28.17	2.5	2.0	2.4	1.8	1.0	0.0	0.0
28.36	1.3	2.0	1.4	1.8	2.0	0.0	0.0
28.49	3.7	0.0	0.0	0.8	0.0	0.0	0.0
28.60	2.0	0.0	0.0	1.2	0.0	0.0	0.0
28.63	2.0	2.2	2.0	0.0	1.0	0.0	0.0
28.73	1.0	1.0	1.0	2.0	0.0	0.0	0.0
28.80	2.3	2.8	2.2	2.3	2.0	1.0	0.0
31.12	3.0	4.0	2.0	2.4	2.0	0.0	0.0
DUG 1[a]	4.4	3.0	2.5	2.6	1.7	1.7	2.0
DUG 2[b]	2.0	2.0	1.5	1.5	2.2	2.2	2.0
Ganjam	3.0	0.0	1.0	1.0	1.0	1.0	0.0

Neutralization titres are the highest dilution of antibody ($\log_{10}$ reciprocal dilution) producing 50% reduction of plaque numbers

[a] DUG 1 was a hyperimmune rabbit antiserum raised against Dugbe 547 virus

[b] DUG 2 was a hyperimmune rabbit antiserum raised against Dugbe Ar 1792 virus (the prototype)

Use of complement or rabbit anti-mouse immunoglobulin to mediate neutralization did not significantly increase the titre of neutralizing MAb and the non-neutralizing MAb still failed to neutralize any of the viruses in the presence of complement or anti-mouse immunoglobulin. Moreover the poorly neutralized viruses (IbH 11480 and Ar 1792) showed no increase in neutralization after either complement or antiglobulin addition.

Polyclonal rabbit antisera were made against the well neutralized Dugbe virus 547 (DUG 1) and the non-neutralized Dugbe virus Ar 1792 (DUG 2). The results in Table 1 show that there was no significant difference between the two antisera, both viruses induced antibodies that neutralized all of the Dugbe strains. Also included was a rabbit polyclonal antiserum raised against Ganjam virus which also neutralized most of the Dugbe virus strains, showing that there are cross-reactive neutralizing epitopes present on DUG and GAN virus.

Comparison of binding capacity of each neutralizing MAb with each virus

Each virus strain was analysed in ELISA binding tests with each antibody that had shown neutralizing activity with Dugbe 547 virus. The results are summarized in Table 2 which presents the readings (O.D. at 490 nm) obtained for each antigen at its optimal concentration, with antibody at 1/100 dilution.

Table 2. Comparison of antibody–antigen binding in ELISA test

Antibody	Dugbe virus strain						
	547	ArD 16095	ArD 16769	ArD 44313	KT 281/75	IbH 11480	Ar 1792
28.5	0.38[a]	0.29	0.31	0.42	0.26	0.31	0.20
28.7	0.43	0.46	0.35	0.47	0.27	0.31	0.20
28.17	0.06	0.00	0.22	0.26	0.08	0.12	0.09
28.36	0.02	0.02	0.05	0.04	0.03	0.05	0.05
28.49	0.24	0.24	0.52	0.45	0.25	0.43	0.21
28.60	0.13	0.10	0.22	0.30	0.08	0.24	0.06
28.63	0.38	0.25	0.41	0.41	0.29	0.35	0.25
28.73	0.04	0.07	0.10	0.10	0.03	0.08	0.04
28.80	0.09	0.11	0.29	0.38	0.19	0.22	0.12
31.12	0.15	0.12	0.32	0.41	0.16	0.26	0.19
DUG 1[b]	0.90	0.98	>2	>2	1.37	>2	1.06

[a] Number is O.D. at 490 nm after correction for control antibody
[b] See footnote Tables 1 and 4

Neurovirulence of the seven Dugbe virus strains in mice

Each Dugbe virus strain was titrated in newborn, 3-week old and 6-week old mice using intracerebral or intraperitoneal administration of virus. Data presented in Table 3 show the relative infectivities (LD_{50}/ml), average survival times and titre of recovered virus in the brains of mice 4 days after infection, i.e., just before the first mice were expected to die. Six-week old mice showed no signs of infection by either route of inoculation and so were not included in further studies. Neurovirulence of all Dugbe virus strains was comparable in newborn mice but when results for three week old mice were compared the viruses differed markedly. In 3-week old mice viruses could be divided roughly into the same two groups as determined using neutralization data. Dugbe strains IbH 11480 and Ar 1792 killed 3-week mice relatively quickly and showed higher infectivity titres than did the other five strains whereas strains ArD 16095, ArD 16769 and KT 281/75 killed these mice more slowly (more than 2 days longer to death) and showed relatively lower infectivity titres. Strains ArD 16769 and ArD 44313 failed to kill 3-week old mice. Nevertheless, these two strains replicated in the brains of the mice and produced titres as high as some of the other viruses that killed mice.

Protection of mice against Dugbe virus infection by passive immunisation with monoclonal antibodies

Each of the neutralizing MAb was given intraperitoneally to groups of 10 mice and after 2 h the mice were challenged intracerebrally with either Dugbe strain 547 (readily neutralized), IbH 11480 (poorly neutralized), or

Table 3. Analysis of Dugbe virus virulence for mice

Dugbe virus	Average survival time		LD_{50} titres[a]		Titres (pfu) in brains ($\times 10^6$)	
	newborn	3 week	newborn	3 week	newborn	3 week
547	6.10	7.20	6.2	4.5	12.0	3.6
ArD 16095	5.20	7.92	5.5	2.1	3.6	0.2
ArD 16769	6.50	NK	4.5	NK	8.4	0.1
ArD 44313	7.50	NK	5.0	NK	9.6	0.2
KT 281/75	6.60	7.40	6.1	2.5	9.0	0.2
IbH 11480	5.30	5.33	7.3	6.5	2.0	0.8
Ar 1792	5.50	4.65	5.7	6.5	11.0	1.6

[a] Titre $\log_{10}$ reciprocal dilution

NK Not killed; in all other cases, all mice died

Table 4. Passive protection of mice using monoclonal antibodies

Antibody	Dugbe virus strain		
	547	IbH 11480	Ar 1792
28.5	0[a]	0	0
28.7	100	0	0
28.17	100	20	0
28.36	0	0	0
28.49	0	0	0
28.60	60	0	0
28.63	60	0	0
28.73	70	0	0
28.80	90	20	0
31.12	100	100	0
DUG 1[b]	100	40	0
Ganjam	50	0	0

[a] Figures show percentage mice surviving intracranial challenge by virus

[b] DUG 1 was a hyperimmune rabbit antiserum raised against Dugbe 547 virus

Ar 1792 (not neutralized). The results (Table 4) show that most of the MAb provided protection of the mice against infection with strain 547 virus but only very limited protection of mice against strain IbH 11480. None of the MAb protected mice against strain 1792 virus, thus linking neutralization directly with the capacity of the MAb to protect mice.

Dugbe virus virulence comparison in hydrocortisone-treated mice

In an attempt to test the effect of host immune responses on Dugbe virus virulence, the AST of mice treated with hydrocortisone acetate was compared with that of non-treated mice challenged intracranially with one of three strains of Dugbe virus (547, lbH 11480, or Ar 1792). The AST for strain 547 was not altered significantly, although this virus killed only 70% of the hydrocortisone-treated mice. In the cases of strain IbH 11480 and Ar 1792, all of the hydrocortisone-treated mice were killed but the time to death was increased significantly (1.2 and 0.85 days, respectively). Despite this increased time to death, these viruses remained more virulent, on the basis of AST, than Dugbe 547.

Discussion

Thirty MAb were produced against Dugbe virus strain 547. Their antigenic specificities and molecular characteristics will be described elsewhere. All MAb reacted with the seven Dugbe isolates when tested in indirect immunofluorescence tests indicating the antigenic similarity of these viruses. Of these 30 MAb, 10 neutralized the Dugbe virus from which the MAb were derived and are therefore presumed to react with the envelope glycoprotein, which is thought to stimulate neutralising antibody responses [8]. Although the seven different Dugbe virus isolates were antigenically indistinguishable by indirect immunofluorescence tests, they differed in their abilities to be neutralized by the MAb that showed neutralization with Dugbe 547. One strain, Ar 1792, was not neutralized by any of the MAb, even if complement or anti-mouse immunoglobulin was incorporated as a promoter of neutralization. Differences in the capacity of different strains of a single virus to be neutralized were reported previously with yellow fever virus [3]. It was proposed that this was due to differences in the conformational arrangement of the epitopes on the viral surface glycoprotein. This was further substantiated in topological studies of envelope glycoproteins of yellow fever virus by Cammack and Gould [4], who showed that epitopes can adopt different relative positions in different strains of the same virus.

In general, the results with Dugbe viruses showed that there was a correlation between neutralization of infectivity and protection of mice by antibody. However, this was not absolute because MAb 31.12 totally protected mice against infection with strain IbH 11480 but did not neutralize this virus. It seems reasonable to suggest that in those cases where neutralization was demonstrated in vitro, protection in vivo was afforded by neutralization of at least some of the input virus. On the other hand, MAb 31.12 presumably protects the mice against challenge with strain IbH 11480 virus by a different mechanism; for example the formation of antigen/antibody complexes or hindrance of virus spread through the host. Failure to neutralize the infectivity of specific strains of Dugbe virus in vitro was not due to differences in the antibody binding capacity of the viruses because there was no correlation between antibody binding, as measured by ELISA and by in vitro neutralization.

Although we have not yet confirmed the molecular specificities of all the MAb, some of the 10 neutralizing MAb have been shown by radioimmunoprecipitation to be specific for the G1 surface glycoprotein of Dugbe virus (Gould, unpubl. data). This is consistent with observations for other viruses of the family *Bunyaviridae*, for which it has been shown that neutralization activity is located on the G1 glycoprotein [8, 9].

Our results suggest that changes in the conformation of the viral surface glycoprotein of Dugbe virus lead to changes in the biological properties seen as neutralization, protection and virulence. The neutralization and protection data showed that strains 547 and Ar 1792 differed markedly in their capacity to be neutralized and in the ease with which antibody could protect mice against these viruses. The difference in virulence was not the result of more rapid replication by Ar 1792 because 24 h before the mice would have died, virus titres in the brains of mice infected with strain Ar 1792 were no higher than those in the brains of mice infected with strain 547. Hydrocortisone treatment, which suppresses immune responses, extended the AST by about one day but did not assist in differentiating between virulent or less virulent viruses. Therefore, it seems likely that the difference in virulence between these Dugbe virus isolates might be explained either by their capacities to target specific cells in the mouse brain or their capacities to exhibit different tropisms in other tissues of the mouse. This is currently being investigated.

It is important to note that whilst the most virulent strains were not neutralized by any MAb, hyperimmune rabbit antisera prepared against this virus (DUG 2) and against the readily neutralized virus (DUG 1) showed the capacity to neutralize all viruses. These results demonstrate how MAb can reveal differences in neutralization domains that are not detected with polyclonal antisera.

The results showed that there is a narrow host age range in which differences between strains can be detected. This possibly reflects the developmental stages of the immune system; none of the strains were virulent for 6-week old mice.

Differences in the presentation of similar epitopes on a protein are likely to arise as the result of changes in the amino acid and, ultimately, the nucleotide composition of the gene that encodes the G1 glycoprotein. Studies are currently being undertaken to determine these sequences.

References

1. Bishop DHL, Shope RE (1979) Bunyaviridae. In: Fraenkel-Conrat H, Wagner RR (eds) Comprehensive virology, vol 14. Plenum, New York, pp 1–156

2. Bishop DHL, Calisher CH, Casals J, Chumakov MP, Gaidamovich SYa, Hannoun C, Lvov DK, Marshall I, Oker-Blom N, Pettersson R, Porterfield JS, Russell PK, Shope RE, Westaway EG (1980) Bunyaviridae. Intervirology 14: 125–143
3. Buckley A, Gould EA (1985) Neutralization of yellow fever virus studied using monoclonal and polyclonal antibodies. J Gen Virol 66: 2523–2531
4. Cammack N, Gould EA (1986) Topographical analysis of epitope relationships on the envelope glycoprotein of yellow fever 17D vaccine and the wild type Asibi parent virus. Virology 150: 333–341
5. Casals J, Tignor GH (1980) The *Nairovirus* genus: serological relationships. Intervirology 14: 144–147
6. Clerx JPM. Bishop DHL (1981) Qalyub virus, a member of the newly proposed *Nairovirus* genus (*Bunyaviridae*). Virology 108: 361–372
7. Foulke RS, Rosato RR, French GR (1981) Structural polypeptides of Hazara virus. J Gen Virol 53: 169–172
8. Gentsch JT, Rozhon EJ, Klimas RA, El Said LH, Shope RE, Bishop DHL (1980) Evidence from recombinant bunyavirus studies that the M RNA gene products elicit neutralizing antibodies. Virology 102: 190–204
9. Gonzalez-Scarano F, Shope RE, Calisher CH, Nathanson N (1982) Characterization of monoclonal antibodies against the G1 and N proteins of La Crosse and Tahyna, two California serogroup bunyaviruses. Virology 120: 42–53
10. Gould EA, Buckley A, Cammack N (1985). Use of the biotin-streptavidin interaction to improve flavivirus detection by immunofluorescence and ELISA tests. J Virol Methods 11: 41–48
11. Gould EA, Buckley A, Cammack N, Barrett ADT, Clegg JCS, Ishak R, Varma MGR (1985) Examination of the immunological relationships between flaviviruses using yellow fever virus monoclonal antibodies. J Gen Virol 66: 1369–1382
12. Gould EA, Buckley A, Groeger BK, Cane PA, Doenhoff M (1987) Immune enhancement of yellow fever virus neurovirulence for mice: studies of mechanisms involved. J Gen Virol 68: 3105–3112

Authors' address: A. Buckley, NERC Institute of Virology and Environmental Microbiology, Mansfield Road, Oxford OX1 3SR, England.

Arch Virol (1990) [Suppl 1]: 207–218

Dugbe virus in ticks:
histological localization studies using light and electron microscopy

T. F. Booth, A. C. Marriott, G. M. Steele, and **P. A. Nuttall**

NERC Institute of Virology and Environmental Microbiology, Oxford, U.K.

Accepted March 1, 1990

Summary. Different methods for the detection and histological localization of Dugbe (DUG) virus infection in vertebrate and tick cells and tissues were compared. Immunohistochemistry with DUG-specific antibody and hybridocytochemistry with a DUG-specific riboprobe were used to investigate the role of different tick cell types in the replication of DUG virus and its tissue tropisms during internal dissemination. DUG virus was localized in both unfed and feeding adults inoculated as nymphs or orally infected by capillary feeding. In non-feeding ticks the main sites of DUG virus replication were the epidermis, hemocytes associated with loose connective tissue and a small number of phagocytic digestive cells in the gut lumen. DUG virus invaded the salivary glands during feeding and was not detected until after 7–10 days feeding on the host. Virus infection increased in the gut during feeding from about 1% to about 40% of cells. The primary site of transstadial persistence of DUG virus is the hemocytes. In the electron microscope, no virus particles or pathological effects were observed in ticks. Tick hemocytes and other motile cells may be important in the transmission of persistent virus infection from one cell or organ to another by diapedesis.

Introduction

It has been demonstrated that the tick *Amblyomma variegatum* is a competent vector of Dugbe (DUG) virus under laboratory conditions, although *Rhipicephalus appendiculatus*, a tick with a geographic distribution overlapping that of *A. variegatum*, is not [18]. The demonstration of vector competence in an ixodid tick requires that virus that is imbibed orally, traverses the 'gut barrier' and replicates, persists trans-stadially during the moulting process, and is subsequently transmitted to another vertebrate host in the tick's next blood feeding stage. DUG virus, a member of the Nairobi sheep disease serogroup of the family *Bunyaviridae*, genus *Nairovirus* [3], is a common tick-borne virus throughout western Africa and has frequently been isolated from domestic cattle and wild populations of *A. variegatum* [9]. Recent work has established that DUG virus is both antigenically and genetically related to Crimean-Congo hemorrhagic fever (CCHF) virus [3, 12], a major health risk to humans. Nucleic acid probes generated to DUG RNA cross-react with CCHF RNA, and CCHF sera cross react with DUG antigen expressed in a baculovirus vector. We therefore use DUG virus as a "safe" model for biological transmission studies with tick vectors and to provide diagnostic reagents that should prove useful in epidemiological studies of CCHF-related infections.

Tick-borne viruses may persist and replicate for many months in quiescent ticks without any obvious pathological effects. Virus that is imbided in the blood meal by an ixodid tick must persist trans-stadially during the moulting process in order to be passed on to a new vertebrate host in the tick's next blood meal. Virus is presumed to be delivered in the tick's salivary secretions over the course of the several days required to complete growth and development whilst attached to the host. We have developed histological localization techniques both as a means to screen ticks for infection and to study the mechanisms of internal dissemination, persistence and transmission of virus in ticks. For light microscopy, the use of immuno-histochemistry to detect DUG antigen was compared with hybridocyto-chemistry with a riboprobe complementary to DUG S-segment RNA. Due to the absence of any virus-like particles or other pathological effects that can be distinguished by electron microscopy (EM) in infected ticks during virus transmission, we have used the technique of immunogold labelling of ultrathin sections to detect viral antigens at the ultrastructural level. These studies allow us to address questions about the timing of virus delivery by the tick and the factors determining vector competence and provide a comparison between the non-pathogenic replication strategies in the tick with the pathologic replication encountered in vertebrate hosts.

Materials and methods

Cells, virus, and antisera

The continuous porcine kidney cell line (PS cells) obtained from Dr. J. S. Porterfield (Sir William Dunn School of Pathology, Oxford) was grown in Leibowitz L15 medium supplemented with 10% tryptose broth and 3% fetal bovine serum (FBS). DUG virus isolate ArD 44313 (chosen as being the closest isolate to the original tick homogenate) was kindly supplied by Dr. J.-P. Digoutte (Institut Pasteur, Dakar, Senegal) as third passage infective suckling mouse brain. Viral stocks were produced by inoculation of PS cells. Antisera to DUG virus was raised in rabbits and hyperimmune ascitic fluid was raised in mice [6, 17].

Virus and antibody assays

Virus titers were determined by plaque titration in PS cells [7]. Dissected tissue samples were homogenized in micro-tissue grinders with 1 ml of L15 medium containing 3% FBS and 2 mg/ml penicillin and streptomycin. The samples were centrifuged and the supernatant fluid assayed for virus by intracranial inoculation of 0.01 ml into 2-day old mice and daily examination of the mice for clinical signs of infection. Neutralizing antibodies were detected by the plaque reduction test [6]. Seroconversion of guinea-pigs following exposure to infected ticks was determined by immunofluorescence tests using DUG virus infected PS cells [1, 7].

Tick rearing and inoculation

A colony of *A. variegatum* established from specimens supplied by Professor M. G. R. Varma (London School of Hygiene and Tropical Medicine) was reared as previously described [11]. Larvae and nymphs were fed on guinea pigs (Dunkin Hartley, average weight 400 g), and adult ticks were fed on Dutch rabbits (approx. 2 kg). Female ticks were introduced onto the rabbits three days after the males, as they do not feed unless males are already present. Replete ticks were removed daily and stored at 28 °C and 85% relative humidity until required for histology. Feeding times were 4–6 days for larvae, 5–9 days for nymphs and 12–14 days for adult females. For some experiments partially fed ticks were removed after 7–12 days of feeding, weighed and examined by histochemistry for DUG infection. Engorged *A. variegatum* nymphs were inoculated with DUG virus on the same day that they completed engorgement on guinea pigs (previous experiments had shown that delay in the time of inoculation dramatically increased tick mortality). The engorged ticks were inoculated via a microsyringe with approximately 20 μl of DUG virus-infected cell culture supernatant fluid, titer 10^5–10^6 PFU/ml. The ticks were stored and specimens were collected and dissected at various times post-infection. Samples were stored at -70 °C and then assayed for DUG virus. Serum was collected from the guinea pigs on which the inoculated ticks engorged and were tested for neutralizing antibodies. Infected specimens of both unfed and partially fed ticks were used for histochemistry.

Capillary feeding of ticks

Capillary feeding with DUG virus was done by modifications of the method of Burgdorfer [2]. Specimens were processed for histochemistry in order to compare viral replication in orally infected ticks with those that were inoculated intracoelomically. Capillary-fed ticks were sectioned immediately after capillary feeding and during their subsequent development, and after development to the adult stage was complete. Questing nymphs were attached ventral-surface upwards to plasticine blocks (Harbutts Ltd., Bath, U.K.) so that they were held down with thin plasticine strips allowing the mouthparts (hypostomes) to protrude. All specimens were contained inside a humidified plastic box during capillary-feeding. Glass capillary tubes (original diameter 3.5 mm) were drawn out over a flame. They were filled with supernatant fluid from DUG virus-infected cell cultures supplemented with 3% new-born bovine serum with a titer of approximately 10^5 PFU per ml. The capillary tubes were placed over the capitulum of the ticks so that the hypostome and palps were enclosed in the neck of the tube. Each tube was held in place by a block of plasticine. The nymphs were left overnight to feed and then removed from the plasticine and placed immediately on a guinea pig to complete engorgement. To ensure that the majority of the specimens were successfully infected only ticks which had visibly distended bodies were selected. If completion of feeding on a host was delayed, a large number of ticks died. We also found that feeding about 50 *Rhipicephalus appendiculatus* nymphs together with a batch of 10–20 capillary-fed *A. variegatum* greatly improved the feeding success of the latter.

Light and electron microscope immunocytochemistry

For immunohistochemistry DUG virus infected PS cells or cryostat sections of infected tissues were acetone-fixed and stained with DUG antibody using a biotinylated secondary antibody with either a streptavidin-conjugated fluorescein isothiocyanate detection system (Amersham International, Amersham, U.K.) or a streptavidin-peroxidase conjugate with amino-ethylcarbazole as the substrate (Sigma, Poole, U.K.). Methylene blue was used as a counterstain for the red peroxidase reaction product, and, for fluorescence, slides were mounted in 'Citifluor' (Agar Scientific, Stansted, U.K.) to enhance specific fluorescence and to minimize and differentiate autofluorescence of some tick cell components.

Initial electron microscopic (EM) investigations were carried out on DUG virus infected PS cells. Cultures were glutaraldehyde-fixed at 48 h p.i. for surface immunogold labelling and processed for electron microscopy as previously described [1]. For post-embedding labelling of DUG virus antigens for EM, tissues and cells were fixed for 1 h at 20 °C with 0.1% glutaraldehyde in 0.1 M phosphate buffer pH 7.4, dehydrated in an ethanol series and embedded in LR White resin (London Resin Co., Basingstoke, U.K.) as described by Newman et al. [14].

Ultrathin sections were cut and mounted on uncoated nickel grids and screened for DUG virus antigen by immunogold labelling. For gold labelling experiments, grids were pre-incubated in 0.1% bovine serum albumin (BSA) in phosphate buffered saline (PBS) for 10 min followed by 2 to 3 h with antibody diluted in PBS/BSA. The dilution of antibody that gave the maximum specific labelling with the minimum of non-specific labelled background was determined for optimum labelling. After washing three times by floating grids on PBS, followed by another 5 min blocking with PBS/BSA, sites of antibody binding were detected by incubating with 10 nm colloidal gold particles conjugated to goat anti-rabbit IgG (Bio Cell Labs, Cardiff, U.K.). Controls included omitting the primary antiserum or substituting pre-immune serum, or an irrelevant antiserum (such as one raised to components of Bluetongue virus), and using uninfected cells or tissue as a replacement for infected ones.

Riboprobe preparation and hybridocytochemistry

Cryosections (6 µm) on glass slides coated with PhotoFlo (Kodak, Hemel Hempstead, U.K.) were fixed in 4% paraformaldehyde in PBS for 10 min. Slides were then treated with proteinase-K, 2.5 µg/ml for 10 min at room temperature. After washing in 0.2 M Tris, 0.1 M glycine the slides were prehybridized for 1 h in a solution containing 50% formamide, 0.3 M NaCl, 30 mM sodium citrate, pH 7.0, 10% dextran sulphate and 0.2 mg/ml single-stranded salmon sperm DNA (Sigma, Poole, U.K.). Probe was prepared by transcription of plasmid BS-36. This plasmid contains the DUG S-segment derived cDNA insert from pDUG36 [19] cloned into the transcription vector pBS (Stratagene, La Jolla, California, U.S.A.). After cleavage with *Sac* l, BS-36 was transcribed by T3 RNA polymerase in the presence of 0.4 mM Biotin-11-UTP (BRL-Gibco, Paisley, U.K.) to produce a 251 nucleotide biotinylated RNA probe, complementary to the viral S-segment. Hybridization was in 50% formamide, 0.3 M NaCl, 30 mM sodium citrate, pH 7.0, 10% dextran sulphate, at 50 °C for 18 hours. After washing off unbound probe and blocking in 3% BSA, bound probe was detected using the streptavidin-alkaline phosphatase protocol (Gibco-BRL, DNA Detection System, Paisley, U.K.) as recommended by the manufacturer.

Results

Immunofluorescent staining of DUG virus-infected cells showed an initial signal in the Golgi region and a strong cytoplasmic reaction at 48 h p.i. (Fig. 1a). Similar results were obtained with the immunoperoxidase method (not shown). The riboprobe also gave satisfactory results for detecting DUG virus infection cytochemically, although the topological localization was less distinct and histology less distinct than with either antibody method. In non-feeding adult ticks, the distribution of DUG virus antigen was essentially the same in both intra-coelomically-inoculated and orally infected ticks. During feeding, several different organs and tissue types that appeared to be non-infected in the non-feeding, questing stage were invaded. Most notably the salivary glands and ovary only became infected after commencement of feeding, and the results from both histochemistry and infectivity titration of dissected organs gave the same conclusions, summarized in Fig. 3. The hemocytes associated with loose connective tissue, and loose connective cells themselves, were major foci of DUG virus infection in both unfed and feeding ticks as determined by antigen detection and S-segment probe hybridization (Fig. 1c and d). The basal laminae, that separate most tissues from the hemolymph (except for hemocytes), and the extracellular connective tissue matrix also reacted strongly for antigen but showed no hybridization with the DUG complementary riboprobe. Certain digestive cells in the gut lumen and hemocytes associated with the gut basement membrane were also infected (Fig. 1b). However, in non-feeding adults infected in the nymphal stage both orally and intracoelomically most of the digestive cells were negative for DUG virus infection by cytochemistry. During feeding, the gut proliferated and larger numbers of digestive cells, including apparently motile phagocytic cells were infected. DUG-positive hemocytes were also embedded among fat body tissues and around the basement membranes of some salivary gland acini (Fig. 1e). The connective tissue sheath around muscles was also positive for DUG virus antigen and there were often DUG-positive hemocytes adhering to muscle surfaces (Fig. 1c). No DUG antigen or hybridization was detected in the salivary gland secretory cells of ticks until after the start of feeding, when small foci of antigen appeared in the e-cells. After about 10–12 days of feeding, antigen was detected throughout the e-cells but DUG-specific RNA could not be detected by the riboprobe (Fig. 1f). Similarly DUG virus was not isolated from dissected salivary glands of non-feeding ticks but was abundant in glands from infected feeding ticks. Guinea pigs on which the ticks fed seroconverted giving a sensitive test of virus delivery by the ticks. Throughout the tick tissues the basement membranes and extracellular matrix was negative for DUG-specific hybridization. In the EM, 90–100 nm virus particles were seen in infected PS cells and adhering to the cell surface (Fig. 2). DUG antigen on virus particles adhering to the cell surface was detected by immunogold labelling and the

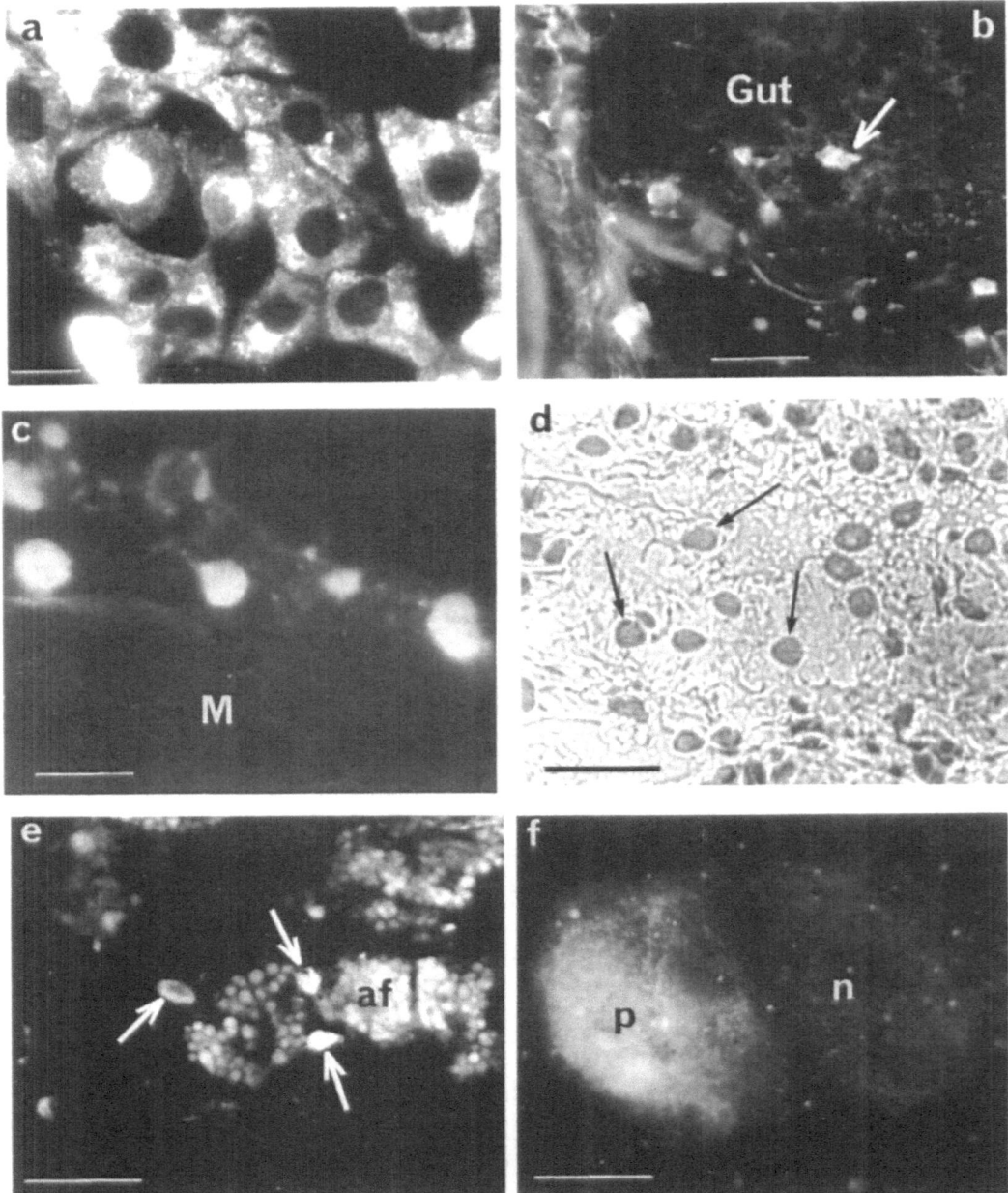

Fig. 1. **a** DUG-virus infected porcine kidney cells stained by immunofluorescence 48 h p.i. Scale: 10 μm. **b** A digestive cell (arrow) in the midgut of a non-feeding trans-stadially infected tick showing a strong DUG immunofluorescent reaction. Most of the gut cells contain no cytoplasmic antigen although the basement membranes surrounding the hemocoel and the hemocytes are positive for antigen. Scale: 20 μm. **c** Immunofluorescence of a trans-stadially infected tick shows hemocytes (arrows) attached to the surface of a muscle (*M*). Antigen is also present in the connective tissue surrounding the muscle. Scale: 10 μm. **d** Positive reaction for DUG-virus by in situ hybridocytochemistry in tick hemocytes (arrows) within tick connective tissue (non-feeding trans-stadially infected adult female). Scale: 20 μm.

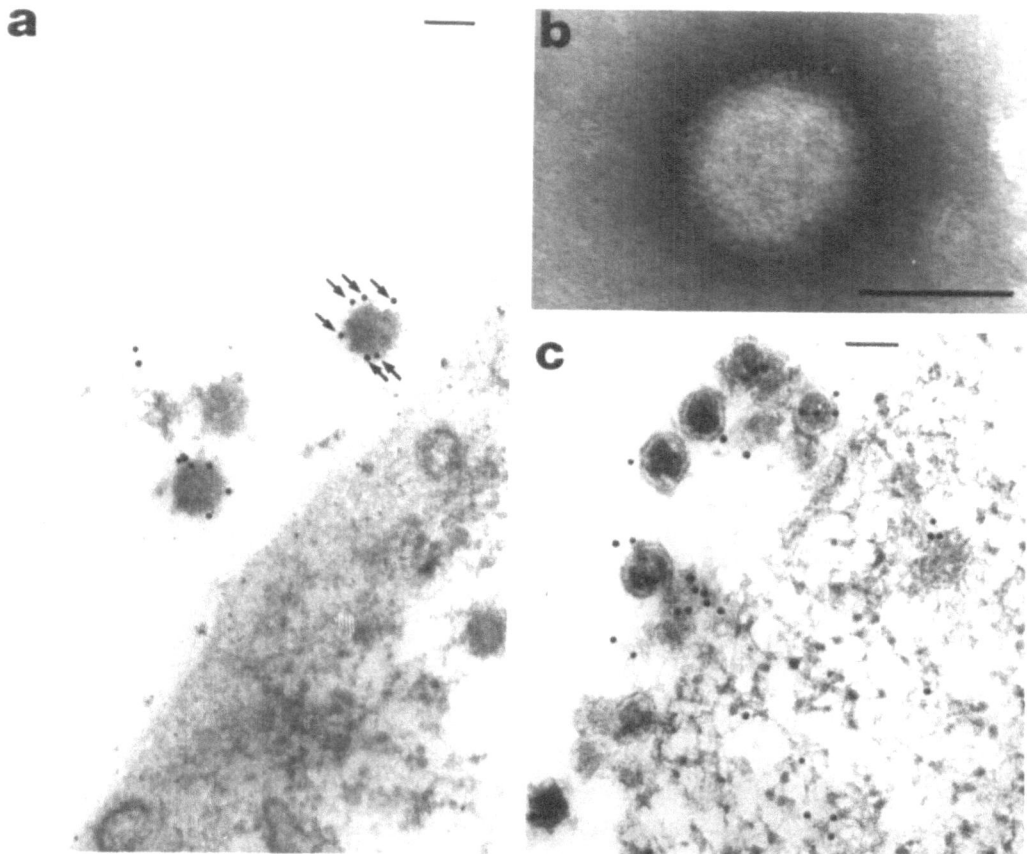

Fig 2. a Pre-embedding immunogold labelling of DUG virus-infected PS cells fixed 48 h p.i. and immunogold labelled. Gold particles (arrows) bind to the surface of virus particles released from infected cells and adsorbed to the cell surface. No labelling is apparent on the cell surface. Scale: 200 μm. b Specimen of DUG virus-infected PS cell culture supernatant negatively stained with ammonium molybdate, showing characteristic glycoprotein spikes on envelope surface. Scale: 100 nm. c Ultrathin section of LR White-embedded DUG-infected PS cell fixed 48 h p.i. and immunogold labelled with polyclonal rabbit antiserum and 10 nm immunogold. Gold particles are mostly associated with virus particles adsorbed to the cell surface. Scale: 100 nm

e Salivary glands of a non-feeding trans-stadially DUG virus infected adult female *A. variegatum* stained by immunofluorescence cytochemistry. A type III acinus is seen surrounded by three DUG antigen positive hemocytes (arrows) in this section, adhering to the basal surface of the salivary gland secretory cells. The a- and d-type secretory cells contain secretory granules that autofluoresce (*af*) strongly but with a slightly different color from that of fluorescein. This autofluorescence varies at different stages of the tick. Scale: 20 μm. f During feeding some secretory cells of the salivary gland become invaded by DUG virus. This infected adult specimen was sectioned after 12 days of feeding on a host and shows an antigen positive (*p*) cell, possibly an e-cell, whilst an adjacent cell is antigen negative (*n*). Scale: 20 μm.

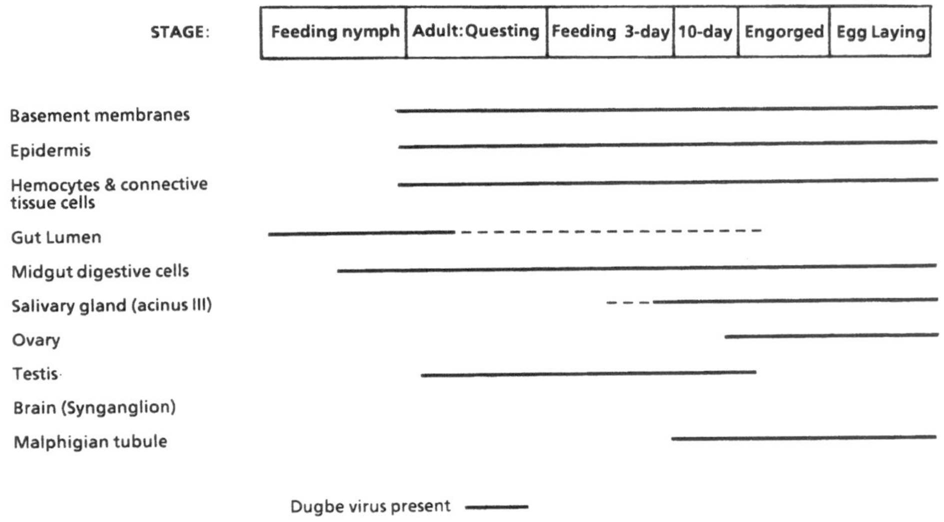

STAGE:	Feeding nymph	Adult: Questing	Feeding 3-day	10-day	Engorged	Egg Laying

Basement membranes

Epidermis

Hemocytes & connective tissue cells

Gut Lumen

Midgut digestive cells

Salivary gland (acinus III)

Ovary

Testis

Brain (Synganglion)

Malphigian tubule

Dugbe virus present ⎯⎯⎯

Fig. 3. Diagram showing changes in Dugbe virus infection of different cell and tissue types of *Amblyomma variegatum* at different developmental stages localized by histochemistry and infectivity titration

cells themselves appeared not to have viral antigen exposed (Fig. 2a). Virus particles had 5–10 nm surface envelope projections which gave an indication of possible icosahedral arrangement (Fig. 2b). Immunogold labelling of ultrathin sections also demonstrated antigen in virus particles in cell-culture (Fig. 2c). In ticks, no virus particles or inclusion bodies were seen by EM, although antigen could be detected by antibody labelling.

Discussion

The results of this study show that DUG virus infection in *A. variegatum* predominates in the hemocytes and, because many tick hemocytes are motile, or loosely attached in the hemocoel, these cells probably play an important role in internal dissemination in the tick and in trans-stadial persistence during moulting. As ixodids only feed once per stadium, this persistence is essential for the tick to be a competent vector of the virus. Arthropod hemocytes (some of which are motile) are capable of invading other tissues and are involved in arthropod immunity, removal of invading microorganisms and rejection of foreign tissues [15, 16]. The salivary glands, which undergo necrosis and regeneration during moulting, were not invaded by DUG virus until the commencement of feeding on a host. The close presence of infected hemocytes to uninfected tick salivary gland secretory cells suggests that virus may be transferred via hemocytes by cell to cell contact or by invasion of salivary tissue by a process of diapedesis (the passage of cells through an epithelial layer without its rupture). This is supported by the fact that hemolymph from infected ticks contains no virus once the hemocytes are removed by low speed centrifugation (data not shown) showing that most virus is cell-associated rather than released freely in the hemolymph. Similarly, DUG RNA could not be detected by polymerase chain reaction in acellular hemolymph, whereas it was detected in organs from the same ticks (unpubl. results). In two previous studies of tick-borne viruses, it was also observed that infection of the salivary glands was not detected until several days after the start of feeding [1, 4]. However, in contrast to both of these previous studies, DUG virus was not detected in the brain or nervous system.

DUG virus infection of the gut in orally infected ticks appears to be limited to a few cells: the 'gut barrier' is rapidly overcome and virus disseminates via the hemocytes. During feeding, DUG virus infection in the gut increases in proportion to cellular proliferation during intracellular digestion and expansion of cells to accomodate the bloodmeal. The results are consistent with the overall dynamics of DUG virus infection in both inoculated and orally infected *A. vareigatum* [18], i.e., a short eclipse phase followed by an increase in viral titer with a reduction and increase during and following moulting. Once virus is released from the gut, dissemination occurred in the same way in the orally-infected ticks as in specimens which were inoculated intra-coelomically with virus. The scarcity of DUG-infected gut cells in newly moulted specimens, and the presence of viral antigen mostly on the basal side of the gut indicates that the gut plays little role in trans-stadial persistence of DUG virus, although it is important in acting as a potential barrier to virus infection [18]. Previous work in our laboratory showed that DUG virus infection was limited to the gut only in orally infected *Rhipicephalus appendiculatus* ticks and under these circumstances this species could not transmit the virus to another host and so must be considered to be a non-competent vector of this virus. However, if the gut was by-passed by intra-coelomic injection with virus, *Rhipicephalus appendiculatus* was able to transmit virus to a vertebrate host by tick feeding. The results from the present study also show that gut cells could be infected both from the lumenal side and from the hemocoel. This is not surprising because the gut has both a phagocytic and secretory role during feeding and virus may be moved into the tick gut via phagocytic cells and out of the tick via secretory cells. In addition to the injection of infected salivary secretions, regurgitation during feeding thus may play a part in transmission to a vertebrate host during feeding.

The presence of DUG virus in the ovary indicates that potential exists for trans-ovarial transmission; however, attempts to isolate DUG virus from the hatched larvae of infected females in our laboratory were unsuccessful [18]. Previous reports of isolations of virus from egg masses [5, 10] may be due to contamination of the egg surface with contaminated secretions from Gené's organ since we observed in this study that the secretory cells of Gené's organ (which secrete the surface wax of the eggs) contained DUG virus antigen. These results suggest that the cycling of virus between infected ticks and vertebrate hosts is more important in maintaining virus infection than vertical transmission from one generation to the next. A previous study reported no evidence for transovarial transmission with Qalyub virus in *Ornithodoros erraticus* [13]. Similarly there is potential for trans-venereal transmission from males to females via infected testis but this has not yet been experimentally demonstrated for DUG virus in ticks.

We have shown that both immunohistochemistry and hybridocytochemistry are useful for detecting DUG virus infections in ticks and that these techniques can be applied in epidemiological studies of nairoviruses. For immunolabelling and observation by light microscopy, immunofluorescence was the most sensitive method, although the immunoperoxidase technique had advantages: phase contrast and counterstaining could be used and autofluorescence problems avoided. The disadvantage of the immunoperoxidase method was that some tick tissues contain endogenous peroxidase which cannot be fully blocked by preincubation in peroxide. Hybridization was the least sensitive method but positive reactions do show that cells contain replicating virus and are not merely presenting antigen produced, for example, in other cells.

The structure of the DUG virus particle appears to be similar to that of CCHF virus [8], although the DUG virion is slightly smaller in diameter (90 nm) than CCHF (120 nm). The DUG viral envelope contains a larger number of smaller surface spikes than CCHF but their arrangement also seems to follow icosahedral symmetry. Results from immunogold labelling assays provide the basis for further high resolution studies to answer questions about DUG virus morphogenesis in ticks. This will allow more precise identification of cell types involved in DUG virus replication and reveal the nature of the association of infected hemocytes with other tissues.

When organs from infected *A. variegatum* were examined by EM, no clearly defined intracellular vacuoles containing virus particles or viral inclusion bodies (such as are present in infected cell culture) were observed. This suggests that viral morphogenesis in ticks differs from that in a cytolytic infection established in cell culture. Although viral antigen, replicative viral sense RNA, and virus infectivity are all present to a high degree in infected ticks, morphogenesis of DUG virus into characteristic bunyavirus particles does not occur. A similar effect was obtained in persistently infected DUG carrier cell cultures which produced antigen and infective virus but which apparently contained no virus particles that could be observed in the EM [7]. These cells were able to grow and divide and continued to produce infectious DUG virus after many passages. This alternative mode of infection may be an important strategy for the maintenance of DUG virus infections in avoiding cytopathological effects that might be deleterious for the tick vector.

Acknowledgement

This work was supported by the U.S. Army Medical Research and Development Command under contract DAMD 17-87-C-7176 and United Kingdom contract No. CB/SL/23c/1831.

References

1. Booth TF, Davies CR, Jones LD, Staunton D, Nuttall PA (1989) Anatomical basis of Thogoto virus infection in BHK cell culture and in the ixodid tick vector, *Rhipicephalus appendiculatus*. J Gen Virol 70: 1093–1104

2. Burgdorfer W (1957) Artificial feeding of ixodid ticks for studies on the transmission of disease agents. J Infect Dis 100: 212–214

3. Casals J, Tignor GH (1980) The *Nairovirus* genus: serological relationships. Intervirology 14: 144–147

4. Chernesky MA, McLean DM (1969) Localization of Powassan virus in *Dermacentor andersoni* ticks by immunofluorescence. Can J Microbiol 15: 1399–1408

5. Cornet J-P, Georges AJ, Gonzalez JP (1987) Passage transovarien 'in natura' du virus Dugbe chez la tique *Amblyomma variegatum*. Ann Inst Pasteur 138: 269–271

6. Davies CR, Jones LD, Nuttall, PA (1986) Experimental studies on the transmission of Thogoto virus, a candidate orthomyxovirus, in *Rhipicephalus appendiculatus* ticks. Am J Trop Med Hyg 35: 1256–1262

7. David-West TS, Porterfield JS (1974) Dugbe virus: a tickborne arbovirus from Nigeria. J Gen Virol 23: 297–307

8. Ellis DS, Southee T, Lloyd G, Platt GS, Jones N, Stamford S, Bowen ETW, Simpson DIH (1981) Congo-Crimean haemorrhagic fever virus from Iraq 1979. I. Morphology in BHK21 cells. Arch Virol 70: 189–198

9. Hoogstraal H (1979) The epidemiology of tick-borne Crimean-Congo haemorrhagic fever in Asia, Europe and Africa. J Med Entomol 15: 304–417

10. Huard M, Cornet J-P, Germain M, Camicas JL (1978) Passage transovarien du virus Dugbe chez la tique *Amblyomma variegatum* (Fabricius). Bull Soc Pathol Exot Fil 71: 19–22

11. Jones LD, Davies CR, Steele GM, Nuttall PA (1988) The rearing and maintenance of Ixodid and Argasid ticks in the laboratory. J Anim Tech 39: 99–106

12. Marriott AC, Ward VK, Higgs S, Nuttall PA (1990) RNA probes detect nucleotide sequence homology between members of two different nairovirus serogroups. Virus Res 16: 77–82

13. Miller BR, Loomis R, Dejean A, Hoogstraal H (1985) Experimental studies on the replication and dissemination of Qalyub virus (*Bunyaviridae: Nairovirus*) in the putative tick vector *Ornithodoros* (Pavlovskyella) *erraticus*. Am J Trop Med Hyg 34: 180–187

14. Newman GR, Jasani B, Williams ED (1983) A simple post-embedding system for the rapid demonstration of tissue antigens under the electron microscope. Histochem J 15: 543–555

15. Ratcliffe NA, Rowley AF (1979) Role of hemocytes in defense against biological agents. In: Gupta AP (ed) Insect hemocytes. Development, forms, functions and techniques. Cambridge University Press, New York, pp 331–414

16. Ratcliffe NA, Leonard C, Rowley AF (1984) Prophenoloxidase activation: nonself recognition and cell co-operation in insect immunity. Science 226: 557–559

17. Shope RE, Sather GE (1979) Arboviruses. In: Lennette EH, Schmidt NJ (eds) Diagnostic procedures for viral, rickettsial and chlamydial infections. American Public Health Association, Washington, DC, pp 767–814

18. Steele GM, Nuttall PA (1989) Difference in vector competence of two species of sympatric tick, *Amblyomma variegatum* and *Rhipicephalus appendiculatus* for Dugbe virus (*Nairovirus: Bunyaviridae*). Virus Res 14: 73–84
19. Ward VK, Marriott AC, El-Ghorr AA, Nuttall PA (1990) Coding strategy of the S RNA segment of Dugbe virus (*Nairovirus: Bunyaviridae*). Virology 175: 518–524

Authors' address: T. F. Booth, NERC Institute of Virology and Environmental Microbiology, Mansfield Road, Oxford OX1 3SR, England.

Arch Virol (1990) [Suppl 1]: 219–225

Biological and molecular characteristics of orbiviruses and orthomyxoviruses isolated from ticks

Patricia A. Nuttall, T. F. Booth, Dorothy Carey, C. R. Davies, Linda D. Jones, Mary A. Morse, and **S. R. Moss**

NERC Institute of Virology and Environmental Microbiology, Oxford, U.K.

Accepted December 12, 1989

Summary. Tick-borne orbiviruses and tick-borne "orthomyxoviruses" possess segmented RNA genomes. Consequently, they have the potential to undergo reassortment following co-infection of a cell with related viruses. Most orbiviruses (family *Reoviridae*) isolated from ticks are members of the Kemerovo (KEM) serogroup, possessing a genome comprised of 10 segments of double-stranded RNA. In vitro studies have demonstrated reassortment between Great Island (GI) subgroup viruses (associated with seabirds) and KEM virus (a human pathogen), but not between GI subgroup viruses and other members of the group that are associated with birds. Such genetic relationships presumably reflect the evolutionary pathways followed by the virus as it circulates in a vertebrate-arthropod infection cycle. Evidence that rapid evolution, via genetic reassortment, can occur in both the vertebrate and tick hosts of an arbovirus has been demonstrated with Thogoto (THO) virus, an unclassified tick-borne virus that shows structural and morphogenetic properties characteristic of orthomyxoviruses. The genome of THO virus is composed of 6 segments of single-stranded, negative-sense RNA. Using *ts* mutants, reassortment of THO virus has been demonstrated in naturally infected ticks and in a vertebrate host. However, in these studies the potential for reassortment was shown to be limited by interference. This paper discusses the relative roles of the vertebrate host and the tick vector in terms of virus evolution resulting from genetic reassortment.

Introduction

Of the 116 viruses isolated from ticks and listed in the International Catalogue of Arboviruses [14], 12 are considered significant to humans or animals. Two of the latter, Kemerovo (and its antigenic relatives) and Thogoto, are described in this brief review.

The genus *Orbivirus* currently comprises 13 virus serogroups. Orbiviruses are arboviruses; all members of the Kemerovo (KEM) serogroup are transmitted by ticks [10]. The 4 antigenic subgroups hitherto distinguished within the KEM serogroup [10] reflect the different ecological associations of these viruses: KEM subgroup viruses are widely distributed in the habitats of *Ixodes* ticks in Eurasia; Chenuda (CNU) subgroup viruses are associated with soft tick species that parasitize pigeons and seabirds; Great Island (GI) subgroup viruses are transmitted by *Ixodes* ticks that feed on seabirds; and the 2 members of the Wad Medani (WM) subgroup have been isolated from hard tick species from tropical/sub-tropical regions [12]. Most of these viruses have not been associated with disease in their vertebrate hosts. However, members of the KEM subgroup are considered potential human pathogens. KEM virus has been isolated from the cerebrospinal fluids of human patients suffering from encephalitis [2], and antibodies to KEM subgroup viruses have been detected in humans [8, 19, 20].

Genetic studies with members of the GI subgroup have demonstrated that the gene carried by segment 5 (which migrates as segment 6 of some isolates) is the major genetic determinant of serotype specificity [25]. In contrast, genomic segment 2 determines the serotype of bluetongue virus (BTV), the type species of the genus *Orbivirus*, while the function of BTV segment 5 is unclear [13, 22]. Comparison of the gene sequence of segment 5 of Broadhaven (BRD) virus (a member of the GI subgroup) with the protein encoded by segment 5 of BTV serotype 10, revealed 30% nucleotide homology and 31% amino acid homology [27]. Significant homology was not shown between BRD segment 5 and segment 2 of BTV. Furthermore, sequences of the 3' and 5' ends determined for BRD segment 5 were similar to the respective 3' and 5' regions of BTV. These data provide evidence of an evolutionary relationship between two ecologically distinct groups of orbiviruses and the changes that have occurred in the functions of equivalent (i.e., at the sequence level) genomic segments.

Thogoto virus originally was isolated from a mixed pool of *Boophilus* and *Rhipicephalus* ticks collected from cattle in Thogoto forest near Nairobi, Kenya [11]. Subsequent isolations have been made from a variety of ixodid ticks species collected in Central Africa, Egypt, Iran, Sicily, and Portugal (reviewed, in [4]). In the laboratory, replication and transmission of THO virus have been demonstrated using *Rhipicephalus appendiculatus* ticks [4]. The natural transmission cycle of THO virus probably involves various species of small mammals, with nymphal and adult tick stages transmitting the virus to larger mammals, including domestic animals. THO virus is of veterinary and possibly medical significance. Haig and associates [11] reported that intravenous inoculation of THO virus produced a febrile reaction in sheep (temperatures of up to 40.5 °C), while Davies et al. [7] described evidence suggesting that THO virus infection was the cause of widespread abortion in sheep in the Masailand region of Kenya. Two isolations of THO virus were made from humans at Ibadan, Nigeria, in 1966 [23]. In one case the virus was isolated from cerebrospinal fluid and was associated with bilateral optic neuritis. In the other, virus was isolated from blood and was associated with a fatal meningitis, although this latter case was complicated by a terminal sickle-cell crisis.

The biological and molecular properties of THO and Dhori (DHO) viruses have several features in common with orthomyxoviruses [3, 9, 29]. However, unlike the recognized members of the *Orthomyxoviridae* (influenza A, B, C virus strains), THO and DHO are arboviruses. THO and DHO virions (ca. 100 nm diameter) acquire a lipid envelope during the processes of virus morphogenesis from the cell plasma membrane [3]. The virions possess a segmented genome of single-stranded, negative-sense RNA (size range 3–7×10^5 Da). Highly purified THO virions have only 6 major RNA species [29]. The influenza viruses have genomes consisting of 8 (influenza A and B) or 7 (influenza C) RNA species [21]. Analyses of THO segment 3 RNA have shown that it codes for a major 68.6-kDa gene product that exhibits some sequence similarity to the segment 3 gene products (PA protein) of the influenza A and B viruses [29]. Whether the 68.6-kDa protein is a transcriptase/replicase component of THO virus is not known. The data are supportive of, but do not prove, an evolutionary relationship between THO virus and the influenza viruses.

In vitro reassortment of Kemerovo group orbiviruses

To examine the evolutionary relationships between KEM viruses, in vitro reassortment assays were performed using spontaneous *ts* mutants. Evidence of reassortment was based on the determination of reassortment frequency or genome profile by polyacrylamide gel electrophoresis. Seven groups of mutants were defined (groups I to VII), presumably corresponding to *ts* lesions in 7 of the 10 viral RNA segments (group I *ts* mutants having lesions in segment 9 of the viral RNA genome, group II in segment 1, group III in segment 5, group IV in segment 6, group V in segment 3, group VI in segment 4, and group VII in segment 2) [24–26].

The GI subgroup appears to constitute a single gene pool. High frequency reassortment was detected following dual infection of cell cultures with 5 geographical representatives of the sub-group [26]. Thus, a member of the GI subgroup introduced into a seabird colony (via an infected seabird or an infected tick carried by a seabird) has the potential to reassort with other members of the group resident in the colony. This explains the remarkable genotypic variation found in isolates from a single seabird colony.

Members of the GI subgroup are related genetically to the KEM subgroup [28]. The frequency of reassortment between the two was significantly lower than that generally found between GI subgroup viruses, possibly reflecting the evolutionary divergence of the two groups. Such divergence may also be reflected in the apparent inability of KEM virus to reassort with GI subgroup mutants having *ts* lesions in either segments 1, 2, or 9. Possibly, the functions of these genes are such that any significant changes in their sequence give rise to incompatibility, for example, in packaging or inter-actions of the particular gene products. Reassortment was not demonstrated between KEM or GI subgroup viruses and representatives of the CNU subgroup. Furthermore, Mono Lake (ML) virus, hitherto classified within the CNU subgroup, did not reassort with CNU virus.

The extant KEM serogroup presumably constitutes a group of viruses that share the same evolutionary origin. This premise is based on the demonstration of group specific reactivity detected by complement fixation and immunofluorescence tests [1]. Reassortment studies indicate that, within the group, the GI subgroup of seabird-associated viruses are more closely related to the KEM subgroup of viruses circulating in a terrestrial ecosystem than to other avian-associated viruses, such as CNU and ML. These relationships presumably reflect the evolutionary pathways followed by the respective subgroups of viruses. Divergence of these pathways may result from selective pressures exerted at the level of infection in the vertebrate host or arthropod vector. What evidence is there that such evolutionary events can occur in the vertebrate or invertebrate hosts of an arbovirus?

In vivo reassortment of Thogoto virus

Thogoto virus provides an ideal model for studying reassortment in vivo. The virus produces high levels of viremia in hamsters. Uninfected *Rhipicephalus appendiculatus* ticks become infected when allowed to feed on viremic hamsters. After moulting to the next development stage, the infected ticks transmit THO virus to uninfected hamsters while feeding on them [4].

Reassortant virus was detected in hamsters that were dually infected by direct inoculation or by oral transmission from infected ticks [17]. Evidence of reassortment was based on detection of wild-type (ts^+) virus in the blood of hamsters dually infected with temperature sensitive mutants belonging to two different reassortant groups. Viral dose and time of co-infection of the hamsters affected the incidence of reassortment. When ticks were dually infected with the *ts* mutants, by interrupted feeding on viremic hamsters, ts^+ reassortant virus was detected in the ticks 12 to 15 days after engorgement [5]. Following moulting, the infected ticks transmitted re-assortant virus to uninfected hamsters.

Opportunities for reassortment in ticks were limited by interference [6]. When ticks were infected by partial feeding on a hamster infected with a *ts* mutant and then, after a delay of several hours or days, super-infected with wild-type (wt) virus, most of the ticks showed complete interference, i.e., wt virus was not detected. Similarly, interference was detected when ticks were infected with a *ts* mutant and then, after moulting, superinfected with wt virus. Interference probably occurs in the gut and does not involve the salivary glands [18]. When ticks infected with the *ts* mutant were later superinfected with wt virus by parenteral inoculation, interference did not occur, i.e., wt virus replicated in the ticks and was delivered by bite to uninfected hamsters. Thus, when the gut is by-passed, the ticks are apparently permissive to dual infection, even when there is a delay in the time of superinfection. The major significance of interference between genetically compatible viruses is that it limits the frequency of reassortment. However, complete interference was not detected in all ticks co-infected with THO virus. Hence the potential for reassortment exists even in ticks orally infected by different viruses in consecutive feeding stages.

Comparative roles of the tick vector and vertebrate host in arbovirus evolution

Arboviruses possessing segmented genomes have the potential to reassort in both their vertebrate hosts and arthropod vectors. What features of these two milieux affect the generation and perpetuation of reassortant viruses?

Vertebrate hosts involved in arbovirus transmission cycles often act as amplifying hosts. This role is characteristic of permissive hosts that develop viraemia of a titre sufficiently high and prolonged to infect arthropod vectors feeding on them. In this context, vertebrate hosts are important sites of reassortant virus formation because they represent a potential source for infecting countless vectors. In contrast to the vertebrate host, the role of an arthropod in perpetuating reassortant viruses may be considered more hazardous. The 'new' virus must rely on the survival of the vector, and transmission horizontally to a susceptible host or vertically to the succeeding vector progeny.

Opportunities for dual infection of ticks are enhanced by long term persistence of arbovirus infections in these arthropods, frequently coupled with vertical transmission. In contrast, viraemia in vertebrates tends to be short-lived, and is often followed by immunity in or death of the host. Thus, compared to arthropod vectors, vertebrates have a relatively short 'time window' in which dual infections can occur. However, vertebrates are often fed on simultaneously by many hundreds of ticks, thereby increasing the probability of multiple infections. An individual tick is less likely to be infected more than once (particularly ixodid species), as it takes only a few bloodmeals throughout its lifetime, and so is less likely to come into contact with more than one infected host.

Interference limits the frequency of reassortment in ticks whereas, in the vertebrate host, immunity [16] is probably a more important constraint on the generation of reassortants. Resistance to tick infestation [15] also may inhibit reassortment in the vertebrate host.

Computer modelling is necessary to evaluate the quantitative roles of the vertebrate host and the arthropod vector in the rapid evolution, via genetic reassortment, of arboviruses that have segmented genomes.

References

1. Carey D, Nuttall PA (1989) Antigenic cross-reactions between tick-borne orbiviruses of the Kemerovo serogroup. Acta Virol 33: 15–23
2. Chumakov MP, Karpovich LG, Sarmanova ES, Sergeeva GI, Bychkova, MB, Tapupere VO, Libikova H, Mayer V, Rehacek J, Kozuch O, Ernek E. (1963) Reports on the isolation from *Ixodes persulcatus* ticks and from patients in western Siberia of a virus differing from the agent of tick-borne encephalitis. Acta Virol 7: 82–83
3. Clerx JPM, Fuller F, Bishop DHL (1983) Tick-borne viruses structurally similar to orthomyxoviruses. Virology 127: 205–219
4. Davies CR, Jones LD, Nuttall PA (1986) Experimental studies on the transmission cycle of Thogoto virus, a candidate orthomyxovirus, in *Rhipicephalus appendiculatus*. Am J Trop Med Hyg 35: 1256–1262
5. Davies CR, Jones LD, Green BM, Nuttall PA (1987) In vivo reassortment of Thogoto virus (a tick-borne influenza-like virus) following oral infection of *Rhipicephalus appendiculatus* ticks. J Gen Virol 68: 2331–2338
6. Davies CR, Jones LD, Nuttall PA (1989) Viral interference in the tick, *Rhipicephalus appendiculatus*. I. Interference to oral superinfection by Thogoto virus. J Gen Virol 70: 2461–2468
7. Davies FG, Soi RK, Wariru BN (1984) Abortion in sheep caused by Thogoto virus. Vet Rec 115: 654
8. Draganescu N, Girjabu E, Iacobescu V, Totescu, Caliniescu C, Petrescu S, Popescu A (1974) Serologic investigations on the presence of antibodies to Kemerovo virus in subjects of several counties in Romania. Rev Roum Virol 25: 211–213
9. Fuller FJ, Freedman-Faulstich EZ, Barnes JA (1987) Complete nucleotide sequence of the tick-borne, orthomyxo-like Dhori/Indian/1313/61 virus nucleoprotein gene. Virology 160: 81–87
10. Gorman BM, Taylor J, Walker PJ (1983) Orbiviruses. In: Joklik WK (ed) The Reoviridae. Plenum, New York, pp 287–357
11. Haig DA, Woodall JP, Danskin D (1965) Thogoto virus: a hitherto undescribed agent isolated from ticks in Kenya. J Gen Microbiol 38: 389–394
12. Hoogstraal H (1973) Viruses and ticks. In: Neuberger A, Tatum AL (eds) Viruses and invertebrates. North Holland, Amsterdam, pp 349–390
13. Inumara S, Roy P. (1987) Production and characterisation of the neutralization antigen VP2 of bluetongue virus serotype 10 using a baculovirus expression vector. Virol 157: 472–479
14. Karabatsos N (ed) (1985) International catalogue of arboviruses including certain other viruses of vertebrates, 3rd edn. American Society of Tropical Medicine and Hygiene, San Antonio, Texas
15. Jones LD, Nuttall PA (1990) The effect of host resistance to tick infestation on the transmission of Thogoto virus by ticks. J Gen Virol 71: 1039–1043
16. Jones LD, Nuttall PA (1989) The effect of virus-immune hosts on Thogoto virus infection of *Rhipicephalus appendiculatus* ticks. Virus Res 14: 129–140
17. Jones LD, Davies CR, Green BM, Nuttall PA (1987) Reassortment of Thogoto virus (a tick-borne influenza-like virus) in a vertebrate host. J Gen Virol 68: 1299–1306
18. Jones LD, Davies CR, Booth TF, Nuttall PA (1989) Viral interference in the tick,

Rhipicephalus appendiculatus. II. Absence of interference with Thogoto virus when the tick gut is by-passed by parenteral inoculation. J Gen Virol 70: 2469–2473

19. Libikova H, Heinz F, Ushazyova D, Stunzer D (1978) Orbiviruses of the Kemerovo complex and neurological diseases. Med Microbiol Immunol. 166: 255–263

20. Malkova D, Holubova J, Kolman JM, Marhoul Z, Hanzal F, Kulkoua H, Markvart K, Simkova L (1980) Antibodies against some arboviruses in persons with various neuropathies. Acta Virol 24: 298

21. Matthews REF (1982) Classification and nomenclature of viruses. Intervirology 17: 1–199

22. Mertens PPC, Pedley S, Cowley J, Burroughs JN, Corteyn AH, Jeggo MH, Jennings DM, Gorman BM (1989) Analysis of the roles of bluetongue virus outer capsid proteins VP2 and VP5 in determination of virus serotype. Virology 170: 561–565

23. Moore DL, Causey OR, Carey DE, Reddy S, Cooke AR, Akinkugbe FM, David-West TS, Kemp GE (1975) Arthropod-borne viral infections of man in Nigeria, 1964–1970. Ann Trop Med Parasitol 69: 49–64

24. Moss SR, Nuttall PA (1986) Isolation and characterization of temperature sensitive mutants of Broadhaven virus, a Kemerovo group orbivirus (family *Reoviridae*).Virus Res 4: 331–336

25. Moss SR, Ayres CM, Nuttall PA (1987) Assignment of the genome segment coding for the neutralising epitope(s) of orbiviruses in the Great Island subgroup (Kemerovo serogroup). Virology 157: 137–144

26. Moss SR, Ayres CM, Nuttall PA (1988) The Great Island sub-group of tick-borne orbiviruses represents a single gene pool. J Gen Virol 69: 2721–2727

27. Moss SR, Fukusho A, Nuttall PA (1990) RNA segment 5 of Broadhaven virus, a tick-borne orbivirus, shows sequence homology with segment 5 of bluetongue virus. Virology (in press)

28. Nuttall PA, Moss SR (1989) Genetic reassortment indicates a new grouping for tick-borne orbiviruses. Virology 171: 156–161

29. Staunton D, Nuttall PA, Bishop DHL (1989) Sequence analyses of Thogoto viral RNA segment 3: evidence for a distant relationship between an arbovirus and members of the *Orthomyxoviridae.* J Gen Virol 70: 2811–2817

Author's address: Patricia A. Nuttall, NERC Institute of Virology and Environmental Microbiology, Mansfield Road, Oxford OX1 3SR, England.

Arch Virol (1990) [Suppl 1]: 227–234

Characterization of tick salivary gland factor(s) that enhance Thogoto virus transmission

Linda D. Jones, Elizabeth Hodgson, and **Patricia A. Nuttall**

NERC Institute of Virology and Environmental Microbiology, Oxford, U.K.

Accepted January 30, 1990

Summary. Thogoto (THO) virus is transmitted from infected to uninfected ticks co-feeding on an uninfected guinea pig, even though the guinea pig does not develop a detectable viremia. We call this "saliva-activated transmission" (SAT). Ten-fold more ticks became infected when feeding on guinea pigs inoculated with THO virus plus salivary gland extract (derived from uninfected, partially fed female *Rhipicephalus appendiculatus* ticks), than those which fed on guinea pigs inoculated with virus alone. To investigate the physicochemical properties of the SAT factor, guinea pigs were infested with 50 uninfected *R. appendiculatus* nymphs and then inoculated with THO virus plus treated or untreated salivary gland extract, or with THO virus alone. Virus enhancement was measured by the number of uninfected ticks that acquired virus. Maximum enhancement of THO virus transmission was observed when salivary glands were derived from female *R. appendiculatus* ticks that had fed for 6 days and inoculated at a concentration of 40 μg of salivary gland protein per guinea pig. The salivary gland extract retained its biological activity after repeated freeze/thaw cycles, at pH 5, 6, and 7, and at temperatures ranging from 4–37 °C. Enhancement of virus transmission was not observed when salivary gland extract was treated with pronase or proteinase-k, indicating that either a protein or a peptide is involved. Complete characterization of the SAT factor will provide insights into the mechanism of this novel mode of tick-borne virus transmission.

Introduction

Arthropod vectors of animal viruses become infected when they feed on the blood of a viremic host [11]. In addition, ticks become infected after feeding on apparently non-viraemic hosts [5]. Investigations of the mechanism of this novel mode of arbovirus transmission indicate that factor(s) potentiating virus transmission are associated with the salivary glands of ticks [7]. Enhancement of *Leishmania* infectivity by salivary glands has been demonstrated with the sandfly, *Lutzomyia longipalpis* [10]. Thus, it appears that the salivary glands of hematophagous arthropods play an integral role in disease transmission, with the parasite or virus utilizing the physiological activities of the vector's feeding mechanism to enhance its own infectivity. In this paper we describe a series of experiments undertaken to establish the physicochemical properties of the factor(s) associated with the salivary glands of ticks that potentiate saliva activated transmission (SAT).

Experiments on SAT were conducted with Thogoto (THO) virus, an orthomyxo-like virus [2, 3, 9] of medical and veterinary significance [3]. The virus was originally isolated from ticks collected in Kenya [4], and has subsequently been isolated throughout central Africa and in parts of the Middle East and southern Europe [3]. Throughout these studies, guinea pigs were used as the vertebrate host and the African three-host ixodid tick species, *Rhipicephalus appendiculatus*, (a competent vector of THO virus [3]) as the arthropod vector.

Materials and methods

Cells and virus

BHK-21 and Vero cell cultures were propagated in modified Eagle's medium (EMEM) supplemented with 10% bovine calf serum (BCS). The Sicilian (SiAr 126) isolate of THO virus [1] was used throughout the study. The virus was passaged seven times in suckling mice, plaque picked three times in Vero cells, and then passaged three times in BHK-21 cells [3].

Ticks

A laboratory colony of *R. appendiculatus* was established by feeding all three stages of the tick on Dunkin Hartley guinea pigs [6]. During the interval between feedings, ticks were maintained in perforated tubes held inside a desiccator at 28 °C and 85% relative humidity.

Virus assay

Nymphs were homogenized individually in 1 ml of EMEM containing 10% BCS and antibiotics appropriate to inhibit bacterial growth. Blood samples were obtained on 4 and 5 days after attachment of ticks, by cardiac puncture from anesthetized Dunkin Hartley guinea pigs. Titration of blood or of tick derived material was undertaken using Vero cells incubated at 35 °C for 4 days, prior to fixation and staining [3].

Time course of virus transmission

Eight uninfected guinea pigs were each infested with 40 uninfected adult *R. appendiculatus* ticks (equal sex ratio). At 1, 4, 6, and 8 days after attachment, female ticks were removed and their salivary glands dissected out, placed in phosphate buffered saline (pH 7.2), extracted by homogenization and low speed centrigugation, and frozen at −20 °C. Guinea pigs (two per salivary gland sample) were infested with 50–70 uninfected *R. appendiculatus* nymphs. Each guinea pig was inoculated subcutaneously with 5000 plaque forming units (PFU) THO virus and salivary gland extract (20 μg protein per animal) from either unfed ticks or from ticks that had fed for a period of 1, 4, 6, or 8 days; control guinea pigs were inoculated with virus alone. Virus transmission was measured by the number of uninfected ticks that became infected. Ticks were assayed for virus by plaque titration 12 days after engorgement, the time of maximum virus titer [3].

Optimum concentration of salivary gland extract

Eight guinea pigs were inoculated with THO virus plus 1 μg, 10 μg, 20 μg, or 40 μg of salivary gland extract; control animals were inoculated with THO virus alone. Protein concentrations were estimated using the Bio-Rad protein estimation kit (Bio-Rad Laboratories, Watford, U.K.)

Physicochemical properties of tick salivary glands

Experiments were undertaken to determine the physicochemical properties of tick salivary glands. Guinea pigs were infested with 50 uninfected *R. appendiculatus* nymphs. Each guinea pig was inoculated subcutaneously with 5000 PFU THO virus with or without treated salivary gland extract. Salivary glands were derived from uninfected female *R. appendiculatus* ticks which had fed for a period of 6 days. The protein concentration was standardized to 40 μg of salivary gland extract per guinea pig; duplicate guinea pigs were used for each test sample. Enhancement of virus transmission was measured by the number of uninfected ticks that acquired virus.

Guinea pigs were inoculated with THO virus and salivary gland extract which had been freeze/thawed one, two or three times; control animals were inoculated with stock salivary gland extract. To assess the affect of pH and temperature on the biological activity of tick salivary gland extract, the extract was incubated at a range of pHs(3, 5, 7, and 9), for 30 min at 21 °C and then readjusted to neutrality prior to inoculation into guinea pigs, or incubated at 4 °C or 21 °C for 24 h, 28 °C for 2 h, 37 °C for 30 min or 60 °C or 100 °C for 10 min prior to inoculation into guinea pigs; control animals were inoculated with virus plus untreated salivary gland extract or with virus alone.

Salivary gland extract was incubated with either pronase (Sigma, Dorset, U.K.) or proteinase-k (Sigma, Dorest, U.K.) at a concentration of 50 μg of enzyme per mg salivary gland extract, for 30 min at 37 °C. Following treatment with these enzymes the salivary gland extract plus THO virus was inoculated into guinea pigs; control guinea pigs either were inoculated with salivary gland extract (incubated for 30 min at 37 °C) plus THO virus, or with THO virus alone.

Results

Inoculation of guinea pigs with THO virus and extracts of salivary glands taken at different stages of feeding

Experiments were conducted to determine whether the enhancing factor(s) associated with the salivary glands of *R. appendiculatus* was present throughout the tick feeding period (Fig. 1). 19/40 (48%) ticks acquired virus when the inoculum included salivary extracts derived from uninfected ticks that had fed for a period of 4 days. Maximum enhancement of virus transmission was

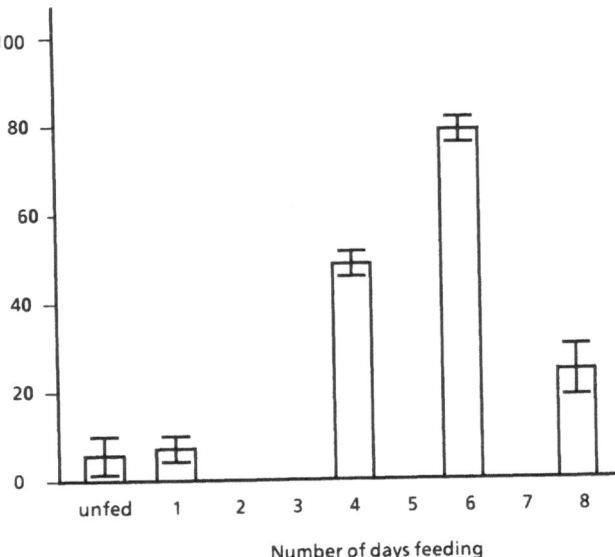

Fig. 1. Guinea pigs were inoculated with THO virus plus extracts derived from salivary glands of uninfected *R. appendiculatus* ticks harvested when unfed or on days 1, 4, 6, and 8 of feeding. Enhancement of virus transmission was measured by the number of uninfected ticks that became infected. The means and ranges were calculated from those ticks found to contain virus

observed (31/40; 78% ticks acquired virus) when the inoculum included salivary gland extracts derived from uninfected ticks that had fed for a period of 6 days. On day 8 the enhancing effect of the salivary gland extracts was significantly reduced (9/40; 23% ticks acquired virus). No enhancement of THO virus transmission was observed when salivary glands were derived from unfed ticks, or ticks that had fed for only one day (2/40, 3/40 ticks respectively, acquired virus).

Inoculation of guinea pigs with THO virus and varying concentrations of salivary gland extract

Further studies were undertaken to determine the optimum concentration of salivary gland extract required for enhancement of THO virus transmission (Fig. 2). When the inoculum included salivary gland extract at a concentration of 40 µg protein, 41/68 (60%) of nymphs tested acquired virus compared to 25/27 (44%) at a concentration of 20 µg. At concentrations below 20 µg the enhancing effect of the salivary gland extracts was significantly reduced.

Stability of salivary gland extract after repeated freeze/thaw cycles

The stability of salivary gland extract after repeated freeze/thaw cycles was assessed (Fig. 3). Salivary gland extract retained its ability to enhance THO virus transmission even after three freeze/thaw cycles with 56/77 (73%) nymphs acquiring virus.

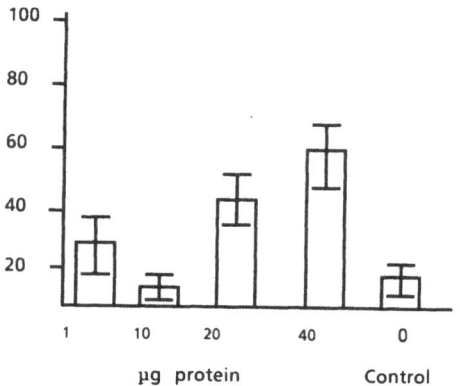

Fig. 2. Guinea pigs were inoculated with THO virus plus various concentrations of salivary gland extract (μg protein); control animals (*0*) were inoculated with THO virus alone. Enhancement of virus transmission was measured by the number of uninfected ticks that became infected when compared to ticks feeding on the control guinea pigs. The means and ranges were calculated from those ticks found to contain virus

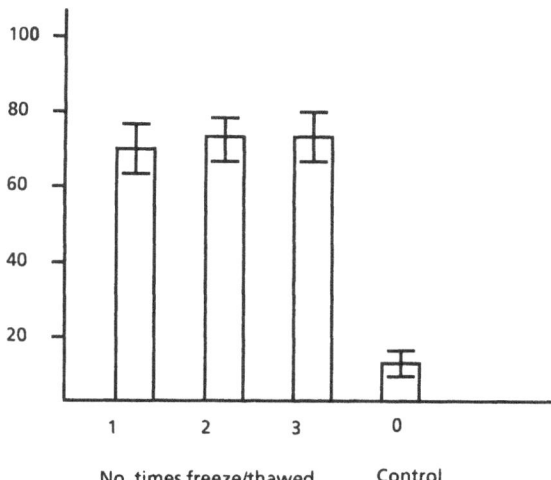

Fig. 3. Guinea pigs were inoculated with THO virus plus salivary gland extract which had been freeze/thawed one, two or three time (*1, 2, 3*); control animals (*0*) were inoculated with THO virus alone. Enhancement of virus transmission was measured by the number of uninfected ticks that became infected when compared to the ticks feeding on the control guinea pigs. The means and ranges were calculated from those ticks found to contain virus

pH sensitivity of tick salivary gland extract

To determine the pH sensitivity of salivary gland extract, salivary gland extract was incubated at various pHs (Fig. 4). At pH 5, 6, and 7 the salivary gland extract retained its ability to enhance virus transmission with 64/118 (54%) of ticks tested acquiring virus. However, at pH 3 the enhancing affect was significantly reduced with only 18/72 (25%) ticks acquiring virus and no enhancement of virus transmission was observed when salivary gland extract was incubated at pH 9.

Incubation of salivary gland extract at selected temperatures

Further experiments were conducted to determine if incubation at various temperatures affected the biological activity of salivary gland extract (Fig. 5). After incubation at 4 and 21 °C for 24 h, 28 °C for 2 h, and 37 °C for 30 min the salivary gland extract retained its biological activity; however, after

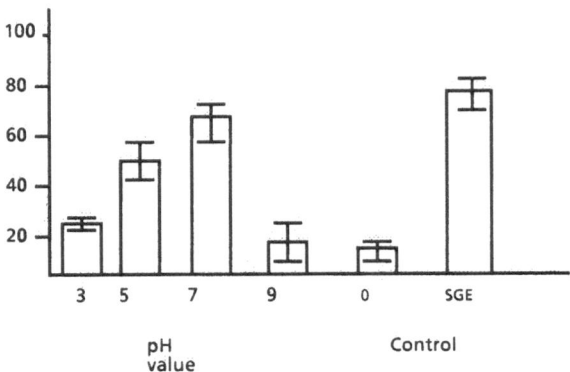

Fig. 4. Guinea pigs were inoculated with THO virus plus salivary gland extract which had been incubated at various pHs; control animals were inoculated with THO virus plus untreated salivary gland extract (*SGE*) or with THO (*0*) virus alone. Enhancement of virus transmission was measured by the number of uninfected ticks that became infected when compared to the ticks feeding on the control guinea pigs. The means and ranges were calculated from those ticks found to contain virus

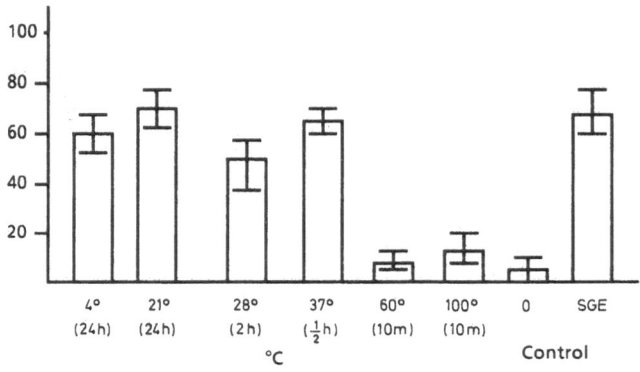

Fig. 5. Guinea pigs were inoculated with THO virus plus salivary gland extract which had been incubated at various temperatures; control animals were inoculated with THO virus plus untreated salivary gland extract (*SGE*) or with THO virus alone (*0*). Enhancement of virus transmission was measured by the number of uninfected ticks that became infected when compared to the ticks feeding on the control guinea pigs. The means and ranges were calculated from those ticks found to contain virus

incubation at 60 and 100 °C for 10 min no enhancement of virus transmission was observed, only 13/115 (11%) uninfected nymphs acquired THO virus.

Effect of pronase and proteinase-k treatment on the biological activity of salivary gland extract

Salivary gland extract was incubated with either pronase or proteinase-k, prior to inoculation into guinea pigs (Fig. 6). Following engorgement, 40/61

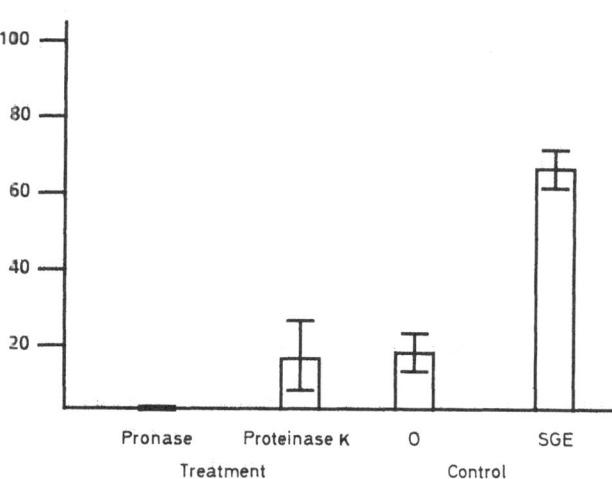

Fig. 6. Guinea pigs were inoculated with THO virus plus salivary gland extract which had been incubated with either pronase or proteinase-k; control animals were inoculated with THO virus plus untreated salivary gland extract (*SGE*) or with THO virus alone (*0*). Enhancement of virus transmission was measured by the number of uninfected ticks that became infected when compared to the ticks feeding on the control guinea pigs. The means and ranges were calculated from thos ticks found to contain virus

(66%) ticks tested acquired virus when feeding on guinea pigs inoculated with virus plus salivary gland extract (incubated at 37 °C), whereas, only 7/118 (6%) ticks acquired virus when feeding on guinea pigs inoculated with virus plus enzyme treated salivary gland extract. Thus, no enhancement of virus transmission was observed when salivary glands were treated with either pronase or proteinase-k.

Discussion

The mechanism by which tick salivary gland constituents enhance virus transmission are unknown. However the inability of the factor(s) to enhance virus transmission in a permissive host or cultured cells suggests that enhancement is not achieved by direct action on the virus, i.e., by protease activity. Previous studies demonstrated that enhancement is a localised response in the host and only occurs when virus and salivary gland are delivered together (as would be the case for virus delivered by infected ticks [7]). Our investigations have focused on the characterization of the tick salivary gland factor(s), which enhances virus transmission.

The process of attachment and feeding of ticks results in changes both in the morphology and in the secretions of their salivary glands [8]. Experiments therefore were conducted to determine if the salivary gland factor(s) that enhance virus transmission was present throughout the tick feeding period. Enhancement of virus transmission was only observed with salivary glands from ticks that had fed for a period of 4–6 days; by day 8 of feeding no enhancement was observed. Thus, the salivary gland factor(s) that enhances virus transmission is not present throughout the tick feeding period.

The tick salivary gland factor(s) is stable after repeated freeze/thaw cycles, at a pH of 5, 6, and 7, and at temperatures of 4–37 °C. Our observation that enhancement of virus transmission was inhibited when salivary gland extract was treated with pronase or proteinase-k, indicates that either a protein(s) or a peptide(s) is responsible (at least in part) for the potentiation of SAT between infected and uninfected ticks.

Determination of the physio-chemical properties of tick salivary gland factor(s) will aid in developing purification procedures. We believe that elucidation of these factor(s) will provide an important insight in to the mechanism of SAT, and enable us to examine how these factor(s) enhance virus transmission. The importance of SAT with regard to the epidemiology of other arthropod-borne viruses is currently being evaluated using other arbovirus-vector-host associations.

Acknowledgements

The authors thank Ms. D. French and Ms. P. Henbest for their help with the tick colony, Ms. S. Clarke for typing the manuscript, and Professor D. H. L. Bishop for his support.

References

1. Albanese M, Bruno-Smiraglia C, Di Di Cuonzo G, Lavagnino A, Srihongse S (1972) Isolation of Thogoto virus from *Rhipicephalus bursa* ticks in Western Sicily. Acta Virol 16: 267
2. Clerx JMP, Fuller F, Bishop DHL (1983) Tick-borne viruses structurally similar to the orthomyxoviruses. Virology 127: 205–219
3. Davies CR, Jones LD, Nuttall PA (1986) Experimental studies on the transmission cycle of Thogoto virus, a candidate orthomyxovirus, in *Rhipicephalus appendiculatus* ticks. Am J Trop Med Hyg 35: 1256–1262
4. Haig DA, Woodall JP, Danskin D (1965) Thogoto virus: a hitherto undescribed agent isolated from ticks in Kenya. J. Gen Microbiol 38: 389–394
5. Jones LD, Davies CR, Steele GM, Nuttall PA (1987) A novel mode of arbovirus transmission involving a nonviraemic host. Science 237: 775–777
6. Jones LD, Davies CR, Steele GM, Nuttall PA (1988) The rearing and maintainance of ixodid and argasid ticks in the laboratory. J Anim Tech 39: 99–106
7. Jones LD, Hodgson E, Nuttall PA (1989) Enhancement of virus transmission by tick salivary glands. J Gen Virol 70: 1895–1989
8. Kemp DH, Stone BF, Binnington KC (1982) Tick attachment and feeding: role of the mouthparts, feeding apparatus, salivary gland secretions and host response. In: Obenchain FD, Galun R (eds) Physiology of ticks. Oxford, Pergamon, pp 119–168
9. Staunton D, Nuttall PA, Bishop, DHL (1989) Sequence analyses of Thogoto viral RNA segment 3: evidence for a distant relationship between an arbovirus and members of the *Orthomyxoviridae*. J Gen Virol 70: 2811–2817
10. Titus RG, Ribeiro JM (1987) Salivary gland lysates from the sand fly *Lutzomyia longipalpis* enhance *Leishmania* infectivity. Science 239: 1306–1308
11. WHO (1985) Arthropod-borne and rodent-borne viral diseases. WHO Tech Rep Ser 719. WHO, Geneva

Arch Virol (1990) [Suppl 1]: 235–241

Serological evidence of the distribution of California serogroup viruses in the U.S.S.R.

A. M. Butenko[1], Irena V. Galkina[1], A. A. Kuznetsov[1], Ludmila V. Kolobukhina[1], S. D. Lvov[2], and Marina S. Nedyalkova[1]

[1]D.I. Ivanovsky Institute of Virology and [2]N.F. Gamaleya Institute of Epidemiology and Microbiology, U.S.S.R. Academy of Medical Sciences, Moscow, U.S.S.R.

Accepted January 18, 1990

Summary. The highest levels of humoral immunity (50%–90%) to California serogroup viruses were found in humans in northern Eurasia in the zone of coniferous forests. In the central regions of European Russia, in the zone of mixed and deciduous forests, the frequency of positive findings was about 30%–40%, and in the forest-steppe and steppe zones and in the arctic regions of Eurasia 10%–20%. Similar results were obtained with serum samples from domestic animals. Other results of these studies provide evidence of sympatric occurrence of Tahyna and Inkoo viruses. The highest activity of Inkoo virus was obtained in central U.S.S.R. and of Tahyna virus in southern U.S.S.R.

Introduction

The California serogroup (family *Bunyaviridae*, genus *Bunyavirus*) at present includes 15 members. Three occur in South America, nine in North America, and one in Africa [2]. In Europe, circulation of Tahyna (TAH) virus in Czechoslovakia, Austria, France, and Norway and of Inkoo (INK) virus in Finland has been established [7]. Data published in the Soviet literature attested to the occurrence in this country of two viruses of the California serogroup: TAH (European part, Caucasus, Central Asia) and INK (Karelia) [4]. In the past three years workers of the Department of Virus Ecology, D.I. Ivanovsky Institute of Virology, also discovered snowshoe hare (SSH) virus (previously known only in Canada and northern U.S.A.) in northern and central areas of Russia, and found an antigenically unusual California serogroup virus (LEIV 12812). Considerable antigenic variability among the recent isolates was demonstrated by comparing them in serological tests with prototype TAH, INK, and SSH virus strains (unpubl. data). Typing of new strains by complement fixation test with antisera to known viruses was shown to be insufficient to accurately identify them. It became evident that demonstration of antibody in human and animal sera by neutralization (N) test or hemagglutination-inhibition test with TAH virus could not be taken as a proof of occurrence of this particular virus in the study area.

In the present work, we studied the prevalence of antibodies to California serogroup viruses in humans living in the Russian Federation and also investigated the distribution and circulation of these viruses.

Materials and methods

Serum samples were collected from humans (5713) and domestic animals (973) in 1986–1988. These were tested by N with TAH (strain 92), INK (strain KN-3641), SSH (original strain) viruses and with strain LEIV 12812 (isolated from mosquitoes of the genus *Aedes* collected in Kalinin region, 1986). Neutralization tests were performed according to a micromethod [6] in continuous culture of pig embryo kidney (SPEV) cells grown in polystyrene panels (Costar, Cambridge, U.K.). The viruses were allowed to remain in contact with 1:10 dilution of sera for 1 h at 37 °C. Working dilutions of viruses contained 100, 1000 or 10,000 CPD_{50}. Results were read on the third day (preliminary) and again 5–6 days later by examining the wells under an inverted microscope. Cytopathic effects were recorded according to a subjective system. Serum samples that neutralized 100 CPD_{50} or more virus were considered to be positive. For determinations of antibody titers, sera diluted 1:10 (heat-inactivated 30 min/56 °C) were tested in 2-fold dilutions against 100 CPD_{50} of virus.

Results

By N tests, the highest frequency of positive responses for TAH virus (50%–80%) was demonstrated in humans living in northern regions (Kirov, Arkhangelsk, Kostroma, Ivanovo) in the zone of coniferous forests of the European U.S.S.R. and Yakut Autonomous Republic, Eastern Siberia (Table 1). In the central part, including Moscow, Vladimir, Gorkiy, and Penza and in the Mordovian A.S.S.R. (zones of mixed and deciduous forests), antibody prevalences were 30%–40%, and in the forest-steppe and steppe zones of the European U.S.S.R. and in Arctic areas (Murmansk, Low-Kolyma district of Yakut Republic) they were 10%–20%.

Similar results were obtained with sera from domestic animals tested for antibody to TAH virus (Table 2). The additional use of INK, SSH viruses, and strain LEIV 12812 in N tests revealed, mostly with INK virus, a considerable number of sera with antibody to these California serogroup viruses (Table 3). Comparative serological studies were made with seropositive specimens from residents of the Volga–Vyatka watershed. Group-specific responses were observed in 60%–80%. A considerable proportion of sera (10%–27%) from Yaroslavl, Kirov, Volgograd, Gorkiy, and Penza and from Mordovian A.S.S.R. selectively neutralized only INK virus or TAH

Table 1. Tests of serum samples from humans residing in the Russian federation by neutralization test with Tahyna virus

Natural zone, administrative unit	No of sera tested	% positive	Natural zone, administrative unit	No of sera tested	% positive
European part			Forest-steppe		
Tundra and coniferous forests			Penza	105	35.2
Murmansk	102	17.6	Mordovian ASSR	224	31.3
			Voronezh	105	20.0
Coniferous forests			Lipetsk	105	20.0
Kirov	90	71.1	Tambov	198	18.2
Arkhangelsk	379	52.8	Belgorod	166	18.1
Komi ASSR	367	44.7			
Karelian ASSR	30	23.6	Steppe		
Vologda	150	21.3	Volgograd	207	15.9
Leningrad	27	7.4			
			Semideserts		
Mixed forests			Kalmyk ASSR	102	14.7
Kostroma	205	81.5			
Ivanovo	230	50.0	**Asiatic part**		
Novgorod	77	44.2	Tundra		
Vladimir	132	43.2	Chukotka Auton.	139	26.6
Yaroslavl	264	43.2			
Kalinin	121	43.0	Tundra and coniferous forests		
Gorkiy	187	42.2	Yakut ASSR	443	58.0
Pskov	102	40.2	Magadan	165	38.8
Moscow	63	38.1			
Bryansk	215	20.9	Coniferous forests		
Kaluga	168	12.5	Kamchatka	201	46.3
Smolensk	173	12.1	Sakhalin	146	21.2
Deciduous forests			Total	5713	36.2
Ryazan	76	27.6			
Kursk	105	21.9			
Orel	59	18.6			
Tula	70	17.1			

ASSR Autonomous Soviet Socialist Republic

virus (6%–12%) (Table 4). Three sera from Gorkiy region reacted only with LEIV 12812 and four sera (one from Yaroslavl and three from Kirov) only with SSH virus.

In the Lower Volga area (Volgograd region and Kalmyk A.S.S.R.), virus-specific antibody to INK virus was detected in 18.6% and 6.3%, respectively. Frequently it was found that sera reacted only with TAH virus or with TAH

Table 2. Tests of serum samples from domestic animals by neutralization test with Tahyna virus

Natural zone administrative unit	Animal species	Number of sera tested	positive	% positive
European part				
Tundra, coniferous forests				
Murmansk	bovine	97	18	18.6
	reindeer	52	11	21.1
	pig	34	3	8.8
Coniferous forests				
Arkhangelsk	bovine	63	43	68.2
	reindeer	27	23	85.2
Kirov	bovine	151	83	55.0
Komi ASSR	bovine	99	48	48.5
	reindeer	24	8	33.3
Forest-steppe				
Belgorod	bovine	37	9	24.3
Voronezh	bovine	32	6	18.7
Lipetsk	bovine	73	7	9.6
Steppe				
Volgograd	bovine	32	3	9.4
Asiatic part				
Tundra				
Chukotka Auton.	reindeer	62	9	14.5
Coniferous forests				
Kamchatka	reindeer	33	3	9.1
Total		973	298	30.6

ASSR Autonomous Soviet Socialist Republic

and SSH virus. Positive reactions with SSH virus may be explained by the presence of antigenic relationships between this virus and TAH virus (Table 5). We also titrated antibody in sera from residents of Kirov and Penza regions and Mordovian A.S.S.R. which had been shown to contain group-specific antibody to all California serogroup viruses. Most frequent findings were fourfold or higher titers to INK virus, fewer with higher titers to TAH virus, and others with similar titers to TAH, INK, and SSH viruses. Five of six sera from residents of Kirov region reacted with TAH, INK, and SSH viruses but did not neutralize the LEIV 12812 strain. In a hyperendemic area including Yaroslavl, Kostroma, and Gorkiy regions, antibodies to California serogroup viruses were found in 74% of men and 67% of women. Immune status was age-dependent: less than 20 years 47.8%, 21–30 years

Table 3. Neutralizing antibodies to California serogroup viruses in humans and bovines in the Volga–Vyatka watershed (eastern European part of the U.S.S.R.) and Yakut Autonomous Republic (eastern Siberia)

Administrative unit	No. of sera examined	% of positive findings with viruses	
		TAH	TAH, INK, SSH, LEIV-12812 (total)
Human			
Kostroma	161	85.1	91.4
Kirov	90	71.1	87.7
Yaroslavl	264	43.2	62.9
Gorkiy	187	42.2	55.6
Penza	105	35.2	44.8
Mordovian ASSR	224	31.3	37.5
Volgograd	207	15.9	20.8
Kalmyk ASSR	107	14.7	15.7
Yakut ASSR	443	58.0	66.0
Bovine			
Kirov	151	55.0	89.4
Mordovian ASSR	157	15.3	41.9
Volgograd	32	9.4	18.7

ASSR Autonomous Soviet Socialist Republic

Table 4. Specificity of neutralizing antibodies to California serogroup viruses Tahyna (TAH), Inkoo (INK), snowshoe hare (SSH) and to strain LEIV-12812 in humans residing in Volga–Vyatka watershed (eastern European part of the U.S.S.R.) and Yakut to Autonomous Republic (eastern Siberia)

Administrative unit	Total no. of positive sera	% virus-specific reactions with				% sera reacting with viruses			
		TAH	INK	SSH	LEIV 12812	TAH, INK	TAH, SSH	INK SSH	ALL
Gorkiy	104	11.5	16.3	0	2.9	18.3	4.8	2.9	43.2
Kalmyk ASSR	16	12.5	6.3	0	n.d.	0	31.3	0	50.0
Kirov	79	6.3	10.1	3.8	0	17.7	2.6	5.1	54.5
Kostroma	128	8.6	6.3	0	0	46.1	3.9	0.8	34.4
Mordovian ASSR	84	11.9	14.3	1.2	n.d.	33.3	3.6	1.2	34.5
Penza	47	4.3	19.1	0	n.d.	21.3	2.1	2.1	51.1
Volgograd	43	2.3	18.6	0	n.d.	2.3	39.5	4.6	32.6
Yaroslavl	166	9.0	27.1	0.6	0	27.7	0.6	3.6	31.3
Yakut ASSR	299	16.4	8.0	6.4	n.d.	15.7	14.1	1.3	38.1

ASSR Autonomous Soviet Socialist Republic
n.d. Not done

Table 5. Titration of antibodies in residents from the Volgograd region and Kalmyk ASSR by neutralization tests with Tahyna, Inkoo, and snowshoe hare viruses

Serum no.	Titer to virus		
	TAH	INK	SSH
Volgograd Reg.			
294	≥320	10	10
303	80	10	<10
327	80	80	20
364	80	40	<10
392	20	40	<10
408	160	10	20
409	160	40	20
443	≥320	80	160
Kalmyk ASSR			
1733	160	80	80
1739	40	<10	20
1836	320	40	160
1963	160	10	40
1991	160	<10	40
1994	80	<10	40
1995	320	<10	20
1998	160	<10	20
2007	160	<10	20

68.2%, 31–40 years 72.5%, 41+ years 65.0%. Those less than 20 years old more frequently had virus-specific N antibody than older subjects (40.9% and 31.5%, respectively).

In 1986–1988, N tests with paired sera from 609 patients with obscure fibrile diseases (mainly, in central European U.S.S.R.) were used to determine 51 illnesses associated with viruses of the California serogroup. In nine of these the etiological role of INK virus was established, in seven TAH virus, but for the remaining 35 no specific determination of the causative agent could be established.

Discussion

The data presented here agree with the results of seroepidemiological studies aimed at the elucidation of TAH virus circulation in Azerbaijan, Uzbekistan, Astrakhan region [5] and in Czechoslovakia [3] as well as with studies of INK virus in Finland [1]. In addition, our data provide the first evidence of the simultaneous circulation of TAH and INK viruses in the same areas of Eurasia. The highest activity of INK virus was observed in central areas of the European U.S.S.R., and that of TAH virus in southern areas. Individual sera from residents of northern regions of the European U.S.S.R. reacted only with SSH virus or only with the LEIV 12812 strain, suggesting infections by these viruses.

Results of N tests with virus-specific sera from naturally infected humans and animals indicate the existence of antigenic differences between TAH, INK, SSH viruses and strain LEIV 12812. It is important to emphasize that a significant number of sera with antibody to TAH virus or with antibody to both TAH and SSH viruses showed no reaction with LEIV 12812 strain. In endemic regions, the proportion of the population immune to viruses of the California serogroup increased with age, with higher probability of infection with one, two, or several related viruses (types) resulting in formation of group-specific antibodies. If high level of humoral immunity in adults is a factor restricting the disease, children may constitute the principal population susceptible to these viruses. In carrying out seroepidemiological surveys and serological diagnosis of the diseases associated with California serogroup viruses in the European and Asian U.S.S.R., at least two endemic viruses, TAH and INK, should be used. In addition, it would be advised to include SSH virus and antigenic variants of all these viruses in future serosurveys and serodiagnostic tests.

References

1. Brummer-Korvenkontio M, Saikku P (1975) Mosquito-borne viruses in Finland. Med Biol 53: 279–281
2. Calisher CH (1983) Taxonomy, classification and geographic distribution of California serogroup bunyaviruses. In: Calisher CH, Thompson WH (eds) California serogroup viruses. AR Liss, New York, pp 1–16
3. Hubalek Z, Bardos N, Medek M, Kania V, Kychler L, Selinck E (1979) Tahyna virus-neutralizing antibodies of patients in Southern Moravia. CS Epidemiol Microbiol Immunol 28: 87–96
4. Lvov DK, Gromashevsky VL, Skvortsova TM, Sidorova SS, Berezina LK, Vladimirtseva EA (1986) Arboviruses and some other natural focality viruses in the USSR. In: Gaidamovich, SYa (ed) Arboviruses. D.I. Ivanovsky Institute of Virology Acad Med Sci, Moscow, pp 15–21
5. Semenov BF, Ismailova ST, Karaseva PS, Yakubov ShKh, Maksumov SS, Chunikhin SP, Butenko AM, Bashkirtsev VN (1968) Serological evidence of Tahyna virus circulation in some regions of the USSR. In: Proc 15th Sci Conf of Institute of Poliomyelitis and Viral Encephalitides of the U.S.S.R. Acad Med Sci 3, pp 260–261
6. Sullivan EJ, Rosenbaum MJ (1967) Methods for preparing tissue culture in disposable microplates and their use in virology. Am J Epidemiol 85: 424–437
7. Karabatsos N (ed) (1985) International catalogue of arboviruses including certain other viruses of vertebrates, 3rd edn. American Society of Tropical Medicine and Hygiene, San Antonio, Texas

Authors' address: A. M. Butenko, Department of Virus Ecology, D.I. Ivanovsky Institute of Virology, U.S.S.R. Academy of Medical Sciences, Gamaleya Street 16, Moscow 123098, U.S.S.R.

Arch Virol (1990) [Suppl 1]: 243–247

Signs and symptoms of infections caused by California serogroup viruses in humans in the U.S.S.R.

Lyudmila V. Kolobukhina, D. K. Lvov, A. M. Butenko, Marina S. Nedyalkova, A. A. Kuznetsov, and Irena V. Galkina

D.I. Ivanovsky Institute of Virology, U.S.S.R. Academy of Medical Sciences, Moscow, U.S.S.R.

Accepted January 23, 1990

Summary. We analyzed 49 of 51 human infections caused by California serogroup viruses. People mainly affected were 14–30 years old. In 32 (65.2%) the disease was influenza-like, without symptoms of central nervous system involvement, 17 (34.8%) developed aseptic meningitis. In the influenza-like disease the principal signs and symptoms included high fever of short duration, headache, weakness. In this group, three patients with general symptoms of infection showed meningeal signs (meningismus) in the first days of the disease: increased intracranial pressure without changes in the cell composition of cerebrospinal fluid. Four patients had paravascular changes in the lungs. Among signs and symptoms in patients with aseptic meningitis severe headache, nausea, vomiting, meningeal signs predominated. In cerebrospinal fluid the number of cells was within several hundreds per ml^3, lymphocytes predominating, the level of protein normal or decreased slightly. In all the cases the disease was of moderate severity and the outcome favourable.

Introduction

California serogroup viruses (family *Bunyaviridae*, genus *Bunyavirus*) were first isolated in the U.S.A. [4]. Later, Tahyna virus was demonstrated to occur in Europe [1], and the role of this virus in infectious diseases has been established [2, 7, 8].

In recent years, circulation of California serogroup viruses Tahyna, Inkoo, and snowshoe hare was shown to occur in the U.S.S.R. The purpose of the work reported here was to establish the role of these viruses in infectious diseases during the season of activity of hematophagous mosquitoes (May–September) and to characterize the clinical picture of the diseases.

Materials and methods

Serum samples were collected from patients with acute febrile diseases and neuroinfection (aseptic meningitis) in May–September 1986–1988 in Moscow City, Kirov, Vologda, Murmansk, Belgorod, Orel regions, and in the Mordovia A.S.S.R. in the European U.S.S.R. Murmansk, Kirov, and Vologda regions are located in the zone of coniferous forests, Moscow region in the zone of mixed forests, Orlov region in the zone of leaved forests, Belgorod region and the Mordovia A.S.S.R. in the forest-steppe zone. Paired acute- and convalescent-phase serum samples from 609 patients (11–60 years old) were examined. The criterion of the laboratory diagnosis of the diseases associated with California serogroup viruses was demonstration of four-fold or greater increase in titer between paired sera detectable in one or several tests with Tahyna (strains 92), Inkoo (strain KN-3641), or snowshoe hare (original strain) viruses: neutralization test in SPEV cell cultures [10], hemagglutination-inhibition test [3]. IgM antibody to CAL serogroup viruses were assayed by MAC ELISA method [6] using Tahyna virus antigen and mouse Tahyna virus-specific antibody conjugated with peroxidase. Assays for IgG antibody were performed by indirect "sandwich" method [5].

Results

Among 609 patients examined, in 51 the etiological role of California serogroup viruses was established. Principal clinical symptoms of the disease were determined from the analysis of 49 case histories.

Young patients predominated: 42 were 11 to 30 years old. The monthly distribution of cases was as follows: 7 in May, 10 in June, 14 in July, 9 in August, 9 in September.

Table 1. Principal signs and symptoms of illnesses associated with California serogroup virus infections of humans in the U.S.S.R., 1986–1988

Frequency of symptoms n = 32 Type 1 (influenza-like)	Signs and symptoms	Frequency of symptoms n = 17 Type 2 (aseptic meningitis)
8	37.0–38.5 °C fever	2
14	38.6–39.5 °C fever	10
10	39.6–40.5 °C fever	5
32	headache	17
32	weakness	17
3	meningeal signs	17
11	nausea, vomiting	17
30	scleritis, conjunctivitis	15
15	dizziness	14
17	retroorbital pain	14
1	rash	2

We observed two types of clinical illness. Type 1 patients were those whose disease was influenza-like, without the involvement of the central nervous system (32 cases), type 2 patients were those with aseptic meningitis (17 cases). In patients of type 1 the disease was characterized by acute onset, chills in all the patients, high febrile response (in 24 patients above 38.5 °C). The main symptoms included frontotemporal headache which in 17 patients was accompanied by retroorbital pains, 11 patients were nauseous and vomited. These symptoms were observed in the first two or three days of the illness and disappeared with the fall to normal of body temperature. All patients showed marked weakness into the convalescence. Objective signs included pale skin, in one case maculopapular rash on the trunk which disappeared within two days without desquamation or pigmentation. Nearly all patients had scleritis, conjunctivitis. Lung auscultation revealed harsh breathing in 16 among 32 patients. The pulse rate corresponded to the fever level, the arterial blood pressure was within the range of 100/60–120/70 mm Hg. Analysis of electrocardiograms done for all the patients showed no significant changes. Five patients had slight enlargement of the liver but biochemical tests showed no functional disorders in any of them. Dyspeptic signs were only in two patients and were limited to loose stools, irregular abdominal pains without deficite localization observed in the first three days of illness. Three patients in the acute period had meningeal signs: nucchal rigidity, Kernig's sign. Studies of their cerebrospinal fluids revealed increased intracranial pressure without changes in the cell composition and normal or slightly reduced (0.264–0.297 g/l) protein levels. Four type 1 patients developed pneumonia. Clinical features of the pneumonia included scanty physical symptoms such as harsh breathing, and local dry rales. X-rays showed paravascular changes against a background of clearly visible vessels. The peripheral blood showed normal or slightly increased (13×10^9/l) number of leukocytes, lack of neutrophil shift and a moderate rise in the erythrocyte sedimentation rate. Rapid regression of pneumonic infiltration was typical: no pathology was found in control X-rays on the 10th day of illness.

Duration of the febrile period in type 1 patients varied from three to seven days, the fever declining within 24–36 h; seven patients, however, had subfebrile temperature in the evening for two-three days. Duration of the illness was from seven to ten days.

Clinical symptoms of type 2 patients (Table 1) were characterized by acute onset without prodromal signs, but with chills and rapid rise of temperature to 38.5–39 °C. The principal complaint was headache, beginning in the first hours of illness, nausea and vomiting. The headache was intense and without definite localization. Vomiting, usually recurrent, appeared either during the first day of illness or on the second or third day. All patients in this group presented with meningeal signs, which were frequently dissociated and of short duration. The cerebrospinal fluid in all the cases was transparent, colorless, with lymphocytic pleocytosis, a normal or slightly lowered level of protein (0.297 g/l), normal sugar, weakly positive Pandy's and Nonne–Apelt's reactions.

Examinations revealed face hyperemia, scleritis in all patients, whereas hyperemia of buccopharyngeal mucosa was weak. We also observed moderate relative bradicardia, occasionally lowered arterial blood pressure, but never below 90/60 mm Hg. The febrile period in type 2 patients varied from five to seven days, the duration of the illness from 20 to 27 days. The criterion for discharge of type 2 patients was subjective and objective improvement, positive dynamics and trend for normalization of cerebrospinal fluid.

The outcomes of all these illnesses were favorable, ending in complete recovery. In two patients, one year (the observation period) after aseptic meningitis astheno-neurotic symptoms such as weakness, undue fatiguability, emotional lability were noted.

Discussion

Our clinical observations agree with the clinical picture of diseases caused by Tahyna viruses as described by others [2, 7, 8] and also suggest that Inkoo virus is an etiological agent in some infectious diseases in the U.S.S.R. Analysis of the clinical picture of illnesses associated with California serogroup viruses (Tahyna, Inkoo) detected in the European U.S.S.R attests to the polymorphism of clinical manifestations and the lack of pathognomonic symptoms. The clinical picture of aseptic meningitis is similar to that of aseptic meningitis associated with other viruses, enteroviruses among them. The illness may have a two-phase course. The first phase is manifested by fever, toxicity symptoms, the second is characterized by meningeal syndrome with changes in cerebrospinal fluid indicating aseptic meningitis [9]. However, a small number of observations and the lack of proper examinations and time-course observations of patients in some cases thus far permit no definite conclusions concerning the clinical pattern of this group of infections in the U.S.S.R. Studies along these lines are underway.

References

1. Bardos V, Danielova VI (1959) The Tahyna virus, a virus isolated from mosquitoes in Czechoslovakia. S Hyg Epidemiol Immunol (Praha) 3: 264–276
2. Bardos V, Medek M, Kania V, Hubalek Z (1975) Isolation of Tahyna virus from the blood of sick children. Acta Virol 19: 447
3. Clarke DH, Casals J (1958) Techniques for hemagglutination and hemagglutination-inhibition with arthropod-borne viruses. Am J Trop Med Hyg 7: 561–573
4. Hammon WMcD, Reeves WC (1952) California encephalitis virus, a newly described agent. Calif Med 77: 303–309
5. Ivanov AP, Rezapkin GV, Dzagurova TK, Tkachenko AE (1984) Indirect solid-phase immunosorbent assay for detection of arenavirus antigen and antibodies. Acta Virol 28: 240–245

6. Jamnback TL, Beaty BJ, Hildreth SW, Brown KL, Gunderson CB (1982) Capture immunoglobulin M system for rapid diagnosis of La Crosse (California encephalitis) virus infections. J Clin Microbiol 16: 577–580

7. Lvov DK, Kostyukov MA, Pak TP (1977) Isolation of Tahyna virus (California antigenic group, family *Bunyaviridae*) from the blood of febrile patients in the Tajik SSR. Vopr Virusol 6: 682–684

8. Lvov DK, Sidorova GA, Gromashevsky VL (1984) New for the U.S.S.R viral natural-focality human diseases. Med Parasitol 2: 86–92

9. Parsonson IM, McPhee DA (1985) Pathogenesis of bunyavirus infection. Adv Virus Res 30: 279–313

10. Sullivan EJ, Rosenbaum MJ (1967) Methods for preparing tissue culture in disposable microplates and their use in virology. Am J Epidemiol 85: 424–437

Authors' address: Lyudmila V. Kolobukhina, D.I. Ivanovsky Institute of Virology, U.S.S.R. Academy of Medical Sciences, 16 Gamaleya Street, Moscow 123098, U.S.S.R.

Arch Virol (1990) [Suppl 1]: 249–258

Arbovirus activity in Canada

H. Artsob

Zoonotic Diseases, Bureau of Microbiology, Laboratory Center for Disease Control, Ottawa, Ontario, Canada

Accepted January 29, 1990

Summary. Nineteen arboviruses have been isolated in Canada, including 11 mosquito-transmitted, 6 tick-transmitted and 2 culicoides-transmitted viruses. Only 6 of these viruses have been documented to have caused symptomatic infections in humans in Canada—western equine encephalitis, St. Louis encephalitis (SLE), snowshoe hare (SSH), Jamestown Canyon (JC), Powassan, and Colorado tick fever viruses. Western equine encephalitis has been the most important arbovirus in Canada, where clinical disease has been recognized, primarily in the prairie provinces, since the 1930's. SLE virus was first isolated in Saskatchewan in 1971 but disease in Canada due to SLE was not documented until 1975 when an outbreak of 66 cases occurred in Ontario and one case each occurred in Manitoba and Quebec. California serogroup viruses have been found throughout Canada; SSH is the most widespread serotype but the JC serotype is also prevalent. Clinical California serogroup virus infections have been recognized in Canada, due primarily to the SSH serotype; JC encephalitis has also been documented. Twelve tick-transmitted symptomatic infections have been diagnosed from 1958–1988. These include 11 Powassan virus infections from Ontario, Quebec, and New Brunswick and a case of Colorado tick fever from Alberta.

Introduction

Nineteen arboviruses were isolated in Canada from 1938–1987. These include 11 mosquito-transmitted, 6 tick-transmitted and 2 culicoides-transmitted viruses (Table 1). The 11 mosquito-transmitted arboviruses in Canada have been isolated from seven provinces as well as the Northwest Territories and the Yukon Territory (Fig. 1); virus isolation has not been attempted from Nova Scotia, New Brunswick or Prince Edward Island. Tick- and culicoides-transmitted arboviruses have been documented in 4 Canadian provinces (Fig. 2).

Table 1. Alphabetical list of arboviruses isolated in Canada

Virus	Taxonomic status[a]	Antigenic group[a]	Principal vector	Reference[b]
Avalon	Nairovirus	Sakhalin	tick	[43]
Bauline	Orbivirus	Kemerovo	tick	[42]
Bluetongue type 11	Orbivirus	Bluetongue	culicoides	[22]
Cache Valley	Bunyavirus	Bunyamwera	mosquito	[28]
California encephalitis	Bunyavirus	California	mosquito	[4]
Colorado tick fever	Orbivirus	Colorado tick fever	tick	[10, 24]
Eastern equine encephalitis	Alphavirus	A	mosquito	[60]
Epizootic hemorrhagic disease	Orbivirus	Epizootic hemorrhagic disease	culicoides	[19]
Flanders	Rhabdoviridae	Hart Park	mosquito	[30]
Great Island	Orbivirus	Kemerovo	tick	[42]
Jamestown Canyon	Bunyavirus	California	mosquito	[32]
Northway	Bunyavirus	Bunyamwera	mosquito	[49]
Powassan	Flavivirus	B	tick	[53]
Silverwater	Bunyavirus-like	Kaisodi	tick	[48]
Snowshoe hare	Bunyavirus	California	mosquito	[47]
St Louis encephalitis	Flavivirus	B	mosquito	[13]
Trivittatus	Bunyavirus	California	mosquito	[67]
Turlock	Bunyavirus	Turlock	mosquito	[28]
Western equine encephalitis	Alphavirus	A	mosquito	[26]

[a] According to International Catalogue of Arboviruses [35]
[b] Refers to earliest published report of isolation in Canada

Arboviruses isolated in Canada of known human disease-causing potential include eastern equine encephalitis, western equine encephalitis (WEE), St. Louis encephalitis (SLE), Powassan (POW), California encephalitis, Jamestown Canyon (JC), snowshoe hare (SSH), and Colorado tick fever (CTF) viruses. In addition, Avalon virus has been implicated as a human pathogen [66]. Dengue 1 and 4 viruses have been isolated from patients in Canada but these individuals had histories of recent travel to the Caribbean [8, 23].

This paper reviews arboviruses that have been documented to cause human disease in Canada.

Western equine encephalitis

Historically, WEE virus has been the arbovirus of greatest human and veterinary importance in Canada. It has been isolated in the four westernmost provinces and in western Ontario (Fig. 1); a WEE seroconversion

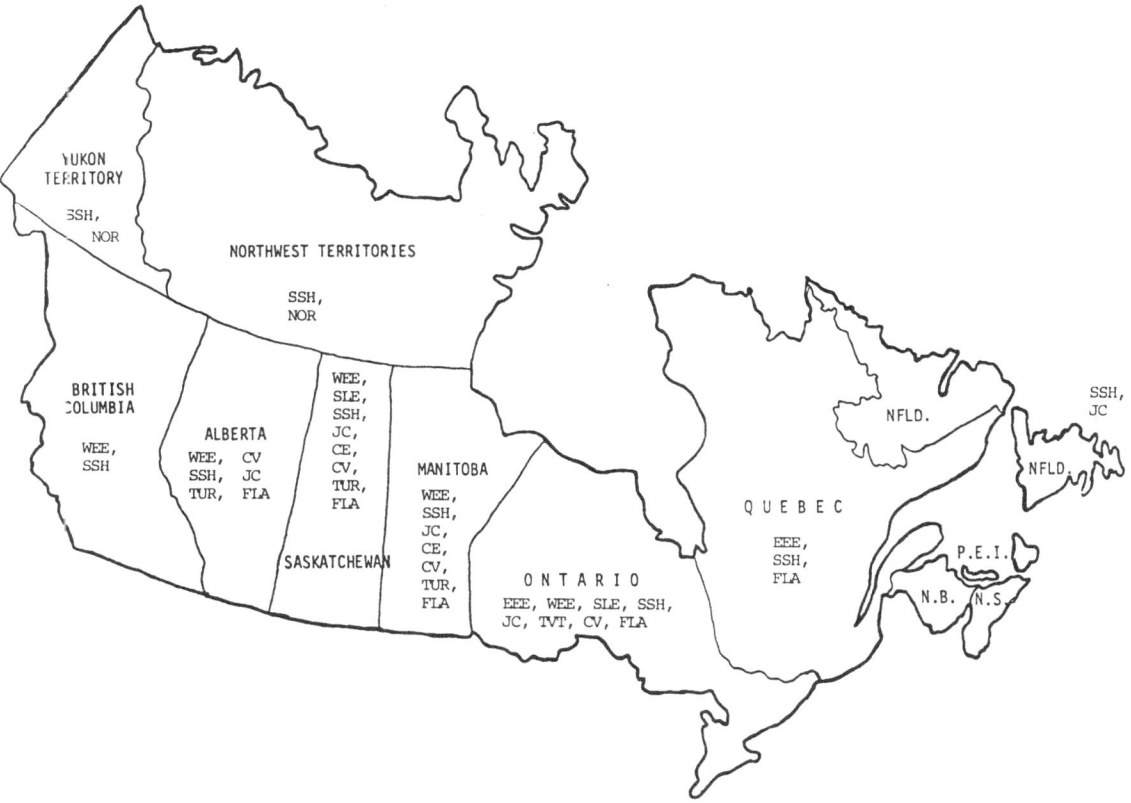

Fig. 1. Mosquito-transmitted arboviruses isolated in Canada. *CE* California encephalitis *CV* Cache Valley, *EEE* eastern equine encephalitis, *FLA* Flanders, *JC* Jamestown Canyon, *NOR* Northway, *SSH* snowshoe hare, *SLE* St. Louis encephalitis, *TUR* turlock, *TVT* trivittatus, *WEE* western equine encephalitis

has been reported in a 20 month old infant in the province of Quebec [58]. Strongest WEE activity traditionally has been in the provinces of Manitoba and Saskatchewan.

Studies to define the natural cycle(s) of WEE virus in the Canadian prairie provinces have included assessments of potential reservoirs. Examination of birds, known amplifying hosts of WEE virus, has resulted in the isolation of WEE virus from English sparrows, Swainson's hawk and mourning dove [15, 56] and in the demonstration of neutralizing antibodies to WEE virus in chickens [33], turkeys [12], wild ducks [11], and 14 other wild bird species [15].

Mammals implicated in the natural cycle of WEE virus in Canada have included the Richardson ground squirrel, *Spermophilus richardsonii*, and the snowshoe hare, *Lepus americanus*. Several WEE virus isolations have been obtained from the Richardson ground squirrel [15, 27, 38] and a high seroprevalence to WEE virus (11.6%) was documented in these rodents during the epidemic year of 1965 [38]. Laboratory studies have indicated that this rodent has the potential to serve as an amplifying host for WEE

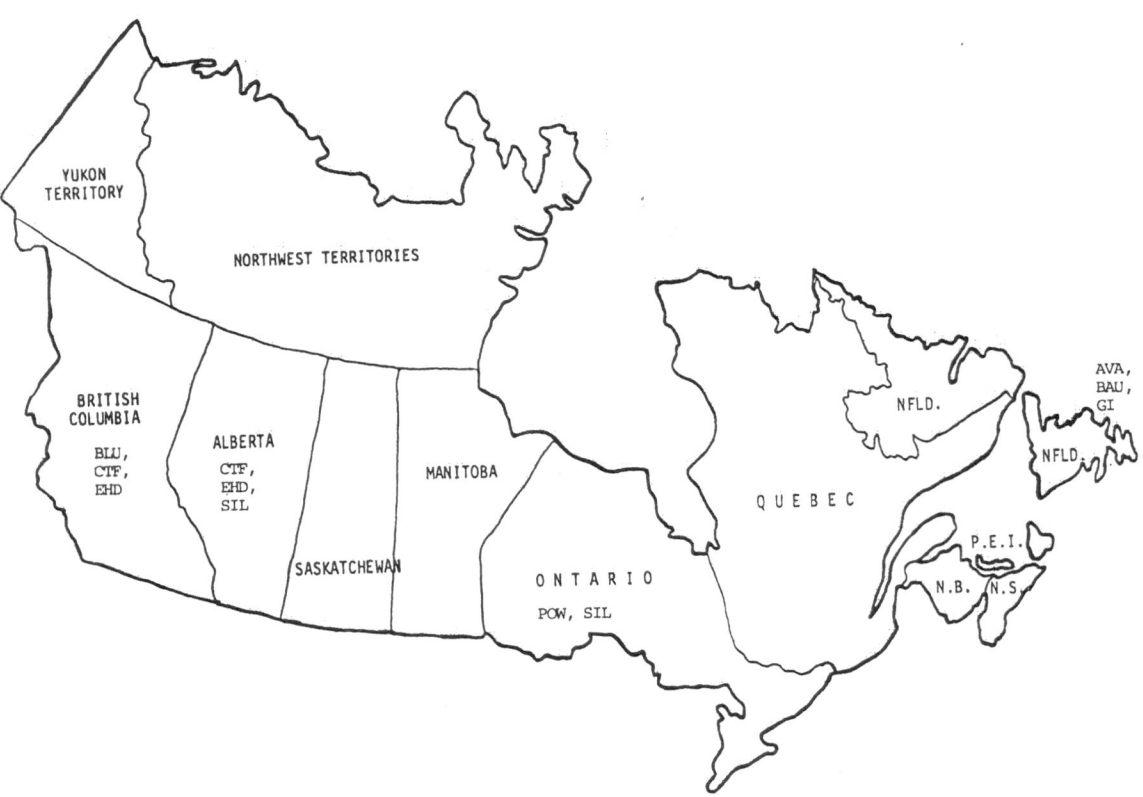

Fig. 2. Tick- and culicoides-transmitted arboviruses isolated in Canada. *AVA* Avalon, *BAU* Bauline, *BLU* Bluetongue, *CTF* Colorado tick fever, *EHD* epizootic hemorrhagic disease, *GI* Great Island, *POW* Powassan, *SIL* Silverwater

virus [39]. Late spring epizootics of WEE in snowshoe hares in Alberta in 1963 and 1965 preceded by two months outbreaks in humans and horses [69, 70] leading to the suggestion that snowshoe hares may have acted as amplifying hosts of WEE virus.

Isolatates of WEE virus have been obtained from garter snakes (*Thamnophis sirtalis parietalis, T. radix haydeni*) and leopard frogs (*Rana pipiens*) in Saskatchewan [14] and neutralizing antibodies to this virus have been demonstrated in them [14, 59, 64]. Their role as possible overwintering hosts of WEE virus in Canada has been postulated.

Five different genera of Canadian mosquitoes have yielded isolates of WEE virus, including *Aedes, Anopheles, Coquillettidia, Culex,* and *Culiseta* [3, 63]. *Culex tarsalis* is clearly the most important mosquito species in Canada for the epidemic transmission of WEE virus [3].

There have been more than 1500 symptomatic infections due to WEE virus in humans in Canada with the earliest diagnosed cases having been noted in 1938 [3]. The most important outbreak year was 1941 when over 1000 cases occurred in the Canadian prairie provinces [17, 20, 34, 46]. Sporadic outbreaks of WEE have been recognized in Canada in every decade since the 1930's. The most recent epidemic occurred in 1983 when 18 cases were diagnosed in the province of Manitoba [61].

St. Louis encephalitis

St. Louis encephalitis virus was first isolated in Canada in 1971 when it was recovered from a pool of *Culex tarsalis* in Saskatchewan during routine surveillance for WEE virus [13]. Subsequently the virus was isolated from human brain in 1975 during a SLE epidemic in Ontario [65] and from *Culex* species mosquitoes collected in Essex county, Ontario in 1976 [67]. Studies in Ontario in 1975 and 1976, following an epidemic of SLE, revealed a high antibody sero-prevalence in English sparrows, likely amplifying hosts during the outbreak, and antibodies to SLE virus were detected in various migratory bird species from 1976–1978 [21].

Human disease caused by SLE virus was first recognized in Canada in 1975 during an unprecedented outbreak of SLE on the North American continent when 66 cases were diagnosed in southwestern Ontario [41] and 1 case each were found in the provinces of Quebec [9] and Manitoba [62]. Four cases of SLE were recognized in southwestern Ontario in 1976 within the focal area of the previous epidemic [41] and 2 cases were reported in 1977 from the province of Manitoba [68].

There has been no clear evidence of SLE virus activity in Canada since 1977 and SLE is not generally considered to be a public health problem in Canada.

California (CAL) serogroup viruses

Snowshoe hare virus has been isolated from mosquitoes collected in the Yukon Territory, Northwest Territories, and 7 Canadian provinces (Fig. 1) and neutralizing antibodies to SSH virus have been demonstrated in animals in the 3 other provinces [1]. SSH virus thus is the most prevalent CAL serogroup virus in Canada.

The JC serotype also is distributed widely. It has been isolated from mosquitoes in 5 provinces (Fig. 1), a human infection due to JC virus has been recognized in the Northwest Territories [6], and neutralizing antibodies to JC virus have been demonstrated in moose in New Brunswick [45] and Nova Scotia [44].

Human disease in Canada due to CAL serogroup viruses was first documented in 1978 when 4 infections were diagnosed—3 from Quebec [25] and 1 from Ontario [5]. In all instances the SSH serotype was implicated as the infecting virus. In 1981 the JC serotype was shown to have caused encephalitis in an Ontario resident [18].

From 1978 to 1989 at least 1 symptomatic infection due to a CAL serogroup virus has been recognized every year in Canada and the total number of humans with encephalitis caused by CAL viruses now exceeds 20. Most of these have been due to the SSH serotype; however at least 2 have been due to JC virus.

Serosurveys have revealed extensive human infection by CAL serogroup viruses throughout Canada with prevalence rates from 0.5 to 31.9% [1]. Most serosurveys have indicated that the SSH serotype infects the greatest number of individuals. An exception was in the province of Newfoundland where human exposure to JC virus was more prevalent [57].

Seroconversions to SSH virus have been documented in 2 Canadian horses with encephalitis [31, 40]; both were yearlings. Perhaps age is as significant a risk factor in horses as it is in humans, in whom SSH virus encephalitis has been documented mainly in children [1].

Powassan

Powassan virus has been isolated in the province of Ontario from several sources, including human brain [53], *Ixodes cookei* and *Ixodes marxi* ticks [7, 50–52, 54] and from the blood of woodchucks, *Marmota monax* [7, 50], and red squirrel, *Tamiasciurus hudsonicus* [51]. Studies in northern Ontario indicated that red squirrels and woodchucks may be important reservoirs of POW virus in Canada [50–52, 54, 55]. Neutralizing antibodies to POW virus have been demonstrated in numerous mammals in Ontario, Alberta, and British Columbia [2].

From 1958–1988, there have been 11 diagnosed cases of POW encephalitis in Canada including 5 from Ontario, 5 from Quebec and 1 from New Brunswick [2; W. M. Fitch and H. Artsob, unpubl. data]. Eight of the 11 Canadian cases occurred in children less than 10 years old and only 1 case of POW encephalitis has been recognized in a Canadian more than 20 years old; 6 have been in males and 5 in females. Two patients expired during the acute phase of the disease and 2 died 1 and 3 years after onset from sequelae. Sequelae have been reported in 4 of the 7 survivors.

Colorado tick fever

Colorado tick fever virus was isolated from *Dermacentor andersoni* ticks collected in Alberta in 1952–54 [10, 24] and in British Columbia in 1965–66 [29]. However, symptomatic infection in humans due to CTF virus was not recognized in Canada until 1986 when a case was diagnosed in a 20 year old woman from southwestern Alberta [16]. Serosurveys revealing exposure of British Columbia residents to CTF virus [36, 37] suggest that human disease due to CTF virus is underrecognized in Canada.

Conclusion

Arboviruses do not pose a major public health hazard in Canada. The only epidemic arbovirus of continuing concern is WEE virus which causes recurrent outbreaks in western Canada, principally Manitoba and Saskatchewan. In addition, CAL serogroup viruses and POW virus have an endemic presence that puts Canadians at slight risk during arbovirus season. However, activity of these viruses is difficult to control so that, for the foreseeable future, Canadians will be at low risk to arbovirus infection.

References

1. Artsob H (1983) Distribution of California serogroup viruses and virus infections in Canada. In: Calisher CH, Thompson WH (eds) California serogroup viruses. Prog Clin Biol Res 123: 277–290
2. Artsob H (1988) Powassan encephalitis. In: Monath TP (ed) The arboviruses: epidemiology and ecology, vol 4. CRC Press, Boca Raton, pp 29–49
3. Artsob H, Spence L (1979) Arboviruses in Canada. In: Kurstak E (ed) Arctic and tropical arboviruses. Academic Press, New York, pp 39–66
4. Artsob H, Spence L, Calisher CH, Sekla LH, Brust RA (1985) Isolation of California encephalitis serotype from mosquitoes collected in Manitoba, Canada. J Am Mosq Control Assoc 1: 257–258
5. Artsob H, Spence L, Caughey WG, Wherrett JR (1981) Aseptic meningitis in Ontario. Can Med Assoc J 125: 958–962
6. Artsob H, Spence L, Sekla L (1985) Arbovirus activity in Canada in 1984. Can Med Assoc J 133: 1026
7. Artsob H, Spence L, Surgeoner G, McCreadie J, Thorsen J, Th'ng C, Lampotang V (1984) Isolation of *Francisella tularensis* and Powassan virus from ticks (Acari: Ixodidae) in Ontario, Canada. J Med Entomol 21: 165–168
8. Artsob H, Spence L, Th'ng C, Gillani A (1978) Dengue fever in travellers returning to Canada. Can Dis Weekl Rep 4: 18–19
9. Artsob H, Spence L, Th'ng C, West R (1980) Serological survey for human arbovirus infections in the province of Quebec. Can J Public Health 71: 341–346
10. Brown JH (1955) Colorado tick fever in Alberta. Can J Zool 33: 389–390
11. Burton AN, Connell R, Rempel JG, Gollop JB (1961) Studies on western equine encephalitis associated with wild ducks in Saskatchewan. Can J Microbiol 7: 295–302
12. Burton AN, McLintock J (1970) Further evidence of western encephalitis infection in Saskatchewan mammals and birds and in reindeer in northern Canada. Can Vet J 11: 232–235
13. Burton AN, McLintock J, Francy DB (1973) Isolation of St Louis encephalitis and Cache Valley viruses from Saskatchewan mosquitoes. Can J Public Health 64: 368–373
14. Burton AN, McLintock J, Rempel JG (1966) Western equine encephalitis virus in Saskatchewan garter snakes and leopard frogs. Science 154: 1029–1031
15. Burton AN, McLintock JR, Spalatin J, Rempel JG (1966) Western equine encephalitis in Saskatchewan birds and mammals 1962–1963. Can J Microbiol 12: 133–141
16. Cimolai N, Anand CM, Gish GJ, Calisher CH, Fishbein DB (1988) Human Colorado tick fever in southern Alberta. Can Med Assoc J 139: 45–46
17. Davison RO (1942) Encephalomyelitis in Saskatchewan, 1941. Can J Public Health 33: 388–398
18. Deibel R, Srihongse S, Grayson MA, Grimstad PR, Mahdy MS, Artsob H, Calisher CH

(1983) Jamestown Canyon virus: the etiologic agent of an emerging human disease? In: Calisher CH, Thompson WH (eds) California serogroup viruses. Prog Clin Biol Res 123: 313–325

19. Ditchfield J, Debbie JG, Karstad LH (1964) The virus of epizootic hemorrhagic disease of deer. Trans North Am Wildl Nat Res Conf 29: 196–201

20. Donovan CR, Bowman M (1942) Epidemiology of encephalitis: western equine type, Manitoba, 1941. Can Med Assoc J 46: 525–530

21. Dorland R, Mahdy MS, Artsob H, Prytula A (1979) Wild bird surveillance in the study of arboviral encephalitis. In: Mahdy MS, Spence L, Joshua JM (eds) Arboviral encephalitides in Ontario with special reference to St. Louis encephalitis. Ontario Ministry of Health, Toronto, pp 208–224

22. Dulac GC, Dubuc C, Afshar A, Myers DJ, Bouttard A, Shapiro J, Shettigara PT (1988) Consecutive outbreaks of epizootic haemorrhagic disease of deer and bluetongue. Vet Rec 122: 340

23. Duperval R, Marcoux JA (1982) Imported dengue type 4 infection from Haiti-Quebec. Can Dis Weekly Rep 8: 9–10

24. Eklund CM, Kohls GM, Brennan JM (1955) Distribution of Colorado tick fever and virus-carrying ticks. JAMA 157: 335–337

25. Fauvel M, Artsob H, Calisher CH, Davignon L, Chagnon A, Skvorc-Ranko R, Belloncik S (1980) California group virus encephalitis in three children from Quebec: clinical and serologic findings. Can Med Assoc J 122: 60–63

26. Fulton JS (1938) A report of two outbreaks of equine encephalomyelitis in Saskatchewan. Can J Comp Med 2: 39–46

27. Gwatkin R, Moynihan IW (1942) Search for sources and carriers of equine encephalomyelitis virus. Can J Res 20: 321–337

28. Hall RR, McKiel JA, Brown JH (1968) Isolation of Turlock virus and a member of the Bunyamwera group, probably Cache Valley virus, from Alberta mosquitoes. Can J Public Health 59: 159–160

29. Hall RR, McKiel JA, Gregson JD (1968) Occurrence of Colorado tick fever virus in *Dermacentor andersoni* ticks in British Columbia. Can J Public Health 59: 273–275

30. Hall RR, McKiel JA, McLintock J, Burton AN (1969) Arboviruses from Saskatchewan mosquitoes-isolation of a member of the Flanders-Hart Park group and of a strain as yet unidentified. Can J Public Health 60: 486–488

31. Heath SE, Artsob H, Bell RJ, Harland J (1989) Equine encephalitis caused by snowshoe hare (California serogroup) virus. Can Vet J 30: 669–671

32. Hoff GL, Yuill TM, Iversen JO, Hanson RP (1969) Snowshoe hares and the California encephalitis virus group in Alberta, 1961–1968. Bull Wildl Dis Assoc 5: 254–259

33. Hoff GL, Yuill TM, Iversen JO, Hanson RP (1970) Selected microbial agents in snowshoe hares and other vertebrates of Alberta. J Wildl Dis 6: 472–478

34. Jackson FW (1942) Encephalitis. Can Med Assoc J 47: 364–365

35. Karabatsos (ed) (1985) International catalogue of arboviruses, 3rd edn. The American Society of Tropical Medicine and Hygiene, San Antonio, Texas

36. Kettyls GD, Verrall VM, Hopper JMH, Kokan P, Schmitt N (1968) Serological survey of human arbovirus infections in southeastern British Columbia. Can Med Assoc J 99: 600–603

37. Kettyls GD, Verrall VM, Wilton LD, Clapp JB, Clarke DA, Rublee JD (1972) Arbovirus infections in man in British Columbia. Can Med Assoc J 106: 1175–1179

38. Leung MK, Burton A, Iversen J, McLintock J (1975) Natural infections of Richardson's ground squirrels with western equine encephalomyelitis virus, Saskatchewan, Canada, 1964–1973. Can J Microbiol 21: 954–958.

39. Leung MK, Iversen J, McLintock J, Saunders JR (1976) Subcutaneous exposure of the

Richardson's ground squirrel (*Spermophilus richardsonii* Sabine) to western equine encephalomyelitis virus. J Wildl Dis 12: 237–246

40. Lynch JA, Binnington BD, Artsob H (1985) California serogroup virus infection in a horse with encephalitis. J. Am Vet Med Assoc 186: 389

41. Mahdy MS, Spence L, Subrahmanyan TP, Joshua JM (1979) Arbovirus encephalitides in Ontario with special reference to St. Louis encephalitis, an overview. In: Mahdy MS, Spence L, Joshua JM (eds) Arboviral encephalitides in Ontario with special reference to St. Louis encephalitis. Ontario Ministry of Health, Toronto, pp 343–355

42. Main AJ, Downs WG, Shope RE, Wallis RC (1973) Great Island and Bauline: two new Kemerovo group arboviruses from *Ixodes uriae* in eastern Canada. J Med Entomol 10: 229–235

43. Main NJ, Downs WG, Shope RE, Wallis RC (1976) Avalon and Clo Mor: two new Sakhalin group viruses from the north Atlantic. J Med Entomol 13: 309–315

44. McFarlane BL, Embil JA, Artsob H, Spence L, Rozee KR (1981) Antibodies to the California group of arboviruses in the moose (*Alces alces americana* Clinton) population of Nova Scotia. Can J Microbiol 27: 1219–1223

45. McFarlane BL, Embree JE, Embil JA, Rozee KR, Artsob H (1982) Antibodies to the California group of arboviruses in animal populations of New Brunswick. Can J Microbiol 28: 200–204

46. McGugan AC (1942) Equine encephalomyelitis (western type) in humans in Alberta, 1941. Can J Public Health 33: 148–151

47. McKiel JA, Hall RR, Newhouse VF (1966) Viruses of the California encephalitis complex in indicator rabbits. Am J Trop Med Hyg 15: 98–102

48. McLean DM (1961) Silverwater virus: characterization of virus isolated from ticks collected in eastern Canada. Fed Proc 20: 443

49. McLean DM (1979) Arbovirus vectors in the Canadian Arctic. In: Kurstak E (ed) Arctic and tropical arboviruses. Academic Press, New York, pp 7–19

50. McLean DM, Best JM, Mahalingam S, Chernesky MA, Wilson WE (1964) Powassan virus: summer infection cycle, 1964. Can Med Assoc J 91: 1360–1362

51. McLean DM, Bryce Larke RP (1963) Powassan and Silverwater viruses: ecology of two Ontario arboviruses. Can Med Assoc J 88: 182–185

52. McLean DM, Cobb C, Gooderham SE, Smart CA, Wilson AG, Wilson WE (1967) Powassan virus: persistence of virus activity during 1966. Can Med Assoc J 96: 660–664

53. McLean DM, Donohue WL (1959) Powassan virus: isolation of virus from a fatal case of encephalitis. Can Med Assoc J 80: 708–711

54. McLean DM, Smith PA, Livingstone SE, Wilson WE, Wilson AG (1966) Powassan virus: vernal spread during 1965. Can Med Assoc J 94: 532–536

55. McLean DM, Vos A, Quantz EJ (1964) Powassan virus: field investigations during the summer of 1963. Am J Trop Med Hyg 13: 747–753

56. McLintock J, Burton AN, Dillenberg H, Rempel JG (1966) Ecological factors in the 1963 outbreaks of western encephalitis in Saskatchewan. Can J Public Health 57: 561–575

57. Mokry J, Artsob H, Butler R (1984) Studies on California serogroup virus activity in Newfoundland, Canada, 1980–83. Mosquito News 44: 310–314

58. Pavilanis V, Wright IL, Silverberg M (1957) Western equine encephalomyelitis: report of a case in Montreal. Can Med Assoc J 77: 128–130

59. Prior MG, Agnew RM (1971) Antibody against western equine encephalitis virus occurring in the serum of garter snakes (Colabridae: Thamnophis) in Saskatchewan. Can J Comp Med 35: 40–43

60. Schofield FW, Labzoffsky N (1938) Report on cases of suspected encephalomyelitis occurring in the vicinity of St. George. Rep Ont Dept Agric Ont Vet Coll 29: 25–29

61. Sekla L, Eadie JA (1984) Manitoba arbovirus surveillance 1983. Can Dis Weekly Rep 10: 94–95

62. Sekla L, Stackiw W (1976) Laboratory diagnosis of western encephalomyelitis. Can J Public Health 67 [Suppl 1]: 33–39

63. Sekla L, Stackiw W, Brust RA (1980) Arbovirus isolations from mosquitoes in Manitoba. Mosquito News 40: 377–380

64. Spalatin J, Connell R, Burton AN, Gollop BJ (1964) Western equine encephalitis in Saskatchewan reptiles and amphibians, 1961–1963. Can J Comp Med 28: 131–142

65. Spence L, Artsob H, Grant L, Th'ng C (1977) St Louis encephalitis in southern Ontario – laboratory studies for arboviruses. Can Med Assoc J 116: 35–36

66. Suinat JL, Leon A, Rendoing J (1985) Polyradiculonevrite apres morsures de tiques. Role possible d'un nouvel agent pathogene chez l'homme: le virus Avalon. La Presse Med 14: 1616

67. Thorsen J, Artsob H, Spence L, Surgeoner G, Helson B, Wright R (1980) Virus isolations from mosquitoes in southern Ontario, 1976 and 1977. Can J Microbiol 26: 436–440

68. Waters JR (1978) Encephalitis surveillance – Manitoba 1977. Can Dis Weekly Rep 4: 46–48

69. Yuill TM, Hanson RP (1964) Serologic evidence of California encephalitis virus and western equine encephalitis virus in snowshoe hares. Zoonoses Res 153–164

70. Yuill TM, Iversen JO, Hanson RP (1969) Evidence for arbovirus infections in a population of snowshoe hares: a possible mortality factor. Bull Wildl Dis Assoc 5: 248–253

Author's address: H. Artsob, Zoonotic Diseases, Bureau of Microbiology, Laboratory Center for Disease Control, Tunney's Pasture, Ottawa, Ontario K1A 0L2, Canada

Arch Virol (1990) [Suppl 1]: 259–266

Surveillance for arboviruses in the Soviet Union: relationships between ecologic zones and virus distribution

[1]D. Lvov, [1]A. Avershin, [1]V. Andreev, [1]Valeriya Aristova, [1]A. Butenko, [1]Olga Voltsit,
[1]Irina Galkina, [1]V. Gromashevsky, [1]G. Dmitriev, [1]Y. Kondaurov,
[1]Nina Kondrashina, [1]A. Kuznetsov, [2]S. Lvov, [1]Tatyana Morozova,
[1]Tatyana Skvortsova, and [1]Marina Shchipanova

[1]D.I. Ivanovsky Institute of Virology and [2]N.F. Gamaleya Institute of Epidemiology
and Microbiology,
U.S.S.R. Academy of Medical Sciences, Moscow, U.S.S.R.

Accepted January 19, 1990

Summary. Results are presented of a systematic search in the territory of the U.S.S.R. for natural foci of arboviruses. We used ecological surveillance of territories through different landscape zones taking into account some climatic factors, determining the distribution of vectors and vertebrate hosts and the reproduction of different arboviruses therein. Surveillance was performed from latitude 72° to 40° North in arctic, subarctic, north taiga, south taiga, leaf-bearing forest, steppe and subtropical zones. Special attention was paid to the specific features of distribution of viruses of the California and Bunyamwera serogroups. For this purpose we collected specimens for virological and serological studies. Collections were made simultaneously in each of nine meridional sections of European and Asian U.S.S.R. covering about 7.5 million km². We collected 1,256,000 mosquitoes, about 20,000 ticks, 3,500 vertebrates, more than 1,000 specimens from patients, and 20,000 serum specimens from asymptomatic humans and domestic animals. In this paper we discuss peculiarities of the distribution of arboviruses in different landscape climatic zones of northern Eurasia.

Introduction

The geographical distribution of arboviruses in nature is determined by a combination of biological and nonbiological factors. In a given area, climate influences the occurrence of arboviruses and weather and other factors affect the intensity of transmission during a given "arbovirus season". We examined some of these parameters in 1984–88 during an extensive surveillance of disparate areas of the U.S.S.R.

Materials and methods

During 1984–88 surveillance was performed by simultaneously collecting material by meridional sections $2-3 \times 10^3$ km long in latitudes from 72° to 40°–50° North. There were nine such sections: five in European and four in Asian areas of the U.S.S.R. In an area covering about 7.5×10^6 km², 1,250,000 mosquitoes, about 27,000 ixodid (Ixodidae and Argasidae) ticks, 3,500 vertebrates, more than 1,000 specimens (paired sera, blood) from patients and about 20,000 asymptomatic human and domestic animal sera have been examined. Ecological surveillance included seven basic landscape-climatic zones: (1) arctic, (2) subarctic (tundra), (3) north taiga, (4) south taiga, (5) leaf-bearing forest, (6) steppe, (7) subtropical (Table 1). The isolation of viruses was performed with the use of suckling mice. Materials of choice were studied by solid phase ELISA. The identification of the isolates was done by electron microscopy, complement fixation and neutralization with reference sera by conventional techniques. Serologic survey with the use of hemagglutination-inhibition (HI) (tick-borne encephalitis, Getah, Karelian fever, Sindbis viruses) and neutralization in tissue culture (California, Uukuniemi serogroups, Batai viruses) were performed by conventional methods.

Results

We isolated 90 viruses from natural sources. Those included 6 togaviruses, 10 flaviviruses, more than 22 bunyaviruses, 6 orbiviruses, 2 each rhabdoviruses,

Table 1. Principal landscape-climatic complexes of nothern Eurasia

Complex	Zone, subzone	Quantity of days $t^0 \geqslant 20\,°C$	$\sum_{t^0} \geqslant 10\,°C$ [a]	Isotherma of July
I	Arctic	0	0	0–2
	1. Central (Inner) arctic	0	0	0
	2. arctic desert	0	0	0–2
II	Subarctic	0–<10	0–600	2–13
	1. arctic tundra	0	0	2–5
	2. moss-lichen tundra	0	0–200	5–8
	3. shrub tundra	0	200–400	8–10
	4. forest tundra	<10	400–600	10–13
III	North taiga	<10–15	800–1600	13–18
	1. northern taiga	<10–10	600–1200	13–16
	2. middle taiga	10–15	1200–1600	16–18
IV	South taiga	15–20	1600–1800	18–19
V	Leaf-bearing forest	20–60	1800–3000	19–21
	2. leaf-bearing forest	20–30	1800–2400	19–20
	2. forest-steppe	30–60	2400–3000	20–21
VI	Steppe	60–120	3000–3800	21–26
	1. steppe	60–90	3000–3200	21–22
	2. semidesert	90–120	3200–3800	24–26
VII	Subtropical (dry and humid)	120–150	3800–5000	26–30

[a] Sum of effective temperatures more than 10 °C

picornaviruses, herpesviruses, and paramyxoviruses, 1 poxvirus, and 37 orthomyxoviruses. Cases of Karelian, Issyk-Kul, Syr-Daria Valley, Tamdy, Powassan, Sindbis, West Nile, and California serogroup virus fevers were found [3, 4]. The activity of arboviruses belonging to different ecological and serological groups in their natural foci is presented in Table 2.

Arctic

Viruses were not isolated.

Subarctic

A low degree of activity of California serogroup, Batai-like virus and Getah-like virus were found. High activity of natural foci of Uukuniemi, Sakhalin, and Kemerovo serogroups connected with *Ixodes uriae* ticks in mass nesting of sea-birds in the ocean islands and sea coast were revealed.

North taiga

High activity of California serogroup viruses was revealed essentially everywhere, and high activity of Karelian fever and Getah-like viruses were found in West and East, respectively. Activity, although low, was determined for Batai-like virus and, in the southern part of middle taiga subzone, tick-borne encephalitis virus.

South taiga

Conditions are optimal for the main vector of tick-borne encephalitis virus, *Ixodes persulcatus*. Evaporation is lower than the annual sum of precipitation, therefore the abundance of marshes, temporary and constant water reservoirs. Weather here ensures favorable conditions for mass propagation of mosquitoes, mainly *Aedes* genus. The sum of effective temperatures is insufficient for the completion of external incubation period for arboviruses transmitted by mosquitoes of equatorial, subequatorial and tropical zones. Among mosquito-borne viruses, moderate activity of California serogroup and low activity of Batai and Getah viruses were determined here.

Leaf-bearing forest

In this zone the conditions for ticks *I. ricinus*, the main vector of Uukuniemi virus and a supplementary vector of tick-borne encephalitis virus, are optimal. According to our data tick-borne encephalitis virus is firmly connected with *I. persulcatus* populations, even in the extreme north-western parts of the area. The infectivity of ticks is high (1% and more). Within this natural zone lies the border of Ixodidae ticks, genera *Haemaphysalis* and

Table 2. Activity of arboviruses in principle landscape-climatic zones of northern Euroasia

Complex	Zone	Activity of the viruses transmitted by									
		Mosquitoes					Ticks				
		Virus serogroups or types					Virus serogroups or types				
		California	Bunyamwera	Sindbis	Semliki	Japanese enceph.	tick-borne enceph.	C-CHF	Uukuniemi	Sakhalin	Kemerovo
I	Arctic	–	–	–	–	–	–	–	–	–	–
II	Subarctic	+	–	–	+	–	–	–	+++	+++	++
III	North taiga	+++	++	++	++	–	+	–	+++	+++	+++
IV	South taiga	++	+	–	+	–	+++	–	+	–	+
V	Leaf-bearing forest	++	++	+	+	+	+	–	++	–	+
VI	Steppe	+	+++	+	+	+	–	+	–	–	–
VII	Subtropics	+	?	+	?	++	–	++	–	–	–

– Not revealed; + weak; ++ temperate; +++ high; ? not revealed needs examination

Dermacentor. The indices of evaporation and moisture are equalized. This adversely affects the places where mosquitoes propagate. Their populations are considerably less in comparison with the northern regions but the number of species is higher. The number of gonotrophic cycles is increased. Populations of Anopheline mosquitoes are increased. Perhaps this is a reason for moderate activity of Batai virus, mainly connected with mosquitoes of the tribe Anophelini, as well as moderate activity of California serogroup viruses, principally connected with mosquitoes of the genus *Aedes*. In the southern parts of the zone the activity of Japanese encephalitis virus has been registered.

Steppe

In this zone evaporation exceeds the annual sum of precipitation. That determines arid climate, absence of sizable marshes and temporary water reservoirs and, as a consequence of it, a deficit of the places of mosquitoes propagation. But high sums of effective temperatures and variable composition of vectors and vertebrates determine wide opportunities for circulation of many viruses ecologically connected with mosquitoes, culicoides and ixodoides (Ixodidae and Argasidae) ticks. Within this landscape zone lies the northern border of the area of *Argas* ticks and of the majority of *Amblyomma* spp. The activity of Batai-like virus of Bunyamwera serogroup is high here. The virus is connected with *Anopheles* mosquitoes and domestic animals. The infectivity for humans has been demonstrated. There are reports of its role in human pathology. A low degree of activity of California serogroup, West Nile, Sindbis, and Getah viruses and, among tick-borne viruses, Congo-Crimean hemorrhagic fever virus has been determined.

Subtropical

This zone covers a sizable territory of the southern part of the country. It is mainly represented by dry arid subtropics of desert type. Only two spots in Transcaucasia may be considered as wet subtropics. The sum of effective temperatures is high enough for replication of those viruses in arthropods, temperature thresholds of which lies within 16–24 °C. Species variability of vectors — mosquitoes, *Culicoides*, *Phlebotomus*, *Argas* and *Ixodes* ticks — is great. The number of *Chiroptera* spp. is considerable and density of bats populations and of *Argas* and *Ixodes vespertilonis* ticks connected with them is high here and there. Sindbis and West Nile viruses circulate in subtropical climate-landscape zone. A sizeable territory is endemic for Congo-Crimean hemorrhagic fever. The etiology of Issyk-Kul, Tamdy, Syr-Daria Valley fever has been disclosed [3, 4].

Some features of the distributions of California and Bunyamwera serogroup viruses in the territory of physico-geographical region of the

Russian Plain should be mentioned. California serogroup viruses predominate in the north, including south-taiga zone; with advance to the south virus activity distinctly decreases (Fig. 1). However, single strains, as well as human illnesses, are seen up to the Subtropical zone. In southern regions the virus may be isolated in the end of May; however, infection rates of mosquitoes increase in July and August. Batai virus has been found in the north taiga zone where it is associated with mosquitoes of the genus *Aedes*. In leaf-bearing forest zone and especially in steppe zone its activity distinctly increases. In the south regions the virus is connected only with Anopheline mosquitoes. The virus was isolated at the beginning of July, a month later than those of the California serogroup. The mosquito infection rate increases by August, achieving high indices at the end of this month (Fig. 2). The results of serologic studies of humans and domestic animals correspond to the basic regularities of mosquito infection rates (Fig. 3). The fraction of the population immune to California serogroup viruses reaches 50 per cent (in some places to 70% in the north taiga complex) reducing to 15% with advance to the south. In humans and domestic animals (cattle, reindeers, pigs) the percentage of individuals with virus neutralizing antibodies is the same (Fig. 3). Immunity to Batai virus in humans does not usually exceed 3–5%, in some instances 10–12%. In domestic animals in south regions the percentage of detection of antibodies is 20–30% (Fig. 4). That fits with the known data of preferential feeding of *Anopheles* mosquitoes on domestic animals. In this case, as in the case of malaria, domestic animals may be biological barriers for transmission of Batai virus to humans. A great variety of antigenic variants is found among more than 100 isolated strains of California serogroup; the biological properties of these strains also differ widely. The majority of these isolates are most closely related to Tahyna, Inkoo, and snowshoe hare viruses. In a preliminary study a dominance of strains related to snowshoe hare virus was registered in the north, to Tahyna virus in central-west regions and to Inkoo virus in south-east regions. Several strains are very unusual in their antigenic properties and are candidates for new representatives of the complex. The general picture is complicated by a great number of antigenically transitional variants among these viruses. Perhaps, they are natural reassortants.

Discussion

The ecology of arbovirus distribution and the interrelations of virus populations and their vertebrate and invertebrate hosts mainly depends on the influence of nonbiological, principally climatic, factors. Understanding these processes, and with due regard for the peculiarities of viral genome organization, allows the use of methods permitting studies of the movement of genetic material within common gene pools of individual viruses (species). Mutation and reassortment can in theory produce a diversity of biological variants

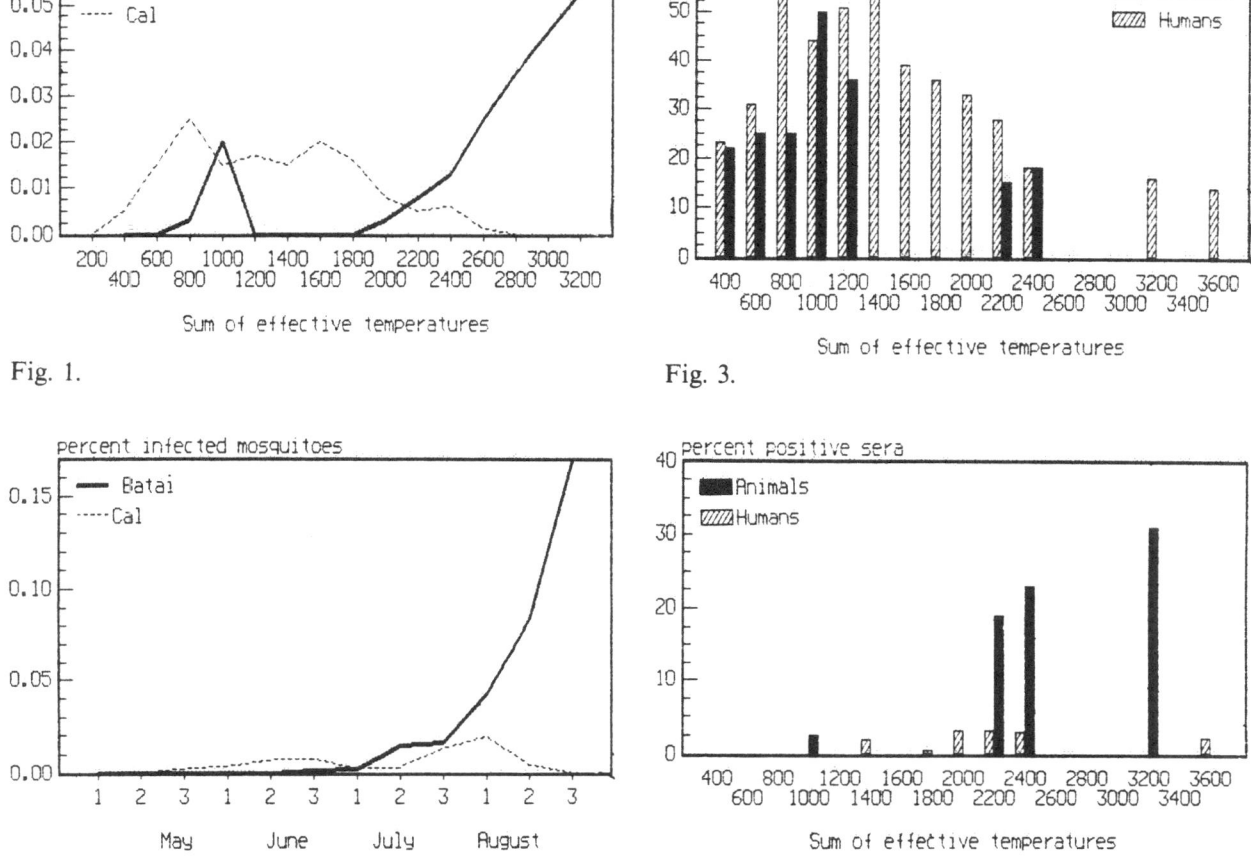

Fig. 1.   Fig. 3.

Fig. 2.   Fig. 4.

Fig. 1. Infection rate of mosquitoes by viruses of the California serogroup and Batai virus in the European part of the U.S.S.R.

Fig. 2. Monthly infection rate of mosquitoes in the south part of the Russian Plain

Fig. 3. Immune structure to California serogroup in different landscapes of the Russian Plain

Fig. 4. Immune structure to Batai virus in different landscapes of the Russian Plain

(biovars). Exchange of genetic material between viruses provides a basis for evolution of new viral populations. Such closely-related viruses can be considered a single species. Further development in field ecology and advances in molecular epidemiology provide approaches to the understanding of the location of ecologic niches of populations, mechanisms by which changes occur in the properties of these ecosystems, and processes by which ecosystems and viruses evolve. Answers to these questions are necessary to explain and to predict conditions leading to the development of epizootic and epidemic situation. In order to get a better understanding of the spread and evolution of arboviruses one has to have more information on fauna, paleogenesis, and landscape zones. For example, the origin of *I. ricinus* as well as of other species of this tick complex is connected with South-Eastern Asia. The formation of its area in the Pliocene era is connected with the genesis of European mesophilic forests. At the end of the Pliocene and in the Pleistocene era it had been pushed towards the west, together with broadleaved and mixed forests, by pressure of taiga from North and steppes from South [2]. Nowadays *I. ricinus* maintains its connection with landscapes including Pliocene flora elements. It is likely that its connection with Uukuniemi virus is equally ancient. In any case, in areas where *I. ricinus* is found, including Talysh mountains in Transcaucasis, this virus was isolated. The formation of *I. persulcatus* area is connected with taiga landscape development. According to Philippova the species area in the upper Pliocene perhaps was limited to Eastern regions and only in the glacial period it has been broadened in a western direction [1].

Concerning California serogroup viruses, one has the impression that it may be time to consider revising its classification. Obviously, serologic studies at this stage are not enough. It is necessary to carry out sequence studies and other methods of molecular epidemiology and to compare strains isolated in different physico-geographic lands of American and Eurasian continents.

References

1. Filipova NA (1971) Species of group *Ixodes persulcatus* (Parasitiformes, Ixodidae). 6. Peculiarities of areas of *I. pavlovsky* Pomerantsev and *I. persulcatus* Schulze in connection with Paleogenesis. Parasitology 5: 305–391
2. Filipova NA (1973) Species of group *Ixodes persulcatus* (Parasitiformes, Ixodidae). 7. Paleogenesis of south branch of group *Ixodes persulcatus* (Pom.) and relationships with *I. ricinus* (L.). Parasitology 7: 3–13
3. Lvov DK (1987) Natural foci of arboviruses in the USSR. Sov Med Rev Virol 1: 153–196
4. Lvov DK, Klimenko SM, Gaidamovich SYa (1989) Arboviruses and arboviral infections. Meditsina, Moscow

Authors' address: D. K. Lvov, D. I. Ivanovsky Institute of Virology, Academy of Medical Sciences of the U.S.S.R., Gamaleya Str. 16, Moscow 123098, U.S.S.R.

Arch Virol (1990) [Suppl 1]: 267–275

Natural foci of arboviruses in far northern latitudes of Eurasia

S. D. Lvov[1], V. L. Gromashevsky[2], V. P. Andreev[2], Tatyana M. Skvortsova[2],
Nina G. Kondrashina[2], Tatyana N. Morozova[2], A. D. Avershin[2], Valeriya A. Aristova[2],
G. A. Dmitriev[2], Y. K. Kandaurov[2], A. A. Kuznetsov[2], Irina V. Galkina[2],
Svetlana S. Yamnikova[2], and Marina V. Shchipanova[2]

[1]N.F. Gamaleya Institute of Epidemiology and Microbiology and
[2]D.I. Ivanovsky Institute of Virology, Academy of Medical Sciences of the U.S.S.R.,
Moscow, U.S.S.R.

Accepted February 7, 1990

Summary. From 1985–1988 we did virologic and serologic studies of hematophagous arthropods, small mammals, cattle and humans in northeastern Europe and Asia, i.e., in the arctic desert, in the moss-lichen and shrub tundra and forest tundra of the subarctic zones, and in northern and middle taiga of the Boreal region. California and Bunyamwera (*Bunyaviridae, Bunyavirus*) serogroup viruses were isolated from mosquitoes in all locations studied. In the northern Far East Getah-like virus of Semliki Forest antigenic complex (*Togaviridae, Alphavirus*) was also isolated from mosquitoes. Viruses from ticks *Ixodes uriae* were isolated Tyuleniy (*Flaviviridae, Flavivirus*), Zaliv Terpeniya (*Uukuvirus*), Okhotsky (*Reoviridae, Orbivirus*) in the European part (eastern Fennoscandia) and additional Sakhalin, Paramushir, Clo Mor, Rukutama (*Bunyaviridae, Nairovirus*, Sakhalin serogroup), Aniva (*Reoviridae, Orbivirus*, Kemerovo serogroup) in the northern Far East. Tick-borne encephalitis and Tyuleniy virus strains were isolated from rodents. We obtained evidence that Zaliv Terpeniya and Tyuleniy viruses, which usually are tick-borne, "splash" from rookeries to the mainland and we found that mosquitoes are involved in circulation there. We also found evidence for a role of California serogroup viruses in human diseases. Thus, we recognized circumpolar distribution of these viruses. In addition, there was a uniformly high level of antibody to these viruses in humans and cattle. Finally, we noted seasonal and regional characteristics of activity of viruses within these complex natural foci.

Introduction

The term "far northern latitudes" is usually applied to arctic and subarctic areas including the north of taiga landscape zone. In America and in the Far

East of the U.S.S.R. these territories are to the north from latitude 50 °N on the whole and in the other territories of Eurasia – to the north from latitude 60 °N. Among abiotic factors the most influential upon biologic systems of far northern latitudes are temperature, light and precipitation. The most important limiting factor with respect to arboviruses is the temperature deficit. In some cases it limits the ability of most arboviruses to replicate in vectors, in other cases the vectors themselves cannot exist.

The Central Arctic is outside continental land. Here the circulation of influenza viruses and paramyxoviruses connected with birds is possible. The arctic desert includes extreme northern parts of land in Asia and America. It is possible that natural foci of viral infections connected with birds, lemmings, and polar foxes exist here.

Four tundra subzones: arctic, moss-lichen, shrub tundra, and forest-tundra we shall call the subarctic. The majority of biological species (mostly birds) have circumpolar distribution. Obligatory bird parasites including *Ixodes uriae* ticks, are typical in the numerous seashore colonies of birds. The northern limit of the arbovirus area probably lies in the subzone of moss-lichen tundra with the sum of effective temperatures up to 200 °C. The number of vectors rises sharply in subzones of shrub tundra and forest-tundra where species of vertebrates characteristic for this zone begin to prevail over the tundra species.

To the north of the temperate zone we regard northern and middle taiga. Relatively high sums of effective temperatures are combined with abnormal rainfall: this ensures mass reproduction of mosquitoes. The number of species is as high as 30 in the south; however, *Aedes communis* and *A. punctor* complexes prevail, density of human attack by these mosquitoes for an hour being 3,000–10,000 and even more. Many arboviruses circulate in the north of Europe [3, 13], the north of America [1, 2, 4, 7, 11] and the north of Asia [5, 6, 12].

Materials and methods

In far northern Eurasia (U.S.S.R.) from 1985–1988, we investigated eastern Fennoscandia, north of the Russian Plain, the North Trans-Pacific, North-Eastern Siberia and the Yakut Hollow (Fig. 1). These territories located in the arctic desert with sum of effective temperatures more than 10 °C ($\sum t° \geqslant 10$–0 °C) in the moss-lichen and shrub tundra and forest tundra of the subarctic zones ($\sum t° \geqslant 10$–0–600 °C), and in the northern and middle taiga of the Boreal region ($\sum t° \geqslant 10$–800–1600 °C). $\sum t° \geqslant 10$ °C means the sum of temperatures 10 °C and higher during a year. This parameter using in biogeography has a great influence upon the forming of ecosystems and correlates with landscape zones.

All the materials collection in the mentioned territories are shown in Table 1. Among collected mosquitoes, *Aedes communis* (87.91%) group dominated absolutely. *A. punctor* was also in the collection (7–10%) as were *A. hexodontus*, *A. nigripes*, *A. impiger* in tundra and *A. pionips*, *A. excrucians*, *A. dianteus*, *A. pullatus*, *A. riparius*, and others in northern taiga (all less than 1%). Isolation and detection of viruses were carried out in suckling mice and by

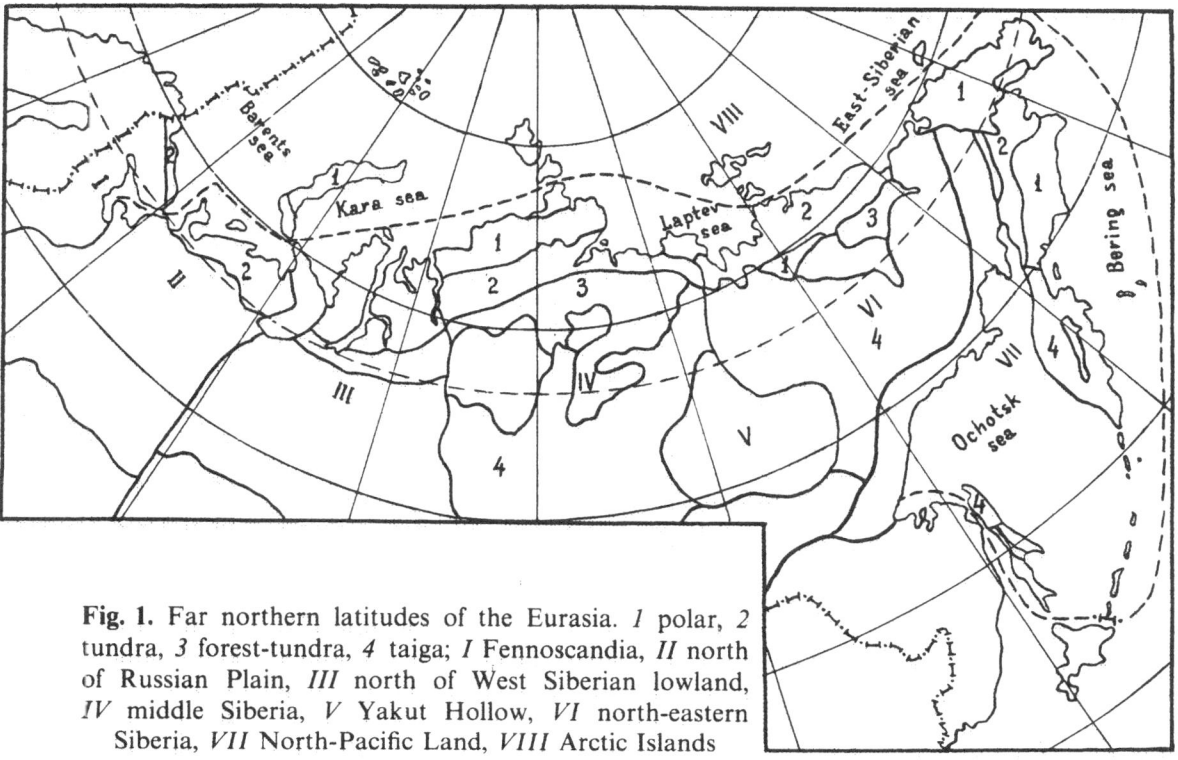

Fig. 1. Far northern latitudes of the Eurasia. *1* polar, *2* tundra, *3* forest-tundra, *4* taiga; *I* Fennoscandia, *II* north of Russian Plain, *III* north of West Siberian lowland, *IV* middle Siberia, *V* Yakut Hollow, *VI* north-eastern Siberia, *VII* North-Pacific Land, *VIII* Arctic Islands

Table 1. Collection of field materials in the Soviet Union, 1985–1988

Physico-geographical lands	Mosquitoes ($\times 10^3$)	Ticks ($\times 10^3$)	Vertebrates	Sera	
				human	domestic animal
Fennoscandia (eastern)	73.1	4.36	225	1132	215
Russian Plain (northern part)	202.4	0.53	437	2270	1124
North-eastern Siberia	164.3	0	390	737	0
Yakut Hollow	130.9	0	51	445	0
North Trans-Pacific	150.1	>10.0[a]	644	1542	136
Total	720.8	>15.0[a]	1747	6127	1475

[a] Including larvae and nymphs

ELISA method (to Tahyna, Inkoo, snowshoe hare, Batai viruses) as described earlier [8, 9, 10]. Identification of strains was carried out by electron microscopy, ELISA, complement fixation and neutralization in tissue culture. Serological tests were performed in using hemagglutination-inhibition (HI) (for Getah and Tyuleniy viruses) and neutralization (for California and Bunyamwera serogroups) as described earlier [8, 9, 10].

Results

In eastern Fennoscandia, besides Karelian fever, connected with a Sindbis-like togavirus, viruses of the California serogroup were isolated from mosquitoes. Zaliv Terpeniya and Uukuniemi viruses (usually connected with ticks) were isolated from mosquitoes in tundra and in the south of Middle taiga zone, respectively.

Viruses of California serogroup are widespread in eastern Fennoscandia. In total, 25 strains of California serogroup have been isolated. The infection rate of mosquitoes is lowest in tundra-forest tundra and highest in Northern taiga and in the north of middle taiga. The isolation of strains from mosquitoes begins from the third week of June in middle taiga, from the second week of July in Northern taiga and forest tundra, with maximum number of isolates obtained at the end of July and in early August.

Isolated strains antigenically are varied. Some of them are related to Tahyna virus, others to Inkoo. There are strains which may be recombinants. Strains from Kola peninsula are related to snowshoe hare virus. Results of serological studies of human and domestic animals correlate with the results of virological testing of mosquitoes.

In the north of the Russian Plain 22 strains of California serogroup and 9 strains of Batai-like virus of Bunyamwera group have been isolated (Fig. 2). The landscape zone and the main parameters of the infection rate of mosquitoes by California serogroup viruses during the season are the same on the whole as in Fennoscandia. However the strains have been isolated only beginning from the second week of July and the maximum isolation rate occurred in August. Strains have been isolated from Shrub tundra in the north to middle taiga in the south. Similar to those from Fennoscandia, strains isolated in tundra are related to snowshoe hare virus. However, there are also isolates of Tahyna and Inkoo and intermediates between them. Antibody prevalence in the human population and in domestic animals is the highest in zones of highest incidence of a virus in mosquitoes in northern taiga and the northern part of middle taiga (Fig. 3). In northern taiga the strains of Batai-like virus have been isolated from mosquitoes. Neutralizing antibodies to the virus have been revealed in humans and domestic animals of northern and middle taiga.

In the extreme north-eastern part of Asia, north Trans-Pacific physico-geographical land, where we conditionally included Sakhalin Island, 7

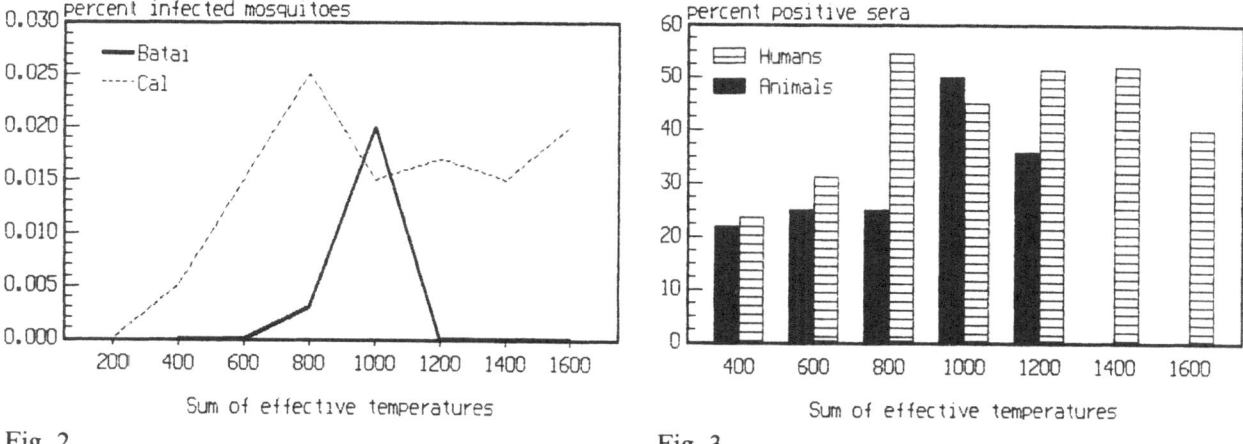

Fig. 2.   Fig. 3.

Fig. 2. Infection rate of mosquitoes by viruses California serogroup and Batai virus on the north of Russian Plain

Fig. 3. Immune structure to California serogroup on the north of Russian Plain

viruses have been isolated from ticks: Sakhalin, Paramushir, and Rukutama viruses of Sakhalin serogroup (*Bunyaviridae, Nairovirus*), Zaliv Terpeniya (*Bunyaviridae, Uukuvirus*) Tyuleniy (*Flaviviridae, Flavivirus*), Okhotsky and Aniva viruses of Kemerovo serogroup (*Reoviridae, Orbivirus*). Of these, Tyuleniy virus penetrates further to north, to moss-lichen tundra. Antibodies to the virus have been revealed in 5–9% of the human population.

Viruses of California serogroup (7 strains), Batai-like virus of Bunyamwera group (4 strains) and Getah virus of Semliki Forest serogroup (1 strain) have been isolated from mosquitoes. Getah virus has been isolated in Chukotka in the landscape subzone of moss-lichen tundra. HI antibodies have been found in 7% of the population. In territories situated further south of this physico-geographical region the antibody prevalence either was not determined or did not exceed 1–2%.

Viruses of the California serogroup were isolated in moss-lichen tundra, forest-tundra and northern taiga (Fig. 4). The infection rate of mosquitoes is essentially lower in comparison with the northern European part. The isolation of strains begins from second half of July and reaches an upper limit in the third week of August. Preliminary studies indicate that at least some of the strains from Chukotka are closely related antigenically to showshoe hare virus. However, these strains require further examination.

Virus-neutralizing antibodies have been revealed in tundra (20%), forest-tundra (53%), northern taiga (40%), middle taiga (25%). Antibody prevalence to arboviruses is lower here in comparison with the corresponding landscape zones of the northern European part.

Strains of Batai-like virus have been isolated in forest tundra and northern taiga. Virus-neutralizing antibodies to Batai virus have been detected among humans and reindeer in tundra-forest tundra-northern taiga (1–2%) and in middle taiga on Sakhalin island (13%).

In north-eastern Siberia and Yakut Hollow there exist unique territories where vestigial steppen landscapes are adjacent to tundra. There we isolated 16 Getah strains, 10 strains of Batai and 26 strains of California serogroup viruses. In Yakut Hollow 2 Batai strains and 25 strains of California serogroup were isolated (Fig. 5). An overwhelming majority of Getah strains were isolated in the eastern part of north-eastern Siberia. Antibodies have been found from tundra to middle taiga.

These virological-serological data demonstrate Getah virus infection in human populations practically everywhere in north-eastern Siberia and Yakut Hollow. One of the strains was isolated from mosquitoes collected in arctic tundra, but the most of the strains were collected in northern taiga. Batai strains have been isolated mainly in the north of northern taiga but some of them were isolated from mosquitoes collected in tundra. In human populations virus-neutralizing antibodies to Batai virus were found in single specimens.

The whole territory of north-eastern Siberia and especially Yakut Hollow is highly endemic with respect to viruses of California serogroup. Even in tundra, including arctic tundra, the infection rate of the mosquitoes is considerable, increasing in northern taiga of north-eastern Siberia and in middle taiga of Yakut Hollow. Antibodies to California serogroup have been found everywhere, but the highest percentage of antibodies has been found in northern taiga (Fig. 6).

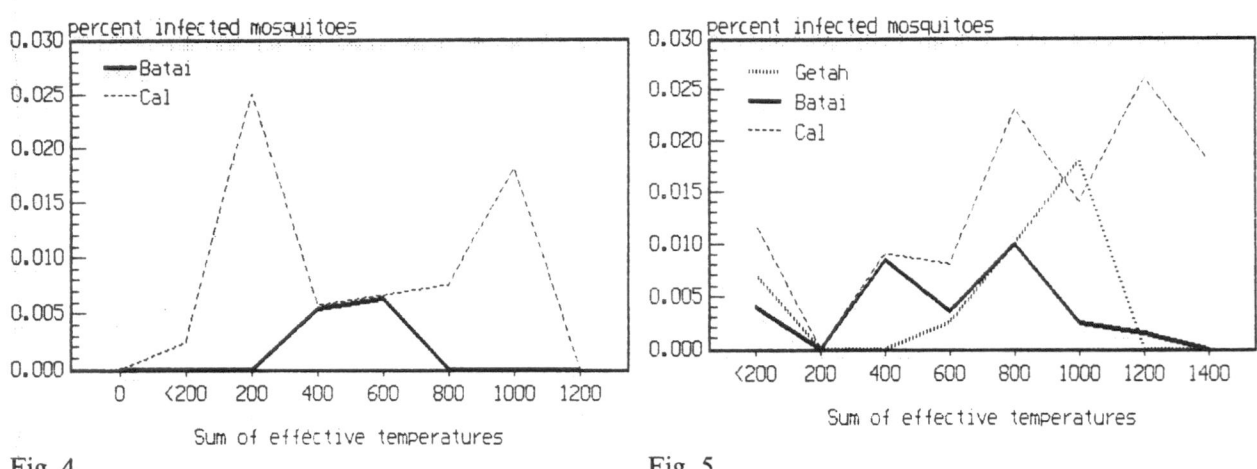

Fig. 4.   Fig. 5.

Fig. 4. Infection rate of mosquitoes by viruses California serogroup and Batai virus in the North Pacific region

Fig. 5. Infection rate of mosquitoes in Siberia and Yakut Hollow

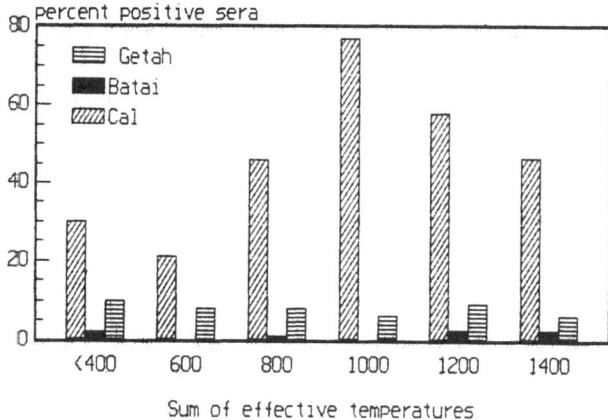

Fig. 6. Immune structure to arboviruses in different landscapes of
north-eastern Siberia and Yakut
Hollow

Discussion

The natural foci of antigenically closely related viruses have different
ecological features. Uukuniemi and Zaliv Terpeniya viruses, are separated in
Fennoscandia by a thousand kilometers. The viruses replicate in mosquitoes
and are associated with birds. This suggests the presence of interpopulational
exchange, recently or in the distant past.

Zaliv Terpeniya virus is adapted to *Ixodes uriae* ticks and Alcidae birds.
We cannot exclude the priority of these very old natural foci of viruses of the
Uukuniemi ecological complex, including Zaliv Terpeniya virus, and the
subsequent introduction of virus into the landscape subzone of middle taiga,
with further southerly spread into areas with *I. ricinus* ticks.

Strains of California serogroup isolated in tundra in the north of
middle Siberia have been related to snowshoe hare virus according to
immunoblotting (pers. comm) and sequencing of S segment (pers. comm).

Data obtained by examining specimens from northern Eurasia without
middle and western Siberia suggest a circumpolar distribution of certain
arboviruses. This applies both to the viruses ecologically related to those
from *I. uriae* ticks and Alcidae birds and to a number of viruses associated
with mosquitoes, particularly viruses of the California serogroup. The
genesis of this complex may be connected with north-eastern Asia, where the
fauna developed in the course of landscape changes, from the tertiary broadleaved forest to the present tundra and taiga. The fauna of eastern Siberia
developed from the Oligocene–Myocene Chinese fauna. The formation of
the fauna in its present condition is related to Golocene (the last 10,000
years). By then the last substantial rise of the ocean level took place, finally
destroying the Beringia land junction between Asia and North America. The
introduction of one or more viruses of the California serogroup to the
American continent from North-East Asia may have occurred at this period
of intensive fauna exchange. Further penetration of these viruses southward

to Central and South America, with the formation of genetically isolated populations and formation of new viruses of the complex, would have been possible. However, the greatest activity of California serogroup viruses was in the center of the area formation – relic steppe and also in taiga and tundra landscapes. Getah virus penetration into North-East Asia probably occurred relatively recently from South-East Asia with birds migrating through East Asia.

The origin of arboviruses associated with *I. uriae* ticks and nesting Alcidae birds has been related to the period of Pliocene transgression, when the direct connection between Atlantic and Polar basins was established. In that period alone the formation of most *Alcidae* spp. birds took place. The bird genus *Plantus* diverged in Miocene era and now is typically found at seashores. Colonies of these birds may have played a role in the formation of foci of arboviruses associated with *I. uriae*. These foci are among the most ancient in far northern latitudes.

References

1. Artsob H (1983) Distribution of California serogroup viruses and virus infections in Canada. In: Calisher CH, Thompson WH (eds) California serogroup viruses. AR Liss, New York, pp 277–290
2. Artsob H, Spence L, Calisher CH, Sekla LH, Brust RA (1985) Isolation of California encephalitis serotype from mosquitoes collected in Manitoba, Canada. J Am Mosq Control Assoc 1: 257–258
3. Brummer-Korvenkontio M, Saikku P, Korhonen P, Ulmanen I, Reunala T, Karvonen J (1973) Arboviruses in Finland. IV. Isolation and characterization of Inkoo virus a Finnish representative of the California group. Am J. Trop Med Hyg 22: 404–413
4. Calisher CH (1983) Taxonomy, classification and geographic distribution of California serogroup bunyaviruses. In: Calisher CH, Thompson WH (eds) California serogroup viruses. AR Liss, New York, pp 1–16
5. Chumakov MP, Moshkin AV, Andreeva EB (1974) Isolation of 5 strains of Getah virus from mosquitoes on a south of Amur district in the U.S.S.R. Med Virol (Moscow) 22: 65–71
6. Gaidamovich SYa, Melnikova EE, Agafonov VI, Lokhova MD, Rodina VY, Goldin RB, Klisenko GA (1974) Identification of group A arboviruses related to Semliki forest virus isolated in the Far East. Arboviruses (Moscow): 93–98
7. Karabatsos N (ed) (1985) International catalogue of arboviruses including certain other viruses of vertebrates. American Society for Tropical Medicine and Hygiene, San Antonio, Texas
8. Lvov SD, Pogorely YuA, Skvortsova TM, Kiseleva LV, Berezina LK, Sharkov AA, Gromashevsky VL, Kadoshnikov YuP, Klimenko SM (1985) Isolation of Tahyna Bunyavirus in the polar circle region. Vopr Virusol 6: 736–740
9. Lvov SD, Gromashevsky VL, Bogoyavlensky GV, Bailuk FN, Skvortsova TM, Kondrashina NG, Kandaurov EK (1987) Isolation of Zaliv Terpeniya, Uukuniemi and Tahyna-like viruses in the tundra, the forest-tundra and the northern taiga of Kola, Taimyr peninsulas and in the middle taiga of Karelia. Med Parazitol 6: 40–43
10. Lvov DK, Scherbin YuD, Zairov GK, Artjukhov NI, Lvov SD, Shilov AD, Gromashevsky VL, Kandaurov EK, Kondrashina NG, Butenko AM, Morozova TN, Kuznetsov

AA, Skvortsova TM (1987) Isolation of a Tahyna-like virus (*Bunyaviridae, Bunyavirus*, California encephalitis complex) in the north of Sakhalin island. Vopr Virusol 5: 588–590

11. McLean D (1985) Bunyaviruses throughout the western Canadian Arctic. In: Proc 6th Int Symp on Circumpolar Health University of Washington Press, Seattle, pp 209–212
12. Takashima I, Hashimoto N (1985) Getah virus in several species of mosquitoes. Trans R Soc Trop Med Hyg 79: 546–550
13. Traavik T, Mehl R, Wiger R. (1984) Mosquito-borne arboviruses in Norway: further isolations and detection of antibodies to California encephalitis viruses in human, sheep, wildlife sera. J Hyg 93: 120–132

Authors' address: S. D. Lvov, The N.F. Gamaleya Institute of Epidemiology and Microbiology, Academy of Medical Sciences of the U.S.S.R., Gamaleya Str. 18, Moscow 123098, U.S.S.R.

Arch Virol (1990) [Suppl 1]: 277–285

Seroepidemiological survey for antibodies to arboviruses in Greece

**A. Antoniadis[1], Stella Alexiou-Daniel[1], N. Malissiovas[1], J. Doutsos[1], Thalia Polyzoni[1],
J. W. LeDuc[2], C. J. Peters[2], and G. Saviolakis[2]**

[1]Department of Microbiology, Arbovirus National Reference Laboratory,
School of Medicine, Aristotelian University of Thessaloniki, Thessaloniki, Greece
[2]U.S. Army Medical Research Institute of Infectious Diseases, Fort Detrick, Frederick,
Maryland, U.S.A.

Accepted March 12, 1990

Summary. Plaque reduction neutralization (PRN) and indirect immuno-fluorescence (IFA) tests were used to detect human antibodies to certain viruses of the families *Flaviviridae* and *Bunyaviridae*. Blood samples for the serosurveys were mainly collected from healthy farmers, wood cutters and shepherds. By PRN test antibodies were found to West Nile, sandfly fever Naples, and sandfly fever Sicilian viruses with seropositives 1.2%, 16.7%, and 2.0%, respectively. Antibodies to Rift Valley fever virus were not detected. By IFA tests, antibodies to the phlebovirus Corfu, tick-borne encephalitis, Crimean-Congo hemorrhagic fever, and Hantaan viruses were found with seropositives 4.0%, 1.7%, 1.0%, and 3.4%, respectively. Epidemiological data concerning the geographic and occupational distributions of the infected individuals are discussed.

Introduction

The first arboviral disease to be reported in Greece was a widespread epidemic of dengue in 1927–1928 [14]. Later, sandfly fever was reported to have affected military personnel during World War II [6]. From 1945 to 1980 no other arboviral disease was diagnosed. In contrast, human and animal serosurveys, mainly using hemagglutination inhibition (HI) tests, were performed to determine the existence of a number of arboviruses in Greece; antibodies to a member of the tick-borne encephalitis (TBE) antigenic complex of flaviviruses, Sindbis, West Nile, dengue types 1 and 2, and Tahyna viruses were found in humans [11]. By agar gel diffusion tests, antibodies to Crimean-Congo hemorrhagic fever (CCHF) virus were found in animals but not in humans [10] and by plaque-reduction neutralization

(PRN) tests, antibodies to sandfly fever (Naples and Sicilian) viruses were found in humans residing on the island of Crete [13]. In 1976, a C-CHF virus strain was isolated from wild-caught ticks in northern Greece [10], and a strain of TBE virus was isolated from a goat in northern Greece in 1969 [9]. Since 1981, the Arbovirus National Reference Laboratory has been carrying out research to identify endemic areas, isolate viruses and diagnose illnesses caused by arboviruses. Several studies have been carried out to assist in our understanding of the epidemiology of hantaviruses in Greece.

So far endemic areas have been identified in northern and central Greece [3], the clinical course of the severe form of the disease as it occurs in Greece has been described [4], a hantavirus has been isolated from a severely ill HFRS patient [2], and the probable rodent host of the virus has been described [7]. In this paper we report results of serosurveys indicating the further existence of certain arboviruses in Greece. Additionally, we present new epidemiological data, concerning the hantavirus.

Materials and methods

Human serosurvey

Sera from 3388 apparently healthy individuals, mainly farmers, wood cutters and shepherds, were obtained in 25 of 54 counties of Greece: 15 in northern Greece (Thrace, Macedonia and Epirus states), five in central Greece (Thessalia state), four in southern Greece (Peloponesos state), one on Corfu island (Ionean Islands state), and two on the island of Crete. Sera were collected annually from 1981 to 1988 and stored at $-20\,°C$ until they could be tested for antibodies.

Individuals were identified by age, sex, occupation, previous travel history (mainly abroad) and location of residence. 2035 of the 3388 sera had been previously examined by IFA test for antibodies to Hantaan virus; the results are presented elsewhere [3].

Serological techniques

Indirect immunofluorescence test

Indirect immunofluorescence (IFA) tests were used to detect antibodies to Hantaan, C-CHF, Corfu, and TBE viruses. Sera were examined by IFA with goat anti-human immunoglobulin conjugate (Gibco Diagnostics, Madison, WI, U.S.A.). Spot slides contained Vero E-6 cells infected separately with Hantaan virus (strain 76–118), TBE (strain Sofin), C-CHF virus (strain Ibr 10200) and Corfu virus (strain Ba Ar 814). These viruses were provided by Dr. R. E. Shope, Yale Arbovirus Research Unit, Yale University, School of Medicine, New Haven, CT. Sera were considered positive if characteristic cytoplasmic fluorescence was detected at a 1:16 dilution for Hantaan virus and at a 1:4 dilution for the other viruses.

Plaque reduction neutralization test

Sera were tested by PRN tests with West Nile virus (strain B 956), Rift Valley fever virus (strain ZH 501), sandfly fever Naples virus (strain Sabin), and sandfly fever Sicilian virus (strain Sabin), following previously described procedures [13].

Results

The results of the serosurveys are summarized in Table 1. With the exception of Rift Valley fever virus, seropositive individuals to all viruses were found. The age distribution is shown in Table 2. Young individuals (to 30 years) were found to be infected with Hantaan, C-CHF, West Nile, TBE and sandfly fever Naples viruses; antibodies to sandfly fever Sicilian virus were found only in individuals 30 to 50 years.

The prevalence of antibodies against seven arboviruses and Hantaan virus in certain human populations of Greece is shown in Table 3. Hantaan and C-CHF viruses are found throughout Greece. In spite of the small number of the individuals examined for antibodies against sandfly fever Naples, sandfly fever Sicilian and Corfu viruses, it appears that these viruses also are found throughout Greece. In contrast, West Nile and TBE viruses,

Table 1. Antibody to seven arboviruses and Hantaan virus in healthy residents of Greece

Virus	No tested	No positive	%	Method
West Nile	245	3	1.2	PRN
Tick-borne encephalitis	475	8	1.7	IFA
Sandfly fever Naples	245	41	16.7	PRN
Sandfly fever Sicilian	245	5	2.0	PRN
Corfu	826	33	4.0	PRN
Rift Valley fever	245	0	0.0	PRN
Crimean-Congo hemorrhagic fever	3388	37	1.0	IFA
Hantaan virus	3388	117	3.4	IFA

Table 2. Antibody to Hantaan, Crimean-Congo hemorrhagic fever, West Nile, tick-borne encephalitis, sandfly fever Naples, sandfly fever Sicilian, Corfu, and Rift Valley fever viruses in residents of Greece, by age

Age (years)	HTN no. (%)	C-CHF no. (%)	WN no. (%)	TBE no. (%)	SFN no. (%)	SFS no. (%)	COR no. (%)	RVF no. (%)	Total
0–10	35 (0.0)	35 (0.0)	7 (0.0)	11 (0.0)	7 (0.0)	7 (0.0)	17 (0.0)	7 (0.0)	126
11–20	578 (1.3)	578 (0.0)	10 (0.0)	46 (0.0)	10 (0.0)	10 (0.0)	37 (8.1)	10 (0.0)	1279
21–30	409 (4.6)	409 (0.4)	45 (2.2)	75 (2.6)	45 (0.4)	45 (0.0)	154 (2.0)	45 (0.0)	1227
31–40	499 (7.2)	499 (1.0)	46 (2.2)	105 (0.0)	46 (1.0)	46 (4.4)	157 (6.4)	46 (0.0)	1444
41–50	612 (4.9)	612 (1.4)	45 (0.0)	151 (0.0)	45 (1.4)	45 (4.4)	151 (6.1)	45 (0.0)	1706
51–60	896 (1.4)	896 (1.0)	42 (0.0)	46 (2.0)	42 (1.0)	42 (2.4)	152 (8.6)	42 (0.0)	2158
61–70	299 (2.3)	299 (2.0)	35 (2.9)	35 (5.7)	35 (2.0)	35 (0.0)	123 (5.7)	35 (0.0)	896
>70	60 (6.6)	60 (10.0)	15 (0.0)	6 (16.6)	15 (10.0)	15 (0.0)	35 (14.3)	15 (0.0)	221
Total	3388	3388	245	475	245	245	826	245	9057

Table 3. Antibody to seven arboviruses and Hantaan virus in healthy residents of Greece: geographic distribution

State County	Hantavirus Hantaan		Nairovirus CCHF		Flavivirus WN		Flavivirus TBE		Phlebovirus							
									Sandfly fever Naples		Sandfly fever Sicilian		Corfu		RVF	
	total tested	% posit- ive	total tested	% posit- ive	total tested	% posit- ive	total tested	% posit- ive	total tested	% posit- ive	total tested	% posit- ive	total tested	% posit- ive	total tested	% posit- ive
Trace																
Evros	246	0.04	246	0.0	58	0.0			58	13.8	58	3.5			58	0.0
Rodopi	202	1.0	202	0.5												
Xanthi	71	1.4	71	1.2												
Macedonia																
Drama	35	0.0	35	0.0												
Serres	42	0.0	42	0.0									107	0.0		
Kilkis	149	4.0	149	2.0									20	5.0		
Thessaloniki	191	0.5	191	0.0			50	0.0					96	0.0		
Chalkidiki	95	6.0	95	0.0									137	0.0		
Pella	93	14.0	93	9.6	36		56	3.57								
Imathia	185	1.6	185	4.3		2.8	44	2.27	36	22.2	36	2.8	44	13.6	36	0.0
Pierria	100	2.0	100	0.0												
Kastoria	62	0.0	62	0.0												
Kozani	153	2.6	153	0.0												
Florina	28	0.0	28	0.0												
Central Greece																
Evritania	94	3.0	94	1.0			33	0.0					57	5.3		
Larisa	103	2.0	103	2.0									43	2.3		

	n	%	n	%	n	%	n	%	n	%	n	%	n	%	n	%
Karditsa	48	2.0	48	6.2	49	0.0	44	0.0	49	20.4	49	2.0			49	0.0
Magnesia	132	3.0	132	0.0												
Trikala	17	0.0	17	0.0												
Epirus																
Ioannina	500	8.0	500	1.6			121	2.47					172	4.1		
Peloponisos																
Messinia	26	0.0	26	0.0												
Laconia	45	2.0	45	1.0	29	6.9	29	13.8	29	0.0	29	0.0			29	0.0
Grete island																
Hania	29	0.0	29	0.0	27	0.0	30	0.0	27	18.5	27	3.1			27	0.0
Iraklio	81	3.7	81	1.2	46	0.0			46	32.6	46	0.0			46	0.0
Corfu island	95	4.2	95	1.0			68	0.0					90	14.4		
Others																
Blood donors	462	4.0	462	0.0												
Veterinarians	24	0.0	24	4.1									60	3.3		
Military	80	0.0	80	0.0												
Total	3388	3.45	3388	1.0	245	1.2	475	1.7	245	16.7	245	2.0	826	4.0	245	0.0

are selectively distributed, i.e, antibodies against these viruses, were not detected in residents of Central Greece. All the individuals tested for antibody to West Nile, TBE, sandfly fever Naples, sandfly fever Sicilian and Corfu viruses were farmers. Those tested for antibody to Hantaan and C-CHF viruses were drawn from farming, wood-cutting, shepherding and other groups, including blood donors, veterinarians and military personnel. It is interesting that of those found seropositive, 56% of farmers were positive to Hantaan virus and 72% positive to C-CHF virus, whereas 37% of wood-cutters had antibody to Hantaan but only 14% of them had antibody to C-CHF virus. Shepherds show a similar patterns to that of farmers, with 11% positive to C-CHF virus and only 3% positive to Hantaan virus.

Discussion

Results of this study indicate a marked difference in the prevalence of antibodies to West Nile, TBE, sandfly fever Naples, sandfly fever Sicilian, Rift Valley fever, C-CHF and Hantaan viruses in residents of Greece. The highest percentage of seropositivity (16.7%) was to sandfly Naples virus, whereas no antibodies were found to Rift Valley fever virus; relatively high seropositivity was found to Corfu (4%) and Hantaan (3.4%) viruses.

Results from the serosurvey in northwestern (Thrace), northeastern (Epirus, Corfu island), central (Thessalia), and southern (Peloponesos, Crete island) states revealed that C-CHF or a virus antigenically related to that virus, and hantavirus(es) are found throughout the country (Table 3). Two counties, one in Epirus state (Ioannina) and another in Macedonia state (Pella), can be characterized as high risk areas for both C-CHF and hanta-viruses (Table 3).

West Nile virus is endemic with a low prevalence (1.2%) in northern (Macedonia) and southern Greece. In a serosurvey conducted in 1974, antibodies against West Nile virus were found in humans in a high percentage (21.7%) [11]. In both previous and recent, serosurveys blood samples were taken from residents of northern Greece but HI was the method used to detect antibodies in the serosurvey conducted in 1974 and PRN was used for the recent survey. The high prevalence of HI antibodies may be either due to cross reactions with other viruses in the same serogroups or to ecologic changes (such as industrialization), better standard of living, and better public hygiene; the latter two have been improved in Greece during the last 15 years. A member of the TBE antigenic complex of flaviviruses causes human disease in Greece [9]. Results of our serosurvey indicate that this virus is endemic in northern states (Macedonia, Epirus) and southern states (Peloponisos) of Greece.

Our serological data indicate a marked difference in prevalence of sandfly fever Naples, sandfly fever Sicilian and Corfu viruses, although these viruses seem to have nearly the same geographical distribution (Table 3). Only 2% and 4% of the individuals had antibodies against Naples and Corfu viruses, respectively. In spite of the low percentage of seropositives (4%), in some areas (Corfu island, Imathia county) the percentage of seropositives to Corfu virus was very high (14.4%, 13.6%, respectively) whereas, in contrast, the percentage of seropositivity to sandfly fever Sicilian virus did not exceed 3.7%. Greece is at the geographic cross-roads between Asia and Europe. Therefore it is at potential risk for Rift Valley fever which is endemic in Asia [1]. In spite of the limited number of individuals examined for antibodies to Rift Valley fever virus, there is no evidence that Rift Valley fever virus occurs in Greece.

From analysis of the age distribution of individuals with antibody we suggest that C-CHF, Hantaan, sandfly fever Naples and Corfu viruses are still active in Greece and infect humans here.

A member of the tick-borne encephalitis antigenic complex of flaviviruses appears to be active in Greece, as supported by the observation that a human case of TBE recently was serologicaly diagnosed [12]. Our sample sizes may have been too small to provide conclusive epidemiological evidence of the existence of West Nile virus.

It is difficult to interpret the significance of antibody prevalence by occupation because the residents of small villages are mainly farmers but sometimes work as shepherds and wood-cutters. However, according to our data, it appears that farmers and wood cutters are at increased risk to infection with Hantaan and C-CHF viruses, whereas shepherds are more likely to acquire C-CHF virus than is the general population. Our data do not allow us to determine the public health importance of C-CHF, West Nile, sandfly fever Naples, sandfly fever Sicilian and Corfu viruses; demonstration of antibodies in humans indicates previous viral infection but does not indicate past illnesses. Thus far, only hemorrhagic fever with renal syndrome and tick-borne encephalitis have been serologicaly diagnosed in Greece.

In a separate study, 1025 blood samples (single or paired) from patients with suspected C-CHF or from patients with pyrexia of unknown origin, influenza-like diseases, and pyrexia with elevated liver enzymes (SGOT-SGPT) were sent to this laboratory for serodiagnosis. None of these patients was found to be infected with C-CHF virus. In contrast, C-CHF virus is endemic in Yugoslavia, Bulgaria and Albania in areas bordering northern Greece and causes human disease [8, 15; E. Eltari, pers. comm. 1988].

Several authors have suggested that infection with sandfly fever viruses causes milder disease in children than in adults [5], and in areas where sandfly fever is endemic most of the population is probably infected during childhood [13]. The fact that such patients are not usually hospitalized may explain why the disease has not been serologically diagnosed thus far in Greece.

Acknowledgements

The authors express their sincere thanks to Drs. R. E. Shope and R. B. Tesh, Yale Arbovirus Research Unit, Yale School of Medicine, New Haven, CT, U.S.A., for performing PRN test against sandfly fever Naples and sandfly fever Sicilian viruses in some of the sera. We also thank Miss Aleka Varna for excellent technical assistance. This research was funded in part by research grant No. 17-87-G-7019 from the U.S. Army Medical Research and Development Command, Fort Detrick, Frederick, MD. The views of the authors do not purport to reflect the positions of the Department of the Army or the Department of Defense.

References

1. Abdel-Wahab KSE, El Baz LM, El Tayeb EM, Omar H, Yasim W (1978) Rift Valley fever virus infection in Egypt: pathological and virological findings in man. Trans R Soc Trop Med Hyg 72: 392–396

2. Antoniadis A, Grekas D, Rossi CA, LeDuc JW (1987) Isolation of a hantavirus from a severely ill patient with hemorrhagic fever with renal syndrome. J Infect Dis 156: 1010–1013

3. Antoniadis A, LeDuc JW, Daniel-Alexiou S (1987) Clinical and epidemiological aspects of hemorrhagic fever with renal syndrome (HFRS) in Greece. Eur J Epidemiol 3: 295–301.

4. Antoniadis A, LeDuc JW, Acritides N, Alexiou-Daniel S, Kyrarissi A, Saviolakis G (1989) Hemorrhagic fever with renal syndrome in Greece: clinical and laboratory characteristics. Rev Infect Dis II [Suppl 4]: 891–896

5. Guelmino DJ, Jevtic M (1955) An epidemiological and hematological study of sandfly fever in Serbia. Acta Tropica 12: 179–182

6. Herting M, Sabin AB (1964) Sandfly fever (Pappataci, Phlebotomus, three-day fever). In: Coats JB (ed) Preventive medicine in World War II, vol 7, communicable diseases. U.S. Government Printing Office, Washington, DC, pp 109–174

7. LeDuc JW, Antoniadis A, Siamopoulos K (1986) Epidemiological investigation following an outbreak of hemorrhagic fever with renal syndrome in Greece. Am J Trop Med Hyg 35: 654–659

8. Obradovic M, Gligic A (1980) Specific antibodies in the sera of patients formerly affected by Crimean-Congo hemorrhagic fever. In: Vesenjak-Hirjan J, Porterfield JS, Arslanagic E (eds) Arboviruses in the Mediterranean countries. Zentralbl Bakteriol Mikrobiol Hyg [A] [Suppl 9]: 267–271

9. Papadopoulos O, Paschaleri-Papadopoulou E, Deligaris N, Doukas G (1971) Isolation of tick-borne encephalitis virus from a flock of goats with abortions and fatal disease (a preliminary report). Vet News Greece 3: 112–114

10. Papadopoulos O, Koptopoulos G (1980) Crimean-Congo hemorrhagic fever (CCHF) in Greece: isolation of the virus from *Phipicephalus bursa* ticks and a preliminary serological study. In: Vesenjak-Hirjan J, Porterfield JS, Arslanagic E (eds) Arboviruses in the Mediterranean countries. Zentralbl Bakteriol Mikrobiol Hyg [A] [Suppl 9]: 189–193

11. Papapanagiotou J, Kyriazopoulou V, Antoniadis A, Batikova M, Gresikova M, Sekeyova M (1974) Hemagglutination-inhibiting antibodies to arboviruses in a human population in Greece. Zentralbl Bakteriol Mikrobiol Hyg [A] 228: 443–446

12. Petridis S, Antoniadis A, Andreadis Ch, Samouilidis I, Alexiou S, Dimitriadis A (1989) A case of meningo-encephalitis characterized as tick-borne. Iatriki 56: 76–79

13. Tesh RB, Saidi S, Gaidamovic SYa, Fodhain F, Vesenjak-Hirjan J (1976) Serological studies on the epidemiology of sandfly fever in the Old World. Bull WHO 54: 663–674

14. Theiler M, Casals J, Moutousses C (1960) Etiology of the 1927–1928 epidemic of dengue in Greece. Proc Soc Exp Biol Med 103: 244–248
15. Vasilenko SM, Katsarov G, Mikhailov M (1972) Crimean hemorrhagic fever in Bulgaria. Trudi Inst Poliomiel Virus Entsephal Akad Med Nauk SSR 19: 100–111 (English summary)

Authors' address: A. Antoniadis, Arbovirus National Reference Laboratory, Department of Microbiology, School of Medicine, Aristotelian University of Thessaloniki, Thessaloniki 54006, Greece.

Arch Virol (1990) [Suppl 1]: 287–293

Sandfly fever in Central Asia and Afghanistan

**Sophia Ya. Gaidamovich, Natalia V. Khutoretskaya, Y. V. Asyamov,
V. I. Tsyupa,** and **Eugenia E. Melnikova**

D.I. Ivanovsky Institute of Virology, U.S.S.R. Academy of Medical Science,
Moscow, U.S.S.R.

Accepted January 16, 1990

Summary. Sandfly fevers are caused by viruses of the family *Bunyaviridae*, genus *Phlebovirus*. Three strains of sandfly fever Naples (SFN) virus (Af-1008, Af-1038 and Af-130) and 2 strains of sandfly fever Sicilian (SFS) virus (Af-1028 and Af-83) were isolated from febrile Soviet servicemen in Afghanistan in May–August, 1986–1987. The new strains of SFN virus were slightly different antigenically from the original Sabin strain but clearly different from prototype Toscana virus. Forty paired serum samples from patients ill in 1986 and 64 paired serum samples from patients ill in 1987 were tested by neutralization and immunofluorescence. A fourfold or greater rise in antibody to SFN or SFS virus was observed in 87 subjects; 16 others had antibody to both viruses but a rise in titer to only one of them. In addition, 101 serum samples with or without antibody to these viruses were tested by immunofluorescence for antibodies to Toscana and Rift Valley fever viruses; none were positive.

Introduction

Sandfly fevers are caused by viruses of the family *Bunyaviridae*, genus *Phlebovirus*. They are transmitted by diptera of the genus *Phlebotomus*. The disease is characterized by sudden onset with high fever, severe headache, retroorbital pains, conjunctivitis and general malaise. One of the typical symptoms is the so-called Pick's symptom – conjunctival injection in the form of a triangle. The outcome of the disease is favourable, without residua. Two antigenically distant viruses cause infections with similar clinical manifestations: sandfly fever Naples (SFN) and sandfly fever Sicilian (SFS) viruses. These agents were first isolated from blood samples by A. Sabin in 1944 during an epidemic among American soldiers in Naples and Sicily, respectively [11].

The SFN antigenic complex includes SFN, Teheran, Karimabad, and Toscana viruses; SFS virus is not assigned to an antigenic complex [2]. The distribution of sandfly fevers coincides with the distribution of the principal vector, *Phlebotomus papatasi*, including Europe, Asia, and Africa between 20° and 45° North latitudes. Large outbreaks of this disease usually have been associated with social (wars) or natural calamities (such as earthquakes), during which non-immune populations enter endemic natural foci or rodents, natural hosts of sandflies, migrate to new areas.

Outbreaks of sandfly fevers were described in the U.S.S.R. in the republics of Central Asia and in the Caucasus, Crimea, and Moldavia, early in this century. The largest outbreaks occurred in Crimea during the Second World War and after the 1948 earthquake in Ashkhabad, Turkmenia.

Of sandfly fever virus isolates from those years, only one (from Ashkhabad) has been maintained. This was retrospectively identified as SFN virus [6]. Since then, no more laboratory-verified cases of sandfly fevers occurred in those regions. According to the results of serological surveys of the general population in Ashkhabad, conducted in mid-70's, antibodies were found mainly in subjects more than 30 years of age, i.e., those infected in 1940–1950 [7, 13]. Nevertheless, these viruses continue to circulate in nature, as indicated by antibodies to SFS virus in natural hosts of *Phlebotomus papatasi*: great gerbils (*Rhombomys opimus*), long-clawed ground squirrels (*Spermophilopsis leptodactylys*), hedgehogs (*Erinaceus auritus*) in Turkmenia [4].

This paper describes the isolation of SFN and SFS viruses from patients and results of serological tests with blood samples of convalescent patients. The outbreaks occurred among Soviet servicemen May–August 1986–1987, in the village Rukha, Parvan province.

Materials and methods

Collection of specimen and virus isolations

Patients were bled 1–2 days after onset of illness and again on 10–20 days. The blood collected in the acute phase was stored at −40 or −70 °C for several months to one year until tested for virus. Samples for serological tests were stored at −20 °C. Virus isolations were done using 1- to 2-day old albino mice inoculated intracranially with 0.01 ml of whole blood diluted 1:5 with sterile 0.9% sodium chloride; animals were observed for 30 days after inoculation. Blind passages at 7- to 8-day intervals were also used for virus isolation. Surviving animals were exsanguinated at 28–30 days and their sera tested for antibodies. Virus isolations were confirmed by reisolation.

Virus identification and serologic tests

The isolates were identified by indirect immunofluorescence (IF), complement fixation (CF), and neutralization (N) tests. Plaque reduction N was done according to the modified method of de Madrid and Porterfield [9] in 24-well plates with a constant dose of virus corresponding to 25–30 plaques per well and two-fold serum dilutions starting at 1:20. Virus-serum mixtures were maintained for 2 h at 37 °C, following which each mixture was added to 2 wells. To the same wells we added (300,000/ml) 0.5 ml each of suspensions of Vero E6 or SPEV (porcine embrio kidney stable line) cells. After 2 h incubation at 37 °C, 0.5 ml of overlay medium was added into each well. The overlay medium contained 4.0 ml 3% carboxymethylcellulose, 0.8 ml noninactivated bovine fetal serum, 0.8 ml MEM medium, 0.08 ml 3% glutamine, and 0.03 ml gentamycin. Plaques were read at 6 days and the test was considered positive if the number of plaques was reduced by 80%.

Immune ascitic fluids (IAF) and antigens of the following phleboviruses were used for identification: SFN (Sabin and Central Asian strains), SFS (Sabin), Karimabad (I-58), Salehabad (I-81), Teheran (I-47), Toscana (ISS Phl 3) and Rift Valley fever (Entebbe). Patients were tested for antibodies by N, IF, and CF tests.

Results

The diseases were characterized by symptoms typical of sandfly fever. All patients reported having been multiply bitten by sandflies. Bloods from patients in the first day of illness yielded 3 strains of SFN virus (Af-1008, Af-1038, Af-130) and 2 strains of SFS virus (Af-1028, Af-83). These strains were isolated both by primary inoculations of mice with patient's blood and by blind passages. In the first passages not all inoculated mice became sick; the incubation period being 10–14 days. After passage the pathogenicity for mice increased, the incubation period stabilizing at 8–9 days for SFN and 7–8 days for SFS strains. The isolates adapted readily to replication in Vero E6 and SPEV cells. By IF, virus accumulation in some cells could be seen as early as 24 h after infection; within 48–72 h more than 50% of the cells were infected. The cytopathic effect on the cells was manifest only after 5–6 passages in mice; however, all isolates eventually produced plaques in Vero E6 and SPEV cells. Indirect IF was used for primary identification of the isolates. Strains Af-1008, Af-1038 and Af-130 cross-reacted and reacted extensively with IAF for SFN virus. The Af-1008 strain also reacted with IAF to the Central Asia strain of SFN virus, with Toscana virus and with Teheran virus, i.e., with all viruses belonging to the SFN antigenic complex. The Af-1028 and Af-83 strains cross-reacted and reacted extensively with IAF for SFS. A detailed identification of the isolates was done by CF (Table 1), comparing them with the closely related SFN and Toscana viruses. Using IAFs, strain Af-1008 reacted similarly with homologous antigen and within fourfold of the homologous titer with SFN antigen. Toscana IAF reacted somewhat less with antigens of SFN and with strain Af-1008. In tests with sera from infected asymptomatic mice, differences from Toscana virus were marked. Toscana antigen did not react with sera to Af-1008, Af-1038, and Af-130. Slight differences were also noted among the field isolates and with prototype SFN; immune sera to Af-1008 and AF-130 reacted to low titer with the SFN (Sabin) antigen and serum to Af-1038 did not react with it. The Af-1028 and Af-83 strains of SFS were similar to each other but strain Af-1028 differed

Table 1. Cross complement-fixation tests comparing strains Af-1008, Af-1038, and Af-130 from Afghanistan with prototype sandfly fever Naples and Toscana viruses

Antigen	Mouse IAF			Mouse sera		
	SFN	Toscana	Af-1008	Af-1008	Af-1038	Af-130
SFN	160/320[a]	160/320	160/320	20/80	<10/<10	80/160
Toscana	20/640	160/640	40/640	<10/<10	<10/<10	<10/<10
Af-1008	320/80	640/160	640/320	320/640	40/80	160/320
Af-1038	320/80	640/160	640/160	—	640/320	160/80
Af-130	320/40	320/40	640/80	40/40	80/80	320/80

[a] Antibody titer/antigen titer
— Not tested

Table 2. Complement-fixation tests comparing strains Af-1028 and AF-83 with prototype sandfly fever Sicilian virus

Antigen	IAF	
	Af-1028	SFS
Af-1028	320/320[a]	640/320
Af-83	320/320	640/320
SFS	80/640	640/640

[a] Antibody titer/antigen titer

slightly from prototype SFS strain (Table 2). As may be seen in Table 3, serum samples from convalescent patients from whose blood the viruses had been isolated neutralized field or prototype strains of SFN and SFS. Forty paired acute- and convalescent-phase serum samples from patients diagnosed as having sandfly fever during the 1986 outbreak and 64 paired acute- and convalescent-phase serum samples from the 1987 outbreak were tested by N and IF (Table 4). A fourfold or greater rise in antibody titer was found in 87; in 16 other patients antibodies to both viruses were found but rising titer was found to one or another of them. In one case results were negative. Antibody titers by N were 320–640. Some patients had N antibody 20–40 in early acute-phase sera (1–2 days of the disease) with a rise of antibody titers in the second sample; by CF, antibodies could be detected irregularly, mostly in blood sera collected not earlier than 20 days postinfection. Of 28 sera from servicemen in 1987 who had experienced the disease in 1986, 8 had antibody to SFS and 3 to SFN. In addition, we tested 32 single serum samples from people who had experienced fabrile illnesses in 1987 but who had arrived at

Table 3. Neutralization tests for antibody to strains Af-1008 and Af-1028 and with sandfly fever Naples virus in serum samples from 5 convalescent patients from whom an isolate was obtained

Serum no	Virus isolations, strain	Day of illness	NT with strains		
			SFN (Sabin)	SFN (Af-1008)	SFS (Af-1028)
80	Af-1008	1	20[a]	20	—
		10	160	160	—
19	Af-1038	1	20	40	—
		10	320	640	—
130	Af-130	1	—	40	—
		20	—	320	—
51	Af-1028	1	—	—	<10
		10	—	—	160
83	Af-83	1	—	—	<10
		20	—	—	40
SFN IAF			320	640	—
Af-1008 IAF			320	640	—
SFS IAF			—	—	160

[a] Antibody titer
— Not tested

Table 4. Neutralizing and indirect immunofluorescence tests with paired acute- and convalescent-phase serum samples from patients with clinically diagnosed sandfly fever

Year of collection	Number of sera tested	Antibody to			Test
		SFS	SFN	SFS+SFN	
1986	40	15/40[a] (37.5%)	25/40 (62.5%)	0	N
	40	15/40 (37.5%)	24/40 (60.0%)	0	IF
1987	64	25/37 (67.5%)	36/63 (57.1%)	16/37 (43.2%)	N
	64	26/63 (41.2%)	9/55 (16.3%)	3/37 (5.6%)	IF

[a] Number with antibody/number tested

the hospital from areas of Afghanistan other than the epidemic zones. Antibody to SFN virus was found in only one (IF). Of 101 samples, some with and some without antibody to SFN or SFS viruses, none had IF antibody to Toscana or Rift Valley fever viruses.

Discussion

As a result of the studies, 3 strains of SFN virus and 2 strains of SFS virus were isolated from patients during an outbreak of sandfly fevers among Soviet servicemen in Afghanistan. From the materials of the same outbreak Perepyolkin et al. [10] reported the isolation of Rift Valley virus as well as SFN and SFS viruses. Our results for Rift Valley virus were negative. Variations in the duration of the incubation period in mice may have been due to slight differences in their age (24–48 h old). Some mice, particularly those inoculated with patient serum, experienced asymptomatic infection, as indicated by development of CF antibody (titers 40–160) 28–30 days postinfection. The laboratory diagnosis of sandfly fevers has certain features that deserve mention. The current generation of virologists in the U.S.S.R. and elsewhere has no experience with sandfly fever because epidemics have not been reported for nearly 40 years. Meanwhile, *P. papatasi* vectors are common in Central Asia and these viruses circulate in nature, conditions conducive to their transmission to people.

The possibility of importation of sandfly fever from other regions of the world should also be considered. In 1987, Calisher et al. [3] described a disease caused by Toscana virus in an American who had returned to the U.S.A. from Italy. Toscana virus, isolated in Italy by Verani et al. [14], is antigenically close to SFN virus and is also transmitted by sandflies. SFN and SFS viruses are regularly isolated from blood on the first day of illness but extremely rarely on the second day. Blood remains infectious for prolonged periods, in our experience to 1 year at −40 °C and, according to Sabin [11], to several years when stored at −70 °C. Newborn mice not older than 48 h should be used for virus isolation because individual strains vary in their pathogenicity for mice [12]. The most reliable method of serodiagnosis appears to be the plaque reduction N test. According to our observations and those of Bartelloni and Tesh [1], SFS virus produces large and small plaques; reduction in the number of large plaques was the criterion of our N test. That IF produces results comparable to N tests was also reported by Feinsod et al. [5]. Studies with SFN strains Af-1008, Af-1038, and Af-130 using sera from mice with asymptomatic infections showed that these strains are completely different from Toscana virus and slightly different from prototype SFN strain, which indicates antigenic variations. The Af-1038 strain differs from SFN virus and from strains Af-1008 and Af-130. This report extends our knowledge of the geographic distribution of SFN and SFS viruses, previously isolated in countries neighbouring Afghanistan including Pakistan [8], Iran [13], and other Asian countries [7, 13].

Acknowledgement

The authors are grateful to Dr. P. Verani (Italy) for kindly supplying IAF and CF antigen for Toscana virus.

References

1. Bartelloni P, Tesh R (1976) Clinical and serologic responses of volunteers infected with Phlebotomus fever virus (Sicilian type). Am J Trop Med Hyg 25: 456–462
2. Calisher CH, Karabatsos N (1988) Arbovirus serogroups: definition and geographic distribution. In: Monath TP (ed) The arboviruses: epidemiology and ecology, vol 1. CRC Press, Boca Raton, FL, pp 19–57
3. Calisher CH, Weinberg AN, Muth DJ, Lazuick JS (1987) Toscana virus infection in a United States citizen returning from Italy. Lancet i: 165–166
4. Chunikhin SP, Chumakov MP, Karaseva PS, Semenov BF (1968) Immunological structure against ten arboviruses in mammals, birds and reptiles in the southestern regions of the Turkmenian SSR. Trudi Inst Poliomiel Virus Entsephal Akad Med Nauk SSSR 12: 328–336 (English summary)
5. Feinsod FM, Ksiazek TG, Scott AK (1978) Sand fly fever-Naples infection in Egypt. Am J Trop Med Hyg 37: 193–196
6. Gaidamovich SYa, Kurakhmedova ShA, Melnikova EE (1974) Aetiology of phlebotomus fever in Ashkhabad studied in retrospect. Acta Virol 18: 508–511
7. Gaidamovich SYa, Obukhova VR, Sveshnikova NA, Cherednichenko YuN, Kostyukov MA, Skofertsa PG, Yarovoy PI, Kurakhmedova ShA, Khanmamedov HM, Ismailov ASh (1978) Natural foci of viruses borne by *Phlebotomus papatasi* in the USSR according to serological surveys of the human population. Vopr Virusol 5: 556–560
8. George JE (1970) Isolation of Phlebotomus fever virus from *Phlebotomus papatasi* and determination of the host ranges of sandflies (Diptera: Psychodidae) in West Pakistan. J Med Entomol 7: 670–676
9. de Madrid AT, Porterfield JS (1974) The flaviviruses (group B arboviruses): a cross-neutralization study. J Gen Virol 23: 91–96
10. Perepyolkin VS, Raevsky KK, Nikolaev VP, Sheibak VV (1989) Etiology of Phlebotomus fever in Soviet servicemen in Afghanistan. Voen Med Zh 187: 48–49
11. Sabin A (1959) Phlebotomus fever. In: Rivers TM, Horsfall F (ed) Viral and rickettsial infections of man. Pitman, Philadelphia, pp 374–381
12. Schmidt JR, Schmidt ML, Said MJ (1971) Phlebotomus fever in Egypt. Isolation of Phlebotomus fever virus from *Phlebotomus papatasi*. Am J Trop Med Hyg 20: 487–490
13. Tesh RB, Saidi S, Gaidamovich SYa, Rodhain F, Vesenjak-Hirjan J (1976) Serological studies on the epidemiology of sandfly fever in the Old World. Bull WHO 54: 663–673
14. Verani PL, Nicoletti L, Ciufolini NG (1984) Antigenic and biological characterization of Toscana Virus, a new Phlebotomus fever group virus isolated in Italy. Acta Virol 28: 39–47

Authors' address: Sophia Ya. Gaidamovich, D.I. Ivanovsky Institute of Virology, Academy of Medical Sciences of the U.S.S.R., Gamaleya Street 16, Moscow 123098, U.S.S.R.

Arch Virol (1990) [Suppl 1]: 295–301

Isolation of Crimean-Congo hemorrhagic fever virus from patients and from autopsy specimens

A. M. Butenko[1] and **M. P. Chumakov**[2]

[1]Department of Virus Ecology, D.I. Ivanovsky Institute of Virology, and
[2]Institute of Poliomyelitis and Viral Encephalitides, U.S.S.R. Academy of Medical Sciences, Moscow, U.S.S.R.

Accepted January 18, 1990.

Summary. We tested 57 blood specimens from 51 Crimean-Congo hemorrhagic fever (CCHF) patients and organ tissue specimens from three autopsy cases for virus isolation. Blood specimens were inoculated intracerebrally into newborn albino mice within 20 min after collection or stored them at 4 °C for no more than 10 days before inoculation. Viremia was observed within the first seven days of the disease in almost all CCHF patients and the virus often was demonstrable in the blood for 8–12 days after onset of illness. Attempts to isolate virus from urine samples of patients were ineffective. Maximum virus titers (to $6.2 \log_{10} LD_{50}/ml$) were present in patients' blood for the first five days of illness. CCHF virus was detected in hypothalamus, bone marrow, breast lymphatic glands, kidney, adrenals, and in the large intestine wall of autopsy specimens.

Introduction

Results described in this paper had been obtained in 1967–1969 during CCHF outbreaks in Rostov and Astrakhan regions of the U.S.S.R. and were available to a very narrow circle of specialists. These results may be of interest to those familiar with results of virological examination of patients during outbreaks in Bulgaria [11], Pakistan [2], Iraq [1], South African Republic [9], and elsewhere because integration of such data obtained during many years in different regions of Eurasia and Africa provides a set of ideas concerning virological diagnosis and pathogenesis of CCHF.

Materials and methods

For virus isolation we used newborn (1- to 2-day old) albino mice inoculated intracerebrally with blood or urine samples from the patients or with suspensions of organ tissues from subjects who had died with CCHF. The isolates were identified in complement-fixation (CF) test [5] and agar gel precipitation (AGP) test [7, 8] using ascitic fluids (MAF) [10] from mice immunized with Drozdov strain [4] of CCHF virus. Altogether, 57 blood samples from 51 CCHF patients were tested, including 2 patients studied in Bulgaria by Dr S. Vasilenko (Research Institute of Infectious and Parasitic Diseases, Sofia, Bulgaria).

Results and discussion

Mice were inoculated with blood samples obtained at the bedside from each of 16 patients. From 15 (93.4%) of them CCHF virus was isolated. The only negative sample was from a patient on the eighth day of illness (fourth day of normal body temperature). When kept at 4 °C for no more than 10 days before inoculation of mice, CCHF virus was isolated from 22 of 24 (91.7%) blood samples.

After prolonged storage of blood samples the frequency of virus isolation was reduced, but in two instances CCHF virus was isolated after 15 days and in three other instances after 21, 22, and 39 days of storage at 4 °C. CCHF virus was successfully kept at 4 °C (at least 10 days) in heparinized blood, in blood clots and plasma.

Table 1 shows differences in CCHF virus titers in the blood of three patients depending on the length of storage of the samples at 4 °C. The titers dropped considerably (from 6.2 to 4.2 $\log_{10} LD_{50}/ml$) in one blood samples stored for 21 days. Duration of viremia in CCHF patients was determined with 57 blood samples from patients in the first 12 days of their illnesses. The rate of virus isolations during days 1–7 of the disease was 85.1% (40 positive of 47 tested) (Table 2). The seven specimens from which CCHF virus was not isolated had been stored at 4 °C for 11–34 days before testing. We conclude that during the first 7 days of the disease the virus was present in the blood of essentially all patients. Only 4 isolates were made from 10 samples collected

Table 1. Changes of the C-CHF virus titers in patients' blood in relation to the length of storage of blood samples at 4 °C

Patient	Day of illness	Length of storage of blood samples (days)	Virus titers ($\log_{10} LD_{50}/ml$)
A	3	0	6.2
		21	4.2
B	6	0	2.7
		6	2.7
		15	2.2
C	4	0	5.1
		5	4.7

8–12 days after onset of illness and stored for shorter period at 4 °C. This suggests that in general viremia decreases considerably or ends by day 8; however, in these 4 virus was isolated on 8, 9, 10, and 12 days after disease onset.

CCHF virus titers in 16 blood samples collected from 11 patients were determined; highest titers were within the first 5 days of disease (Table 3). Titers began to decrease 6–7 days after onset of illness and by the day 8–9 viremia had essentially disappeared. This coincides with the time of development of detectable complement-fixing, precipitating, and neutralizing antibodies in CCHF patients [3].

It is known that there are three main clinical forms of CCHF: severe, moderate and mild. The severe course is characterised by high temperature

Table 2. Isolation of C-CHF virus in relation to the time of blood collection after the onset of the disease

No. of samples	Days of illness											Total
	1	2	3	4	5	6	7	8	9	10	12	
Tested	2	4	4	10	12	5	10	4	2	1	3	57
Isolated	2	3	4	8	11	4	8	1	1	1	1	44

Table 3. C-CHF virus titers in 16 blood samples collected from 11 patients

Patient	Day of illness	Virus titers ($\log_{10} LD_{50}/ml$)
1	1	6.2
2	3	5.2
3	3	4.2
4	3	6.2
5	3	4.2
6	4	5.1
7	4	3.6
4	4	3.2
6	5	5.1
4	6	3.2
8	6	2.7
9	7	4.2
10	7	2.7
11	7	<1.7
10	7	<1.7
7	9	2.3

(39–41 °C) lasting sometimes for 12 days (average 7.9), sharp general intoxication and pronounced hemorrhagic syndrome. There is severe blood loss from profuse nasal, gastrointestinal, pulmonary and uterine hemorrhages. In the moderate course, hemorrhages are not life-threatening. Frequently, only hemorrhages of oral and nasal cavities are observed; the fever lasts 4–7 days. In the mild course, fever and general intoxication are relatively short (up to 4 days). Hemorrhagic syndrome is manifested mostly by skin hemorrhages (petechial rash or small hematomas) or are absent.

In our studies CCHF virus was isolated from the blood of 22 (88%) of 25 patients with the severe form of CCHF, from 14 (70%) of 20 with moderate form of CCHF, and from 2 of 6 with mild CCHF. Disregarding results with blood specimens stored inappropriately, the frequency of virus isolation from the blood of patients with the severe, the moderate and the mild forms of CCHF were determined as 22 of 23 (95.7%), 14 of 15 (93.3%) and 2 of 2, respectively.

According to data obtained with CCHF patients in Astrakhan and Rostov regions [6], CCHF is characterized mainly by four general types of febrile reactions. In 39% of the patients, body temperature usually fell to normal levels between 2 and 5 days after onset. This decline usually lasted less than 24 h, seldom 48–60 h, but the curves have a biphasic character (Fig. 1, curves A and B). In two other types, temperature curves appear

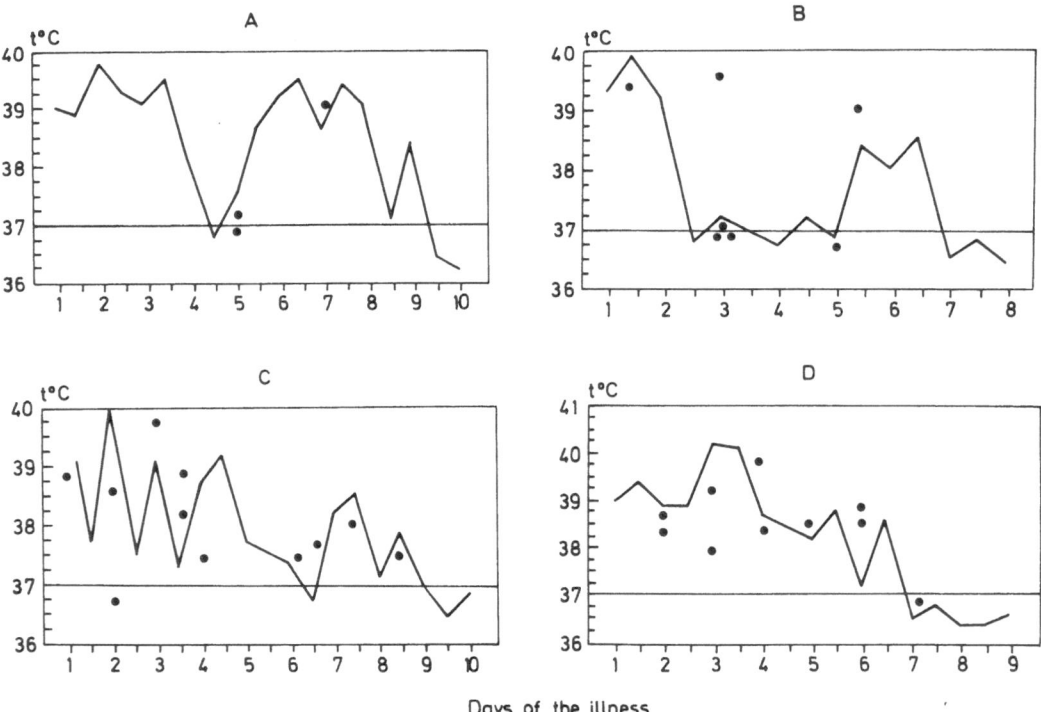

Fig. 1. Isolation of CCHF virus from patient's (**A–D**) blood depending on types of their t °C-reaction. Points designate virus isolation

continuous or irregular with temperature variations of 1.5–2 °C (Fig. 1, curves C and D).

Analysis of virological findings showed that in patients with the first and second types of temperature patterns viremia was observed during the first rise of body temperature, during its decrease, and during the second peak of fever. In patients with the third and fourth types of temperature patterns CCHF virus was isolated from their blood throughout the febrile period.

Urine samples from 10 patients in the acute phase of CCHF were filtered through 220 nm Millipore filter and inoculated into mice undiluted and at dilutions 1:5 and 1:10. CCHF virus was isolated from none of these but was isolated from the blood of these patients.

Organs and tissues from three fatal cases of CCHF in the Rostov region were also tested for virus. According to the records (EV Leschinskaya, VN Lazarev, unpubl. data) these patients had suffered from typical CCHF (fever, acute general toxicity, prostration, multiple hemorrhages into the skin and mucous membranes, massive bleeding mainly from the gastrointestinal tract). Portions of organs and tissues obtained at autopsy were put into

Table 4. Virus isolation from organs and tissues of three fatal C-CHF cases

Materials	Results of examinations		
	patient 12	patient 13	patient 14
Cortex	−	−	ND
Hypothalamus	+	−	ND
Cerebellum	−	−	ND
Spinal cord	−	ND	ND
Bone marrow	+	−	+
Tonsil	−	ND	ND
Lymphoid gland (breast)	−	−	+
Myocardium	−	ND	−
Lung	−	+	ND
Liver	+	−	+
Spleen	−	−	+
Kidney	−	+	ND
Adrenal gland	ND	+	ND
Pancreas	+	ND	ND
Gastric mucosa	−	ND	ND
Large intestine wall	+	ND	ND
Skeletal muscle	−	ND	ND
Skin	−	ND	ND
Blood	−	ND	−

− Virus not isolated
+ Virus isolated
ND Not done

sterile vials and sent to the laboratory on dry ice. The samples were thawed and washed with sterile saline solution before preparation of suspension for virus isolation. From patient 12, who died on the eighth day of illness, CCHF virus was isolated from blood collected 4–5 days after onset of illness before death; autopsy was done 6 h after death (Table 4). Patient 13 died on the ninth day of illness; autopsy was done 10 h after death. CCHF virus had been isolated from blood of this patient on the sixth day of illness. Patient 14 had died on the sixth day of illness; autopsy was done 24 h after death. Virological examination of organs and tissues showed that CCHF virus was present in hypothalamus, bone marrow, lymphatic gland (breast), lung, liver, spleen, kidney, adrenal gland, pancreas, and large intestine wall. Attempts to isolate the virus from post mortem blood of two of these patients were unsuccessful. No virus was isolated from the brain cortex, cerebellum, tonsils, myocardium, skeletal muscles, gastric mucosa, and skin. These results emphasize the role of hematogenous routes of CCHF virus dissemination during the disease and indicate the pantropic nature of this virus.

Acknowledgement

We are most grateful to Dr. E. V. Leschinskaya, Institute of Poliomyelitis and Viral Encephalitides, U.S.S.R. Academy of Medical Sciences, Moscow, and Dr. V. N. Lazarev, Volgograd State Medical Institute, Volgograd, U.S.S.R. for conducting the clinical part of this study.

References

1. Al-Tikriti SK, Al-Ami F, Jurji FJ, Tantavi H, Al-Moslih M, Al-Janabi N, Mahmud MJA, Al-Bana A, Habid H, et al (1981) Congo-Crimean hemorrhagic fever in Iraq. Bull WHO 59:1: 85–90
2. Burney MI, Ghafoor A, Saleen M, Webb PA, Casals J (1980) Nosocomial outbreak of viral hemorrhagic fever caused by Crimean-Congo virus in Pakistan, January 1976. Am J Trop Med Hyg 29: 941–947
3. Butenko AM (1971) Data from studying etiology, laboratory diagnosis and immunology of Crimean hemorrhagic fever; questions of ecology of the viral agent. Avtoref Diss Soisk Uchen Step Dokt Biol Nauk Inst Poliomiel Virus Encephal Acad Med Nauk SSSR, Moscow, (in Russian) (in English, NAMRU3-T-1152)
4. Butenko AM, Chumakov MP, Bashkirtsev VN, Zavodova TI, Tkachenko EA, Rubin SG, Stolbov DN (1968) Isolation and investigation of Astrakhan strain ("Drozdov") of Crimean hemorrhagic fever virus and data on serodiagnosis of this infection. Mater 15 Nauchn Sess Inst Poliomiel Virus Encephal Acad Med Nauk SSSR (October 1968) 3: 88–90 (in Russian) (in English, NAMRU3-T-866)
5. Casey HL (1965) Standardized diagnostic complement fixation method and adaptation to micro test, part II. In: Pub Health Monogr No 74; Adaptation of LBCF method to micro technique, pp 31–34 (P.H.S. Publication No 1228) US Government Printing Office, Washington, DC
6. Leshchinskaya EV (1967) Clinical picture of Crimean hemorrhagic fever and its comparison with hemorrhagic fevers of other types. Avtoref Diss Soisk Uchen Step Dokt Med Nauk Inst Poliomiel Encephal Acad Med Nauk SSSR, Moscow (in Russian) (in English, NAMRU3-T-1180)

7. Murphy FA, Coleman PH (1967) California group arboviruses: immunodiffusion studies. J Immunol 99: 276–284

8. Ouchterlony O (1958) Diffusion in gel methods for immunological analysis. Prog Allergy 5: 1–78

9. Shepherd AI, Swanepoel R, Shepherd SP, Leman PA, Blackburn NK, Haller AF (1985) A nosocomial outbreak of Crimean-Congo hemorrhagic fever at Tygerberg Hospital. V. Virological and serological observations. S Afr Med J 68: 733–736

10. Tikasingh ES, Spence L, Downs WG (1966) The use of adjuvant and sarcoma 180 cells in the production of mouse hyperimmune fluids to arboviruses. Am J Trop Med Hyg 15: 219–226

11. Vasilenko SM (1973) Results of the investigation on etiology, epidemiologic features, and the specific prophylaxis of Crimean hemorrhagic fever in Bulgaria. In: Proc 9 Int Congr Trop Med Malaria, Athens, Greece 1: 32–33

Authors' address: A. M. Butenko, Department of Virus Ecology, D.I. Ivanovsky Institute of Virology, Academy of Medical Sciences of the U.S.S.R., Gamaleya Street 16, Moscow 123098, U.S.S.R.

Arch Virol (1990) [Suppl 1]: 303–322

Ecology of ticks as potential vectors of Crimean-Congo hemorrhagic fever virus in Senegal: epidemiological implications

J.-L. Camicas[1], M. L. Wilson[2, 3], J.-P. Cornet[1], J.-P. Digoutte[2], M.-A. Calvo[2], F. Adam[1], and J.-P. Gonzalez[1, 2]

[1]Institut Francais de Recherche Scientifique pour le Developpement en Cooperation (ORSTOM), Laboratoire ORSTOM de Zoologie medicale, Dakar
[2]Institut Pasteur, Dakar, Senegal
[3]Departments of Population Sciences and Tropical Public Health, Harvard School of Public Health, Boston, Massachusetts, U.S.A.

Accepted February 8, 1990

Summary. At least 30 tick species from 7 genera have been found naturally infected with Crimean-Congo hemorrhagic fever (CCHF) virus worldwide. To this list we add *Rhipicephalus guilhoni*. In the sub-saharan Africa, 17 tick species have been implicated as vectors, of which 12 are present in Senegambia or Mauritania. We studied the five principal species that appear to be the most important in CCHF virus transmission in Senegal, namely *Amblyomma variegatum*, *Hyalomma impeltatum*, *H. marginatum rufipes*, *H. truncatum*, and *Rhipicephalus guilhoni*. We report on the distribution, host associations, seasonal activity patterns and CCHF virus infection of these ticks, as well as the epidemiological implications for human disease. Despite similarities in ecological characteristics, not all of these species are equally likely to be important in the transmission cycle. The most important vectors in enzootic and epidemic transmission throughout Senegal appear to be *Hyalomma truncatum* and *Amblyomma variegatum*.

Introduction

The geographic distribution of Crimean-Congo hemorrhagic fever (CCHF) is astonishingly large for an arboviral zoonosis and includes most of Africa, southern Eurasia, southern Europe, and the Middle East [10]. CCHF virus transmission occurs in ecologically diverse sites of at least 30 different countries in the palearctic, oriental, and afrotropical biogeographic regions

[10, 23]. Equally astonishing is the number and diversity of ticks from which this virus has been isolated. Nearly 30 species from seven genera have been shown to be naturally infected [23]. It is precisely such diversity of potential vectors that represents a major obstacle to our understanding both of the ecology of CCHF virus and of the epidemiology of human outbreaks. In most settings, little is known about the important vector species and the conditions under which enzootic transmission is maintained. Which of these or other ticks are active in epizootic or epidemic transmission? What do the feeding patterns of the three stages of each tick species imply for vertebrate hosts of the virus? Which species competently transmit virus transovarially, transstadially and horizontally? Which tick stages and species can transmit the virus to humans? The diversity of tick species and stages that have been found infected in various faunal regions and habitats poses a formidable task for studies on CCHF virus transmission.

One common denominator appears to be the presence of *Hyalomma* spp. ticks. Despite CCHF virus isolations from numerous species of seven genera, active transmission has been most often linked to one or another species of *Hyalomma* ticks. Evidence of CCHF virus transmission is found only where these ticks are present, and epizootic or epidemic transmission corresponds with periods of increased abundance of *Hyalomma* spp. [23]. Conversely, enzootic transmission may be feeble or nonexistent where vector tick abundance is minimal or nil. However, this global relationship between tick species diversity and the magnitude of CCHF virus transmission has not been systematically studied within a defined geographic region.

Study sites and methods

In sub-saharan West Africa, CCHF virus has been isolated from ticks, domestic and wild vertebrates, and humans in Senegal, Mauritania, and Burkina Faso [3, 16–18]. This region of large ecological diversity is situated in a zone of transition between the Sahara desert in the north and rainforests in the south; the biota changes dramatically from north to south primarily in association with increasing rainfall. Annual precipitation, averaging 200–300 mm annually in the northern sahelian zone, is five times greater in the southern sudano-guinean and sub-guinean zones. The diversity and abundance of ixodid ticks is also varied [3, 6–8]. In this study, we analyse the potential of these ticks as vectors of CCHF virus in these zones and attempt to synthesize the ecological and epidemiological implications of these findings to determine risk of human infection.

Our analysis is based on experience and studies spanning nearly two decades. During this period, casual observations of host associations and CCHF virus infection have been increasingly directed toward systematic field observations of tick population dynamics, tick feeding patterns, and CCHF virus transmission cycles. In addition to anecdotal observations, we report on regular samples of domestic animals and wild rodents from the Senegal study sites in the villages of Bandia, Dahra, and Yonofere (Fig. 1). Many of our results and conclusions should be applicable to ecologically similar sites throughout the region.

Potential vector ticks

Worldwide CCHF virus infection of ticks

A total of 30 species of ticks (Acari: Ixodidae), have been found naturally infected with CCHF virus, including 28 species of ixodids ("hard" ticks) and 2 of argasids ("soft" ticks) [10, 19, 23]. The 28 ixodid species are distributed among 7 genera as follows: 11 *Hyalomma*, 8 *Rhipicephalus*, 4 *Boophilus*, 2 *Dermacentor*, 1 *Amblyomma*, 1 *Haemaphysalis*, and 1 *Ixodes* (Table 1). The 2 species of argasids belong to the genera *Argas* and *Alveonasus* (a subgenus of *Ornithodoros* for some taxonomists). The relevance of infection in these latter two "soft" ticks recently has been questioned by the experimental work of Shepherd et al. [19], who demonstrated CCHF virus replication in numerous ixodids, but could not show replication in 3 Argasina infected by intracoelomic inoculation. They concluded that strains obtained from the two argasid ticks may be merely surviving virus from recent viremic blood-meals, as was probably the case in the isolation from *A. lahorensis* [20]. Thus, CCHF virus appears to be primarily a virus transmitted by hard ticks (Ixodidae).

The biogeographic diversity of CCHF virus-infected ticks is striking. Five Ixodidae belong to the oriental biogeographical region, 17 ixodids and the 2 argasids to the palearctic region, and 18 or 19 ixodids plus 1 argasid are found infected in sub-saharan Africa. Note that the total is greater than 30, because some species are found in several faunal regions.

Infected ticks in Senegambia and Mauritania

Of the 17 ixodids implicated in published reports as potential vectors from sub-saharan Africa, (18 if *H. anatolicum anatolicum*, which may have been introduced into Nigeria, is included), 12 (13?) are present in Senegambia or Mauritania. Among these, 7 have been found naturally infected [3, 15, 18]. These include *Amblyomma variegatum*, *Boophilus decoloratus*, *B. geigyi*, *Hyalomma impeltatum*, *H. impressum*, *H. marginatum rufipes*, *H. truncatum*. To this list we add an eighth potential vector, *Rhipicephalus guilhoni* Morel & Vassiliades, 1963, from which CCHF virus was isolated in August, 1988 (Table 1). That isolation was made from a pool of 20 male *R. guilhoni* removed from sheep in Yonofere, Senegal (Fig. 1). The identity of the virus was confirmed at the WHO Collaborating Center for Arboviruses at the Pasteur Institute in Dakar. Thus, there is a total of 8 tick species from which CCHF virus has been isolated in Senegal or Mauritania (Table 1).

Other potential vectors that have been shown to be infected in other areas, and that exist in this region include *B. annulatus*, *H. dromedarii*, *H. nitidum*, *R. evertsi evertsi* and *R. sanguineus*. However, the low population densities of these species or the peculiarities of their feeding ecology suggest

Table 1. Potential vectors of CCHF virus in relation to geographic region

Genus, species	Pal	Ori	Afr	Sen
Hyalomma				
H. anatolicum anatolicum	○	○	+?	
H. asiaticum	○			
H. detritum	○			
H. dromedarii	○	+	+	+
H. impeltatum	+		○	○
H. impressum			○	○
H. marginatum marginatum	○	+		
H. marginatum rufipes			○	○
H. marginatum turanicum	○		+	
H. nitidum			○	+
H. truncatum			○	○
Rhipicephalus				
R. appendiculatus			○	
R. bursa	○			
R. evertsi evertsi			○	+
R. guilhoni			○	○
R. pulchellus			○	
R. pumilio	○			
R. rossicus	○			
R. sanguineus	○	+	+	+
R. turanicus	○		+	+?
Boophilus				
B. annulatus	○		+	+
B. decoloratus			○	○
B. geigyi			○	○
B. microplus		○	+	
Dermacentor				
D. daghestanicus	○			
D. marginatus	○			
Amblyomma				
A. variegatum			○	○
Haemaphysalis				
H. punctata	○			
Ixodes				
Ixodes ricinus	○			

Pal Palearctic region
Ori Oriental region
Afr Afrotropical region
Sen Senegal
+ Tick species present
○ Tick species present and naturally infected

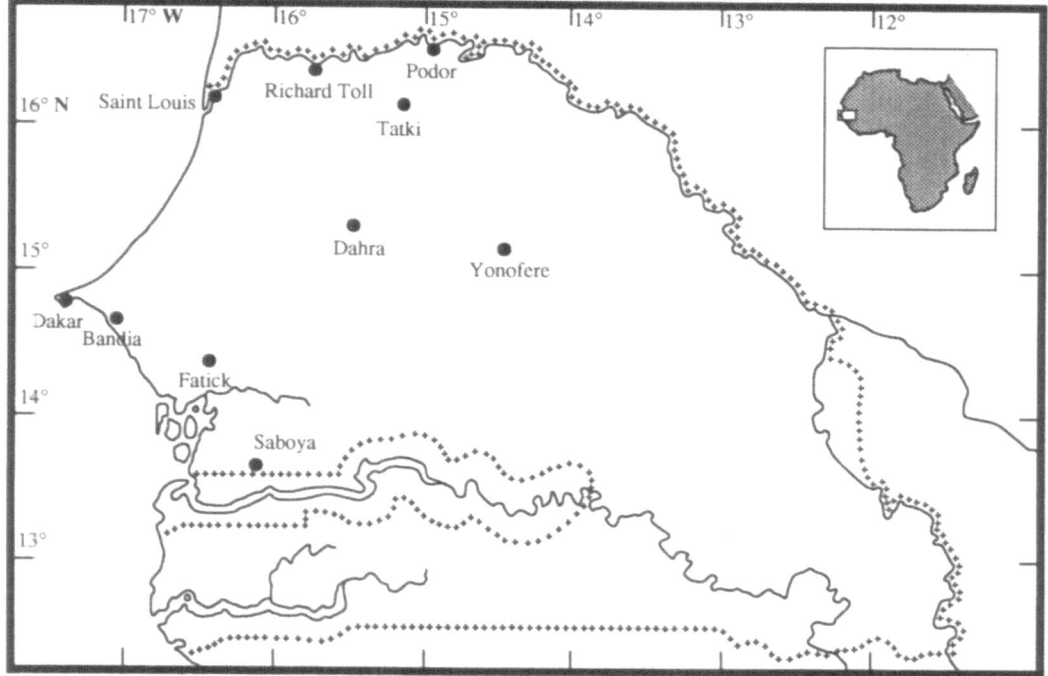

Fig. 1. Map of Senegal indicating major villages and the 3 sites (Bandia, Dahra and Yonofere) in which studies were done

that they are unlikely to play an important epidemiological role in the transmission of CCHF virus. The possibility that *R. sanguineus*, which typically infests canids, may be involved in a domestic cycle involving dogs remains to be ascertained. Thus, we consider there are five principal species (*A. variegatum*, *H. impeltatum*, *H. marginatum rufipes*, *H. truncatum*, and *R. guilhoni*) that are likely to be involved in either enzootic or epizootic transmission of CCHF virus because of their widespread distributions or relative densities.

For each of these ticks, we have analysed ecological factors that influence their potential as vectors of CCHF virus in Senegal, namely their distribution, host associations, and seasonal activity pattern. The classification scheme of Morel [13] has been employed to characterize the feeding ecology of ixodid ticks in terms of the types of hosts that larvae, nymphs, and adults feed upon during each of the three blood-meals, and the number of different individuals utilized while feeding. Ixodid ticks are considered as either mono-, di-, or triphasic signifying that the number of different individual hosts that they feed upon is either one, two, or three. The diversity of host types (e.g., ungulates, ground feeding birds and small mammals, carnivores) that are typically fed upon is characterized by the terms mono- or bitropic which indicate one or two host types, or telotropic for tick species whose

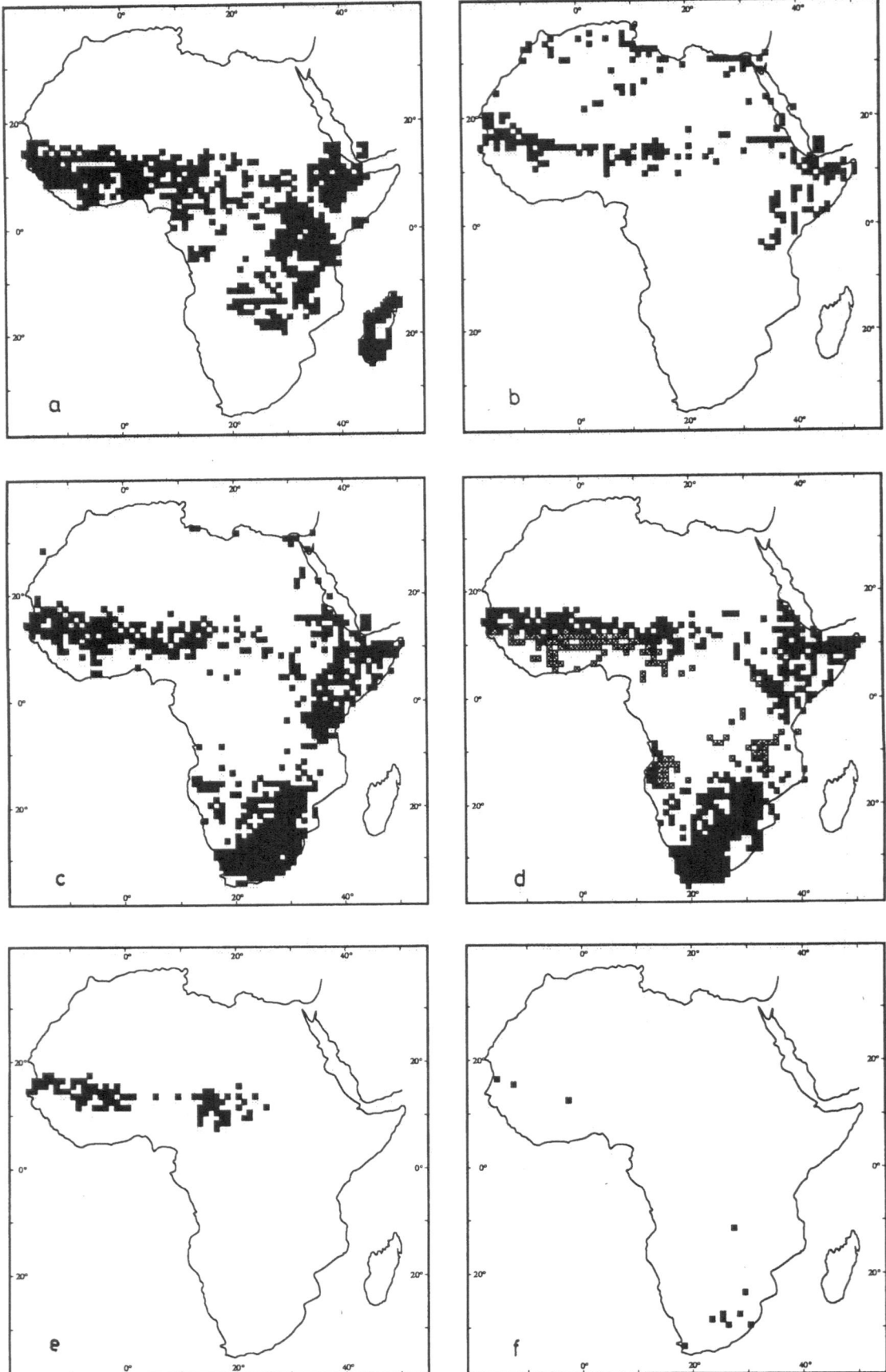

immature stages utilize multiple host types while the adults are host type-specific. These factors influence the population dynamics of the vector and the possibilities for horizontal transmission of CCHF virus. Discussion of the distributions and host associations of these ticks is based on studies by Camicas et al. [1–3], Gueye et al. [6–8], and our unpublished observations.

Amblyomma variegatum

Distribution

This species exists, more or less abundantly, throughout most of Senegal (Fig. 2), its northern limit being the Senegal River basin [3, 7, 8]. It is rare in northeastern Senegal, particularly far from this river. This species tends to predominate where rainfall or humidity is more abundant. For adult ticks, the principal hosts are domestic ungulates, especially cattle which are typically ten times more parasitized than small ruminants (sheep and goats). Larval and nymphal ticks are found on practically all available terrestrial vertebrates including humans [2], although they seem to prefer small ruminants.

Host associations

A. variegatum typically exhibits a triphasic feeding cycle (i.e., each of the three stages engorge on and drop from a different individual host) and a telotropic host association pattern in which immatures are not host-specific and can be found on most available vertebrates, whereas adults feed principally on domestic ungulates.

Larval *A. variegatum* may intensely parasitize a very wide variety of vertebrate hosts. A study performed at Bandia, Senegal on a herd of 10 to 20 goats from 1976 through 1980, has shown a mean of about 500 larvae per animal during November of each year (Fig. 3). Ground-feeding birds are also important hosts for larvae because of their abundance and significant parasitic load. Indeed, we have occasionally found intense infestations of tens or hundreds of larvae on birds such as the red-beaked hornbill (*Tockus erythrorhynchus*), Senegal coucal (*Centropus senegalensis*), grey-breasted helmet guinea-fowl (*Numida meleagris galeata*), double-spurred francolin (*Francolinus bicalcaratus*), and long-tailed glossy starling (*Lamprotornis*

Fig. 2. Geographic distributions of 5 potential CCHF virus vector ticks and of naturally acquired human hemorrhagic CCHF cases in Africa. Tick species include *Amblyomma variegatum* (**a**), *Hyalomma impeltatum* (**b**), *H. marginatum rufipes* (**c**), *H. truncatum* (■) (⊠ *H. truncatum* and/or *H. nitidum*) (**d**), and *Rhipicephalus guilhoni* (**e**), Natural human hemorrhagic cases (**f**)

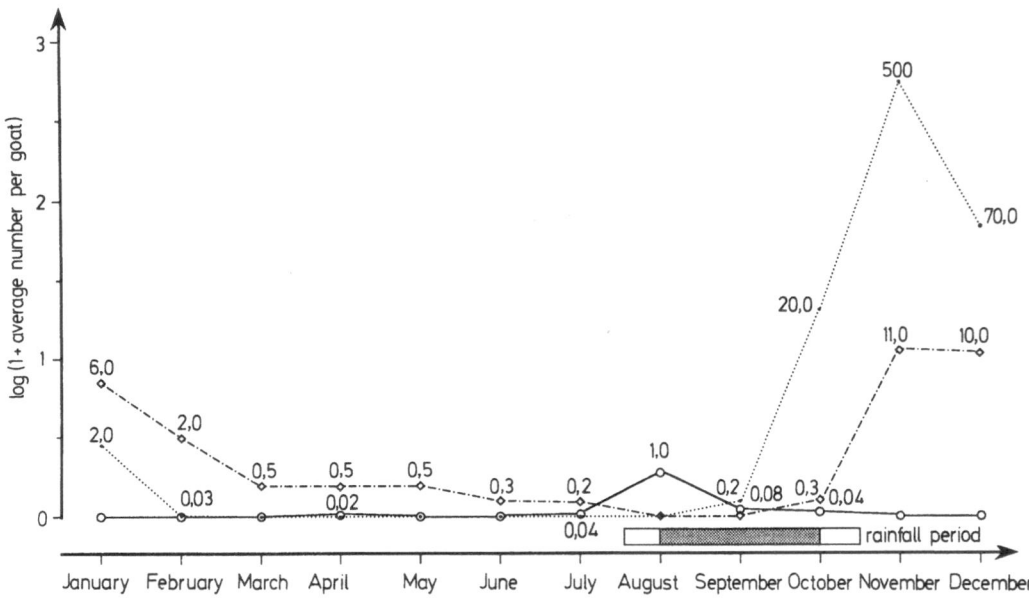

Fig. 3. Seasonal activity cycle of *Amblyomma variegatum* larvae (● ···· ●), nymphs(◇–·–◇), and adults(○—○) at Bandia, Senegal. Average number of ticks per goat per month during 1976 through 1980. Rainy season during the 1972–86 drought (dark bar) was shorter than the typical rainy period (lighter extension)

caudatus). Several species of medium and small mammals may be heavily parasitized by *A. variegatum* larvae. Except for myomorph rodents, which are never parasitized by this tick, larvae can be found on most small mammals. We have found numerous larvae on the common jackal (*Canis aureus*), the serval (*Felis serval*), and Crawshay's hare (*Lepus crawshayi*). Primates may be parasitized, including the green monkey (*Cercopithecus aethiops sabaeus*) and the patas monkey (*Erythrocebus patas*).

Preferred hosts of nymphal *A. variegatum* are large and small ruminants, as well as large rodents, lagomorphs and primates. We have removed nymphal *A. variegatum* from the above-mentioned monkeys, hares, and the ground squirrel *Xerus erythropus*. Some birds have also been found infested.

Adult *A. variegatum* prefer ungulates, and particularly (zebu) cattle. In addition, they occasionally can be found on carnivores, especially domestic dogs. There appears to be a tendency toward monotropism on ungulates and dog.

Seasonal activity

The seasonal pattern of *A. variegatum* in Senegal indicates that it is univoltine [2]. Adults are most active on ungulates during the rainy season, typically from June or July through September and October (Fig. 3). The larval population is at its maximum just after this period, particularly in

November; larvae usually cannot be found in February and March. The nymphal activity peak follows that of larvae by about a month. Nymphs, however, do not disappear abruptly; they persist in low numbers until the next rainy season. Unlike larvae and adults, which are highly hygrophilous, nymphs of this species are much more tolerant of dryness, and are found in reduced numbers on vertebrates throughout most of the dry season.

Epidemiological implications

Because immature A. variegatum exhibit a strong anthropophily (personal observations of 5 people infested with a total of 129 larvae and 12 people with 14 nymphs), they may be able to transmit CCHF virus to humans. Transmission by larvae should be rather exceptional because this would have required that the larva be previously infected by transovarial transmission of the virus, an apparently rare event. Alternatively, nymphs, infected by the transtadial route from larvae fed on viremic hosts, could be important vectors to humans. Adults, emerged from nymphs infected mainly by feeding on ruminants, may transmit CCHF virus to other ungulates, thereby supporting viral circulation and amplification. These adults, however, are of no direct importance in human infection because they almost never feed on humans.

Hyalomma impeltatum

Distribution

The geographic range of H. impeltatum (Fig. 2) was first described by Morel [13] who noted that this tick "does not belong to the Ethiopian fauna but constitutes in Oriental and Occidental Africa a Mediterranean Palearctic component". The southern limit of the distribution of this species in coastal West Africa is at about 14° N. Adult H. impeltatum are abundant on cattle in northern Senegal where rainfall is low, averaging ca. 500 mm or less. This boundary extends east-southeast reaching 10° N in Mali [22] and even 9° N in Nigeria [13]. In Oriental Africa, the phytoclimatic situation is quite different and parallel bands of increasing rainfall and corresponding vegetation do not exist. There the species will be found as far south as the latitude 5° S. In Senegal, this species is rarely found at the Bandia station (14°30′ N); we found immature stages in 1967 and 1970 on *Lepus crawshayi* and again two larvae on the rodent *Arvicanthis niloticus* in 1988.

Host associations

H. impeltatum is typically pholeo-exophilic ditropic, that is, its immatures feed principally on small mammals, while adults most often infest ungulates and carnivores [13]. Specifically, the major hosts of adults are domestic ungulates, particularly zebu cattle, and to a lesser extent sheep and goats. The hosts for both larvae and nymphs appear to be similar, principally myomorph rodents of the genera *Acomys, Arvicanthis, Gerbillus, Psammomys, Jaculus*, as well as hares (*Lepus crawshayi*) and hedgehogs (*Erinaceus*) [1, 9; Morel, pers. comm.].

Seasonal activity

The seasonal population dynamics of *H. impeltatum* has been studied in Dahra (Fig. 1) where, each month, 50 randomly selected sheep were examined, 10 from each of 5 different flocks. Adults were most numerous from the beginning of January through August, showing increased activity during the dry season (Fig. 4). Interestingly, a similar study of sheep in Yonofere, about 80 km due east of Dahra (Fig. 1), indicates that this tick is rarely found there (ML Wilson, unpubl.).

Observations on immature *H. impeltatum* are scarce, although it appears that most immatures are active just after the rainy season. Except for two larvae found in mid-February 1988 in Bandia, the six other findings of immatures on rodents and hares were in November and December. Nymphal activity occurs typically a month or two following that of larvae. Perhaps those eggs that are laid during the rainy season have a greater chance of hatching. Yet larvae, nymphs and adults, which are xerophilous, are active during the dry season. Thus, it may be that members of this species reproduce one generation per year.

Epidemiological implications

H. impeltatum probably is incidental to infection of humans. Although Hoogstraal [9] mentioned that humans are fed upon by adult and immature

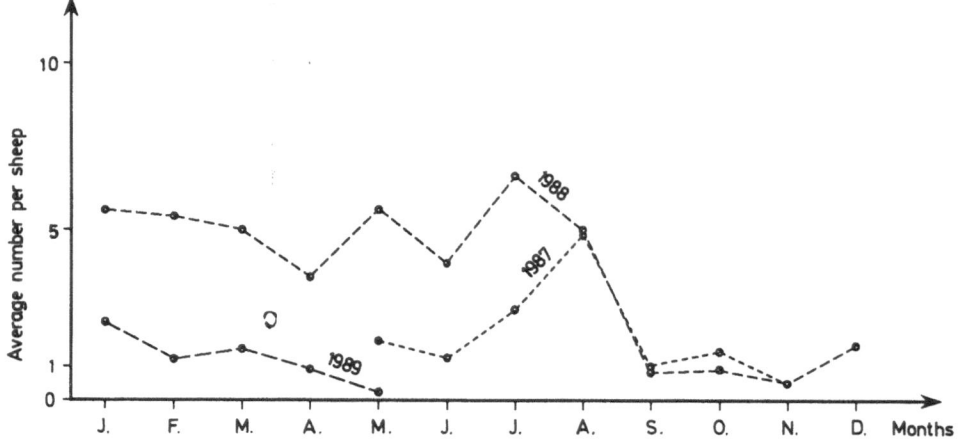

Fig. 4. Seasonal activity cycle of adult *Hyalomma impeltatum* on sheep in Dahra, Senegal. Monthly mean number of ticks from 50 randomly selected sheep (10 from each of 5 herds) examined monthly from May 1987 through May 1989

stages, we have never noticed such occurrence in Senegambia during the past two decades. Infected immature ticks may transmit CCHF virus between small mammals (rodents and lagomorphs). Larval *H. impeltatum*, however, would be able to infect small mammals only if they had been infected transovarially. Nymphs that emerge from larvae engorged on a viremic rodent may be able to infect a naive rodent. Adults that develop from infected nymphs are possible vectors, particularly to ungulates that could become infected at any point during the year, more likely during the dry season. To the extent that *H. impeltatum* adults feed on humans, transmission would occur during the period from November through June.

Hyalomma marginatum rufipes

Distribution

In his study, Morel [13] explained that "the normal habitat of the species is situated between the isohyets 150 and 750 mm of annual rainfall" (Fig. 2). *H. marginatum rufipes* disappears where annual rainfall exceeds 1250 mm, except in southern Senegal and western Guinea. In this area, despite abundant precipitation (ca. 2,000 mm per year), the seven-month long dry season apparently creates acceptable conditions. In Senegal, this tick is present nearly everywhere. Despite widespread distribution, *H. marginatum rufipes* is never very abundant, typically much less abundant than *A. variegatum* and *H. truncatum*.

Host associations

H. marginatum rufipes typically exhibits a biphasic ditropic cycle in which engorged larvae molt on the host producing nymphs that feed on the same individual; adults, however, feed on another type of host. Immatures most often parasitize birds or lagomorphs, while adults usually infest ungulates. A 1969 study showed that 23% of birds were parasitized in Bandia and 4.5% in Saboya (JL Camicas et al., unpubl.). Ground-feeding birds that are often infested by immature *H. marginatum rufipes* include Levaillant's cuckoo (*Clamator cafer*), spotted eagle-owl (*Bubo africanus*), broad-tailed roller (*Eurystomus glaucurus*), red-beaked hornbill (*Tockus erythrorhynchus*), long-tailed glossy starling (*Lamprotornis caudatus*), and the common garden bulbul (*Pycnonotus barbatus*). In addition, terrestrial monkeys (e.g., *Erythrocebus patas*), hedgehogs (*Erinaceus* [*Atelerix*] *albiventris*), and hares (*Lepus crawshayi*) have been found heavily infested. The major hosts of adults are ungulates, particularly zebu cattle.

Seasonal activity

The monthly mean number of adult *H. marginatum rufipes* on natural hosts shows little seasonality; they are present, though not abundant, throughout

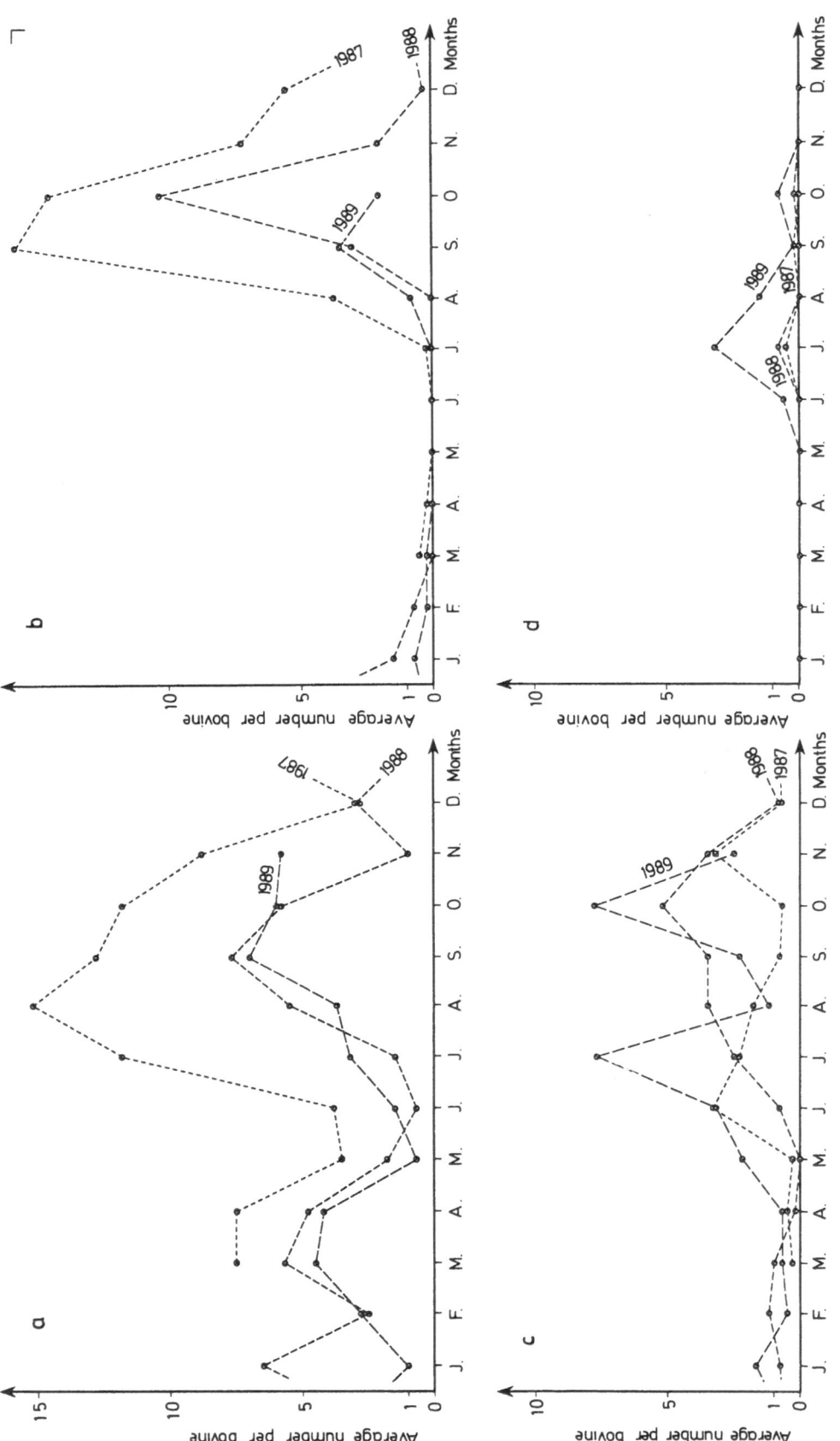

Fig. 5. Seasonal activity cycle of adult *Hyalomma truncatum* (**a**), *Rhipicephalus guilhoni* (**b**), *H. marginatum rufipes* (**c**), and *Amblyomma variegatum* (**d**) in Bandia Senegal. Monthly mean number of ticks from bovines examined three times each month from March 1987 through November 1989

the year (Fig. 5). Similarly, we have not been able to determine a seasonal pattern for immatures, either because infestation levels are low or because of annual variation in weather. Thus, it has been impossible to determine whether *H. marginatum rufipes* has one or more generations per year, or does not follow any seasonal cycle in Senegal.

Epidemiological implications

From an epidemiological point of view, the rarity of human infestation by adult *H. marginatum rufipes* suggests that this species plays a minor role in human infection. Moreover, as birds apparently are unable to develop significant viremias with CCHF virus, and as the proportion of immature ticks that feed on mammals is very small, *H. marginatum rufipes* may be of little epidemiological importance unless transovarial transmission of the virus is a frequent phenomenon.

Hyalomma truncatum

Distribution

In West Africa, *Hyalomma truncatum* is found predominantly in areas with annual rainfall between 400 and 1,000 mm, although its range extends to regions with annual precipitation ranging from 150 mm to 2,000 mm [13]. The southern boundary of the recognized distribution of this species perhaps should be revised in view of the possible confusion with *H. nitidum* Schulze, 1919, a more hygrophilous species. According to our observations, *H. nitidum* is not found in Senegal west of 13° W, whereas *H. truncatum* is found throughout the region southward to areas experiencing 1,500 mm of annual rainfall. Our observations indicate that *H. truncatum* exists throughout Senegal (Fig. 2). In the southeast *H. truncatum* coexists with *H. nitidum*. In Senegal *H. truncatum* is, by far, the most abundant and widespread member of the genus *Hyalomma*.

Host associations

H. truncatum is a species with a triphasic-ditropic cycle. Specifically, hosts of larvae tend to be ground-feeding birds, rodents, hedgehogs, and hares. The resulting nymphs typically feed on a different individual host, but these also tend to be rodents, hares and hedgehogs. The principal hosts of adult *H. truncatum* are cattle and sheep.

Seasonal activity

Our studies of *H. truncatum* infestations of cattle in Bandia suggest that they produce only one generation per year. With the first significant rains of July,

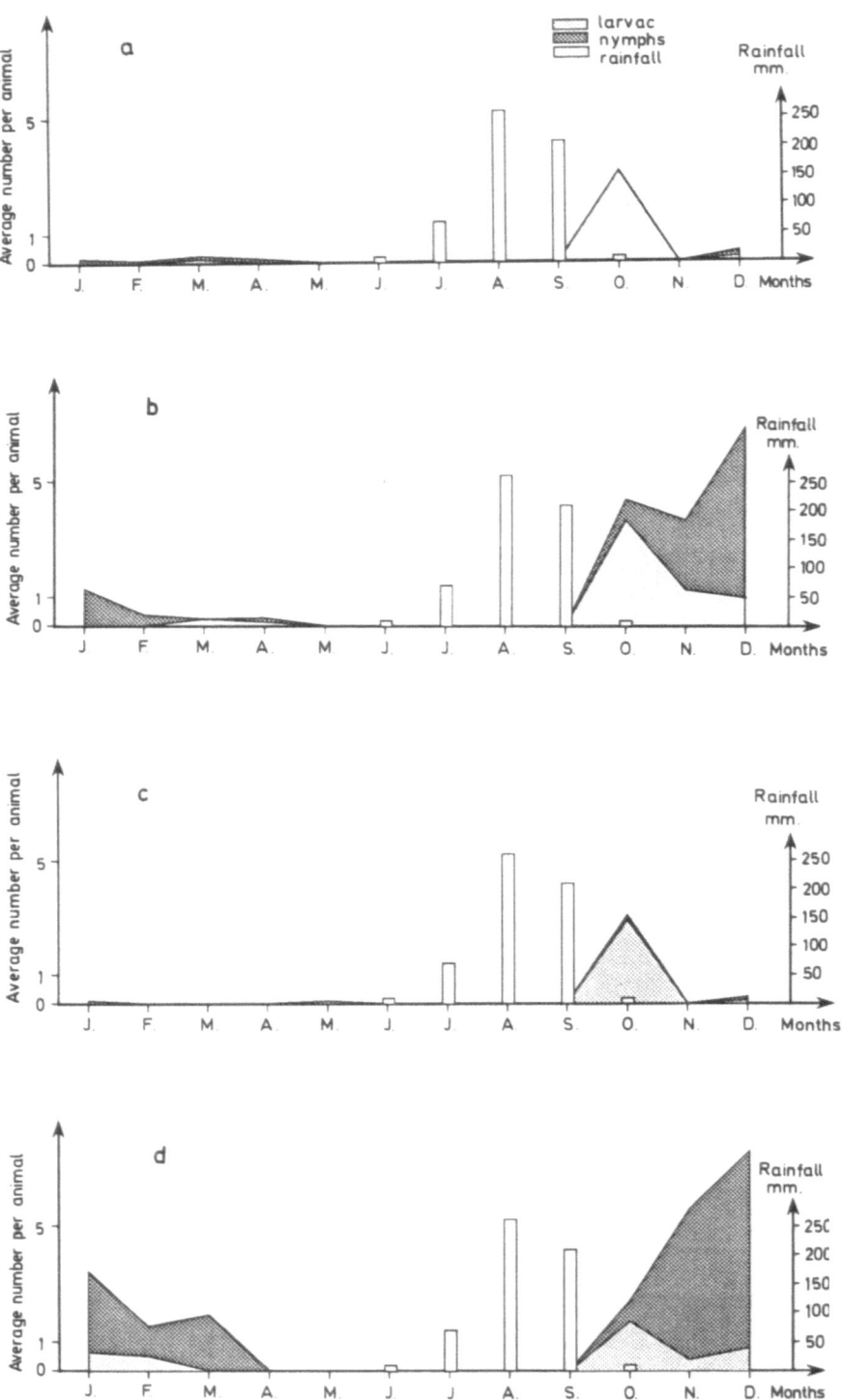

Fig. 6. Seasonal activity cycle of larval and nymphal *Hyalomma truncatum* attached to *Mastomys erythroleucus* (**a**) or *Arvicanthis niloticus* (**b**), and of *Rhipicephalus guilhoni* attached to *M. erythroleucus* (**c**) or *A. niloticus* (**d**) in Bandia, Senegal. Monthly mean number of ticks from rodents were examined monthly from March 1987 through December 1988

adults appear in large numbers and become progressively more scarce through June of the following year (Fig. 5). Parasitism of rodents, determined by holding captured rodents until all ticks have detached, indicates that larvae are in maximum abundance on these hosts in October, and nymphs in December (Fig. 6). Occasionally, we have found a few larvae in March. However, because adult *H. truncatum* feed throughout the year, it may be that engorged females produce eggs that emerge as larvae at that time. Larvae emerging in October from females engorged one or two months earlier find optimal conditions of temperature and relative humidity. Later, as the dry season progresses, the hygrometric conditions become unfavourable and probably lead to a high mortality of laid eggs or emerged larvae and nymphs. The occasional appearance of some larvae on rodents in March and April may correspond to momentarily favorable conditions due to erratic rains that can occur at this period of the year [11].

Epidemiological implications

H. truncatum may be an important vector of CCHF virus to humans in Senegal for several reasons. First, it is among the most abundant and widely distributed of potential vectors. Second, tick infection by horizontal transmission may be relatively more common because a proportion of immatures feed on small mammals that are liable to harbor CCHF virus such as *Erinaceus albiventris* [4] and *Mastomys* sp. [5]. Recent experimental results have shown that larva can be infected by feeding on viremic new-born laboratory mice, and that adults are transstadially infected and transmit virus to guinea pigs [12]. Finally, *H. truncatum* adults have been found to feed on humans (four personal observations). Our studies in Yonofere indicate a strong correlation between IgG antibodies to CCHF virus in humans and their reports of being bitten by adult *H. truncatum*, but not by adult *A. variegatum* or *Rhipicephalus guilhoni* (L. Chapman et al., unpubl.).

Rhipicephalus guilhoni

Distribution

The recognized distribution of *R. guilhoni* is limited to habitats between the isohyets of 250 mm and 750 mm annual rainfall [13] (Fig. 2). In Senegal, this species is found from the Senegal River in the North, south to 14° N. Our collections are consistent with this distribution. Our most southern findings of *R. guilhoni* are in Bandia, where this species is abundant.

Host associations

R. guilhoni shows a triphasic ditropic cycle. The major hosts of its adults are domestic ungulates, particularly cattle, sheep and goats; on goats, it is the dominant *Rhipicephalus*. Adults bite humans having close contact with ruminants (3 personal observations). The immature stages feed essentially on rodents, particularly *Arvicanthis niloticus*.

Seasonal activity

Analysis of the mean number of *R. guilhoni* on cattle and goats of Bandia provides clear evidence for the existence of only one annual generation (Fig. 5). In Bandia, May and June are the only months during which adult *R. guilhoni* are not found. They appear in small numbers after the first significant rains of July. The annual peak abundance is attained two months after the beginning of the rainy season, generally in September. This is different from *A. variegatum* and *Hyalomma* spp., which appear to increase activity immediately following the beginning of the rainy season, usually in July. Then, as with the *Hyalomma* spp., *R. guilhoni* adults are seen during much of the dry season, typically until April.

Studies of the seasonal abundance of immature *R. guilhoni* on rodents *Arvicanthis niloticus*, and *Mastomys erythroleucus* have shown a cycle nearly identical to that observed for immature *H. truncatum* (Fig. 6). For both species, maximum larval activity occurs in October; two months later, nymphs are at their maximum. Both immature stages usually disappear by April.

Epidemiological implications

R. guilhoni is implicated in the natural cycle of CCHF virus transmission in Senegal by our isolation of a strain from adults collected in 1988. The role of this species in the epidemiology of human infection may be similar to that of *H. truncatum*. Although immatures are not believed to bite humans, infected adult *R. guilhoni* may be capable of transmitting CCHF virus to humans.

Conclusions

Of the 17 African hard ticks found infected with CCHF virus, 12 are present in Senegambia or Mauritania and eight have been shown to be infected there. Yet, the widespread and focally prevalent nature of CCHF virus transmission suggests that those tick species that are neither widely distributed nor abundant are unlikely to play a role in enzootic transmission. Furthermore, the particular feeding patterns of some ticks, such as monophasic-monotropism, would limit the possibility of horizontal transmission by such species. Even those ticks that become infected but are rarely anthropophilic would not be important in transmission to humans. Thus we consider only five ticks as potential enzootic vectors.

For each of these ticks, we have compared those ecological factors that may influence their status as vectors of CCHF virus in Senegal. These factors include their population dynamics, seasonal activity, and host associations. Of the five tick vectors of potential importance in the epidemiology of CCHF in Senegal, each, with the possible exception of *H. marginatum rufipes*, has a similar population cycle of one generation per year. For *A. variegatum*, it is the adults that influence the timing of reproduction by becoming active only after the first significant rains (July); adult *A. variegatum* disappear with the beginning of the dry season (November). For *H. impeltatum, H. truncatum,* and *R. guilhoni*, adults, which are more xerophilous, engorge on ruminants during the first half of the dry season. Climatic conditions following the end of the rainy season (October–December) permit development of these species during this period. As the dry season progresses, only adult ticks continue to feed on ruminants, laying eggs that may not develop. Larvae that emerge during this dry period probably die before finding a host.

Considering their seasonal activity patterns, host ranges, proclivity for human-biting, and CCHF virus infection rates, the ticks that appear to be the most likely vectors to humans in Senegal are nymphs of *A. variegatum* and adults of *H. truncatum*. Such human infection should occur predominantly during periods of maximal vector activity: November through February if nymphal *A. variegatum* are vectors, or during July through November if adult *H. truncatum* transmit. Nymphal *A. variegatum* derived from larvae that fed on viremic hosts or that were infected transovarially could transmit to humans. Transovarial transmission of CCHF virus by this tick has not yet been recognized; however, horizontal transmission by infection of vertebrates that also serve as hosts for nymphal *A. variegatum* may occur. Adult *A. variegatum* that emerge from such nymphs could transmit CCHF virus to other ungulates; however, they would not directly influence risk of human infection because they almost never feed on people.

H. truncatum appears to be an important vector of CCHF virus to humans in Senegal because it is among the most abundant and widely distributed of infected ticks, immatures feed readily upon small mammals that have been shown to be infected, experimental evidence has demonstrated horizontal and transstadial transmission [12], and adult ticks bite humans. Furthermore, the distribution of CCHF virus antibody seroprevalence among humans in Africa, and of naturally acquired hemorrhagic cases, fits more closely the distribution of *H. truncatum* than that of *A. variegatum* (Fig. 2). Thus, we suggest that *H. truncatum* is likely to be the principal anthropo-zoonotic vector in Senegal.

The natural cycle of this virus contains numerous links, some of which could change considerably the dynamics of virus maintenance and transmission to humans (Fig. 7). Other hypothetical links remain to be identified. Adding to these complex interactions are temporal and spatial variations in antibody prevalences [24], the unknown protective properties of neutralizing antibodies, variation in the timing and magnitude of viremia among vertebrates, differences in the vector competence of ticks, and changing

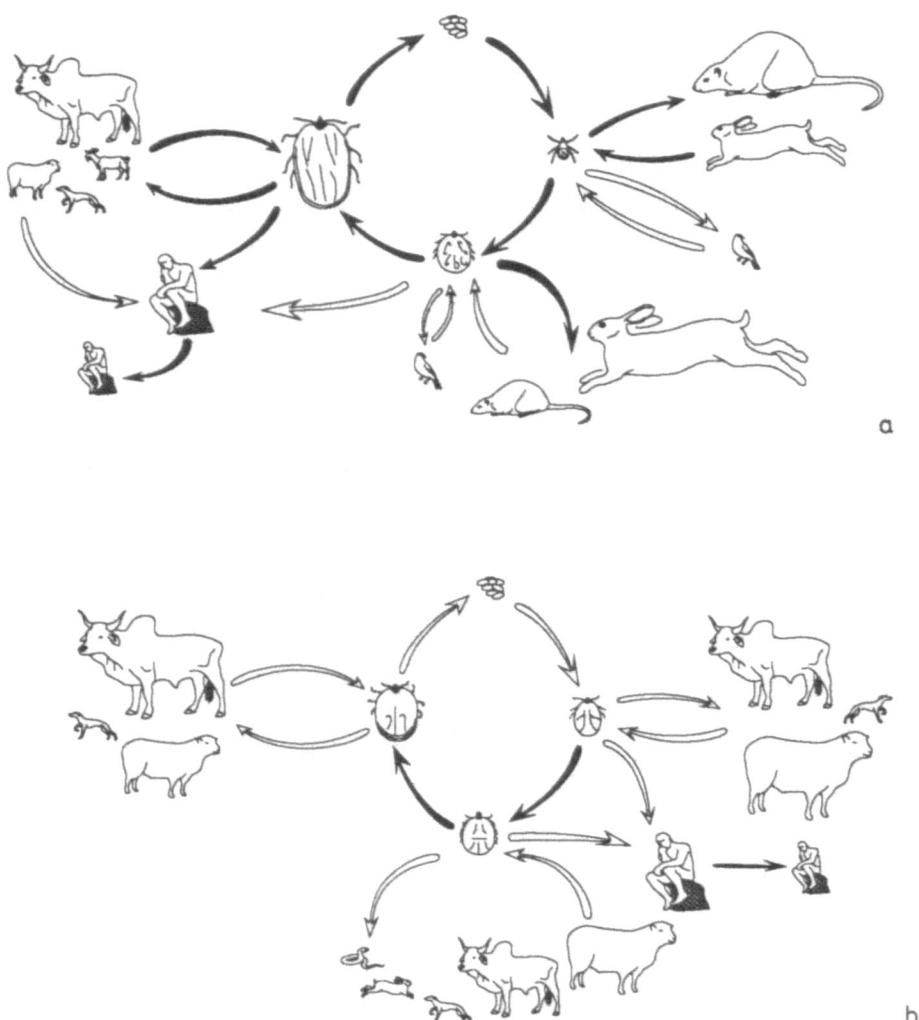

Fig. 7a, b. Scheme of the transmission cycle of Crimean-Congo hemorrhagic fever virus among vector ticks and vertebrates. The sizes of vertebrates represent their relative importance as hosts to the tick. Demonstrated transmission (solid arrows), suggested transmission (open arrows). Virus may be vertically transmitted among eggs, larvae, nymphs and adults, or horizontally transmitted to and from vertebrates. Transmission cycles for *Hyalomma truncatum* (**a**) and *Amblyomma variegatum* (**b**) are illustrated

infection rates. Schematic diagrams of CCHF virus transmission (Fig. 7) are useful tools in conceptualizing the potential pathways and interactions, both deriving from and guiding research on longterm transmission dynamics under natural conditions. We lack, at present, sufficient observations of this sort to paint a satisfactory picture of the ecology of CCHF virus throughout most of its range.

Acknowledgements

Funding of this research was provided through grant DAMD 17-87-G-7003 from the U.S. Army Medical Research Institute of Infectious Diseases (USAMRIID), Ft. Detrick, Frederick, Maryland, U.S.A. We thank Khalilou Ba, Mamoudou Diallo, Abdoulaye Diouf, Elizabeth Dykstra, and Ibrahim Samb for technical assistance. We are grateful to Drs. Arona Gueye and Racine Sow of the Institut Senegalais de Recherches Agricoles (ISRA) for their cooperation and for permission to work at the ISRA research station at Dahra.

References

1. Camicas JL (1970) Contribution a l'etude des tiques du Senegal (Acarina, Ixodoidea). I. Les larves d'*Amblyomma* Koch et de *Hyalomma* Koch. Acarologia 12: 71–102

2. Camicas JL, Cornet JP (1981) Contribution a l'etude des tiques du Senegal (Acarida; Ixodida). III. Biologie et role pathogene d'*Amblyomma variegatum*. Afr Med 20: 335–344

3. Camicas JL, Robin Y, Le Gonidec G, Saluzzo JF, Jouan A, Cornet JP, Chauvancy G, Ba K (1986) Etude ecologique et nosologique des arbovirus transmis par les tiques au Senegal. 3. Les vecteurs potentiels du virus de la fievre hemorragique de Crimee Congo (virus CCHF) au Senegal et en Mauritanie. Cah ORSTOM Ser Entomol Med Parasitol 24: 255–264

4. Causey OR, Kemp GE, Madbouly MH, David-West TS (1970) Congo virus from domestic livestock, african hedgehog and arthropods in Nigeria. Am J Trop Med Hyg 19: 846–850

5. Chunikhin SP, Chumakov MP, Butenko AM, Smirnova SE, Taufflieb R, Camicas JL, Robin Y, Cornet M, Shabon Zh (1969) Results from investigating human and domestic and wild animal blood sera in the Senegal Republic (West Africa) for antibodies to Crimean hemorrhagic fever virus. In: Proc. XVI Scientific Conference of the USSR AMS Institute of Poliomyelitis and Virus Encephalitides, Moscow, October 1969, 2: 158–160 (in Russian) (in English: NAMRU-3, T 810)

6. Gueye A, Mbengue M, Diouf A, Seye M (1986) Tiques et hemoparasitoses du betail au Senegal. I. La region des Niayes. Rev Elev Med Vet Pays Trop 39: 381–393

7. Gueye A, Camicas JL, Diouf A, Mbengue M (1987) Tiques et hemoparasitoses du betail au Senegal. II. La region Sahelienne. Rev Elev Med Vet Pays Trop 40: 119–125

8. Gueye A, Camicas JL (1990) Distribution des tiques du betail. In: Centre Technique de Cooperation Agricole et Rural (eds) Elevage et potentialites pastorales saheliennes syntheses cartographiques—Senegal. IEMVT, Maisons-Alfort, France, p 20

9. Hoogstraal H (1956) African Ixodoidea. I. Ticks of the Sudan (with special reference to Equatoria Province and with preliminary reviews of the genera Boophilus, Margaropus and Hyalomma). Department of the Navy, Bureau of Medical Surgery, Washington, DC

10. Hoogstraal H (1979) The epidemiology of tick-borne Crimean Congo hemorrhagic fever in Asia, Europe and Africa. J Med Entomol 15: 307–417

11. Leroux M (1983) Climat. In: Pelissier P (ed) Atlas du Senegal. Jeune Afrique, Paris, pp 12–17

12. Logan TM, Linthicum KJ, Bailey CL, Watts DM, Moulton JR (1989) Experimental transmission of Crimean-Congo hemorrhagic fever virus by *Hyalomma truncatum* Koch. Am J Trop Med Hyg 40: 207–212

13. Morel PC (1969) Contribution a la connaissance de la distribution des tiques (Acariens, Ixodidae et Amblyommidae) en Afrique ethiopienne continentale. Thesis, DSc, Orsay, France

14. Morel PC (1976) Etude sur les tiques d'Ethiopie (Acariens, Ixodides). IEMVT, Maisons-Alfort

15. Robin Y, Camicas JL, Jan C, Heme G, Cornet M, Valade M (1978) Ecology of tick arboviruses in arid areas of Senegal. In: Cherepanov, AI (ed) Transcontinental connections of migrating birds and their role in distribution of arboviruses. Papers of the Symposium 1976, Novosibirsk (Akademgorodok). Nauka, Novosibirsk, pp 209–211

16. Saluzzo, JF, Digoutte JP, Cornet M, Baudon D, Roux J, Robert V (1984) Isolation of Crimean-Congo haemorrhagic fever and Rift Valley fever viruses in Upper Volta. Lancet 1: 1179

17. Saluzzo JF, Aubry P, McCormick JB, Digoutte JP (1985). Haemorrhagic fever caused by Crimean-Congo haemorrhagic fever virus in Mauritania. Trans R Soc Trop Med Hyg 79: 268

18. Saluzzo JF, Camicas JL, Chartier C, Martinez D, Digoutte JP (1986) Le virus de la fievre hemorragique de Crimee-Congo en Mauritanie. Cah ORSTOM, Ser Entomol Med Parasitol 24: 129–137

19. Shepherd AJ, Swanepoel R, Cornel AJ, Mathee O (1989) Experimental studies on the replication and transmission of Crimean-Congo hemorrhagic fever virus in some African tick species. Am J Trop Med Hyg 40: 326–331

20. Sureau P, Klein JM, Casals J, Digoutte JP, Salaun JJ, Piazak N, Calvo MA (1980) Isolement des virus Thogoto, Wad Medani, Wanowrie et de la fievre hemorragique de Crimee-Congo en Iran partir de tiques d'animaux domestiques. Ann Inst Pasteur Virol 131 E: 185–200

21. Swanepoel R, Shepherd AJ, Leman PA, Shepherd SP, McGillivray GM, Erasmus MJ, Searle LA, Gill DE (1987) Epidemiologic and clinical features of Crimean-Congo hemorrhagic fever in southern Africa. Am J Trop Med Hyg 36: 120–132

22. Teel PD, Bay DE, Ajidagba PA (1988) Ecology, distribution and host relationships of ticks (Acari: Ixodidae) infesting livestock in Mali. Bull Entomol Res 78: 407–424

23. Watts DM, Ksiazek TG, Linthicum KJ, Hoogstraal H (1988) Crimean-Congo Hemorrhagic Fever. In: Monath TP (ed) The arboviruses: epidemiology and ecology, CRC Press, Boca Raton, FL, pp 177–222

24. Wilson ML, Gonzalez JP, LeGuenno B, Cornet JP, Guillaud M, Calvo MA, Digoutte JP, Camicas JL (1990) Epidemiology of Crimean-Congo hemorrhagic fever in Senegal: temporal and spatial patterns. Arch Virol [Suppl 1]: 323–340

Authors' address: J.-L. Camicas, Institut Pasteur, B.P. 220, Dakar, Senegal.

Arch Virol (1990) [Suppl 1]: 323–340

Epidemiology of Crimean-Congo hemorrhagic fever in Senegal: temporal and spatial patterns

M. L. Wilson[1,2], J.-P. Gonzalez[1,3], B. LeGuenno[1], J.-P. Cornet[3], M. Guillaud[4], M.-A. Calvo[1], J.-P. Digoutte[1], and J.-L. Camicas[3]

[1]Institut Pasteur, Dakar, Senegal
[2]Departments of Population Sciences and Tropical Public Health,
Harvard School of Public Health, Boston, Massachusetts, U.S.A.
[3]Institut Francais de Recherche Scientifique pour le Developpement en Cooperation
(ORSTOM), Laboratoire ORSTOM de Zoologie medicale, Dakar, Senegal
[4]Institut d'Elevage et de Medecine Veterinaire des Pays Tropicaux, Maisons Alfort,
France

Accepted April 15, 1990

Summary. Aspects of the spatial and temporal patterns of transmission of Crimean-Congo hemorrhagic fever (CCHF) virus were studied in Senegal, West Africa. A country-wide serological survey of domestic animals indicated that transmission was most intense in the northern dry sahelian zone and least in the southern, more humid guinean zone. Human IgG prevalence, ranging from nearly 20% to <1% among 8 sites throughout the region, also was greatest in the north. A fatal human case of CCHF from Rosso, Mauritania in 1988 was studied and an accompanying serosurvey of human contacts and domestic animals indicated epidemic transmission during that period. Systematic samples of adult ixodid ticks on domestic animals allowed us to analyze the distribution and relative abundance of potential CCHF virus vectors, demonstrating that *Hyalomma* spp. predominated in those biotopes where transmission was most intense. A prospective study of CCHF virus infection and tick infestation in sheep exposed a period of epizootic transmission in 1988 that corresponded temporally with increased abundance of adult *H. truncatum* and *H. impeltatum*. Four strains of CCHF virus were isolated from pools of these ticks and of *Rhipicephalus guilhoni*. Our results suggest that CCHF virus is focally endemic throughout the region, although highly variable in time and space, and that the relative abundance of *Hyalomma* ticks may be the primary determinant of epidemic transmission.

Introduction

The distribution of Crimean-Congo hemorrhagic fever (CCHF), although irregular and focal, is remarkably widespread. CCHF virus circulates throughout much of the southern Soviet Union, central Asia, southern Europe, the Middle East, and Africa [45], covering three of the world's seven major biotic zones [18]. Zoonotic disease in humans has been recognized in about a dozen countries, but serological or virological evidence of CCHF virus transmission has been found throughout these regions. At least 30 species of ticks [3] and more than 20 different vertebrate species [45] have been shown to be infected. Despite the abundance of research documenting the widespread distribution, diversity of possible vectors and numerous potential vertebrate reservoirs for CCHF virus, our understanding of the transmission cycle(s) remains incomplete. We can only speculate, at present, as to the importance of variables that maintain transmission and stimulate epizootics or epidemics.

In sub-saharan West Africa, CCHF virus and antibodies have been detected in ticks, domestic and wild vertebrates, or humans from Benin [10], Burkina Faso [10, 37], Mauritania [34–36], Nigeria [4, 8, 44] and Senegal [9, 29–33]. Initial observations in Senegal during the late 1960's demonstrated indirect evidence of domestic animal and tick infections [5]. During the 1970's and early 1980's, researchers at the Pasteur Institute in Dakar isolated numerous strains of CCHF virus from ticks feeding on cattle and sheep at Senegal slaughterhouses [29–33]. Antibodies were detected at sites throughout the country and a human case and antibodies in animals were documented along the border in Mauritania [36]. Thus, CCHF virus has been circulating in Senegal during at least the past 2 decades.

The factors that influence transmission of CCHF virus include the density of competent vector ticks, particularly of the genus *Hyalomma*, and the relative abundance of vertebrates that serve as both hosts to these ticks and as possible reservoirs of CCHF virus [18]. Sustained transmission is found only where *Hyalomma* ticks are present and epizootic or epidemic transmission is believed to occur during periods of increased abundance of these ticks [45]. However, the general relationship between tick species diversity and the magnitude of CCHF virus transmission has not been systematically studied within a particular geographic region. While the vertebrate host associations of these ticks typically are well-defined [18], the role that most of these vertebrates play in the natural maintenance cycle or amplification of CCHF virus remains enigmatic.

Human cases of CCHF have been recognized from more than a dozen countries on 3 continents [45]. In sub-saharan Africa at least 69 human cases of CCHF have been documented, mostly in southern Africa. In West Africa, however, only two non-fatal human cases have been previously reported, this despite a relatively high seroprevalence indicating a high prevalence of human infections. The first case, in 1983, was from Selibaby, Mauritania near the northeastern border with Senegal [36]. Another case later that year was reported from southeastern Burkina Faso [37]. Thus, the true incidence of human infection and particularly of human disease remains unknown. We report here on various studies designed to investigate the spatial and temporal aspects of CCHF virus transmission within this region of West Africa. Specifically, we present summaries of studies on the seroprevalence of sheep and of humans [47], a recent fatal case [11], the distribution of potential vectors, an epizootic corresponding with changes in tick abundance, and recent virus isolations from ticks.

Study sites and methods

A prospective, multidisciplinary research program based at the Pasteur Institute in Dakar, Senegal was begun in 1987 to investigate various aspects of the ecology and epidemiology of CCHF virus transmission in Senegal, Gambia, and Mauritania. This region, encompassing a variety of habitats and ecological zones, ranges from the Sahara desert in the north to humid forests in the south. The biota changes primarily in association with differences in average annual rainfall ranging from 200 mm in the northern sahelian zone to 1500 mm in the southern sudano-guinean and sub-guinean zones [23]. The diversity and abundance of potential vector ticks is also large [2, 3, 12], as is the vertebrate fauna of potential hosts [21].

Serological studies

In order to define the distribution and intensity of CCHF virus transmission, and the environmental factors associated with such infection in Senegal, a systematic serosurvey of IgG antibodies among sheep was undertaken [47]. These abundant and widespread domestic ungulates are infested by potential tick vectors and express antibodies to CCHF virus infection. Furthermore, because their average age is only a few years, sheep seroprevalence serves as an index of relatively recent transmission. Blood samples were obtained between November 1987 and February 1988 from herds chosen in 26 randomly selected administrative districts throughout Senegal [47]. Age was estimated by dental examination and grouped into four categories: < 14, 14–28, 29–36, and > 36 months. Sera were frozen in the field and later tested at the Pasteur Institute in Dakar for IgG antibody to CCHF virus.

Human serological studies were based on IgG prevalence from samples collected during 1986–88 from 8 different sites in the region. In addition to our samples from Yonofere, Dahra, Bandia, and Kedougou, we tested sera collected as part of other studies from Dagana and Ziguinchor by Dr. Alain Jouan, from Tambacounda by Dr. Jean-Francois Saluzzo, and from Rosso by Dr. Elizabeth Manus. The details of the sampling methods varied among the sites; however, all sera were obtained from apparently healthy people who were asked to donate blood for medical research. People were bled by venipuncture and sera were frozen at −20 °C from 1 to 4 days later and tested as above.

Sera were tested for evidence of anti-CCHF virus IgG using an ELISA test [28] modified slightly by adding a saturating solution of PBS with 0.05% Tween 20 and 1% non-fat bovine milk. In this direct ELISA test, 96-well plates (Immulon II, Dynatech Laboratories, Alexandria, VA) were coated with diluted CCHF virus hyperimmune mouse ascitic fluid. CCHF virus (strains IbAr 10200 from Sokoto, Nigeria and the recently isolated Dak H49199 from a human in Rosso, Mauritania) in crude suckling mouse brain was heat inactivated at

60°C for 1 h. and then added. Test sera, diluted 1:400, followed by test-species specific anti-immunoglobulin conjugated with horse radish peroxidase (Biosys, Compiegne, France) was used to detect the IgG. A chromogenic substrate (orthotoluidine, Sigma, LaVerpilliere, France) was added for colorimetry. All plates included a control of crude suckling mouse brain without CCHF virus antigen. Differences in optical density (OD) between the test and control wells were measured at 450 nm using an automatic reader (Multiscan MCC/340, Flow Laboratories, Irvine, Scotland) coupled to a microcomputer. By iterations of the distribution of OD values, we determined the mean of the population of negatives. Sera were considered positive if the OD was greater than 3 standard deviations above the mean of negatives.

IgM antibodies were detected by immunocapture ELISA [38]. Plates were coated with anti-μ chain specific for the species being tested. The test serum, followed by CCHF viral antigen was then added. The detecting antibody was a high-titered mouse ascitic fluid against CCHF virus antigen. Anti-mouse immunoglobulin conjugated with horse radish peroxidase and the chromogenic substrate were added as above. Evaluation and criteria were as for IgG.

Virus and antigen identification

Virus isolation was attempted by intracranial inoculation of suckling mice and by inoculation of Vero cells using undiluted and 10-fold diluted sera. Virus identification was made by an indirect immunofluorescent test on Vero cells, using polyclonal and monoclonal antibodies. The identity of virus isolates was confirmed by a complement fixation test at the WHO collaborating Center for Arboviruses at the Pasteur Institute in Dakar.

An antigen capture ELISA [38] was also employed to test for presence of CCHF virus antigen in human sera. Plates first were coated with anti-human IgM μ chain specific antibody, followed by human sera with high titer IgM. The test serum was added next and then a high-titered anti-CCHF virus monoclonal IgG was added to bind to any antigen captured from the test serum. An anti-mouse IgG, conjugated with horse radish peroxidase and the chromogenic substrate were used for colorimetry.

Vector tick distribution and abundance

The relative abundance of potential vector ticks was studied by analyzing monthly samples of adult ticks attached to domestic animals in various sites throughout Senegal. Sheep, goats or cattle were deticked monthly throughout at least 1 year in order to avoid bias due to seasonal variation in tick activity. We sampled 3 sites in northern Senegal: Yonofere, Dahra, and Bandia (Fig. 1). In Yonofere, 5 herds of sheep were chosen monthly by chance encounter from May 1987 through May 1989 and 10 randomly selected individuals from each herd were sampled. All ticks were removed with forceps and stored for later identification and virus isolation. The same regime was followed in Dahra from May 1987 through August 1989. In Bandia, 12 sentinel goats and 2 cattle were examined similarly 3 times each month from April 1987 through March 1989.

In addition, published data from other sites in Senegal were also analyzed. Sites were compared by calculating the mean number of ticks per host (cattle, sheep, or goats) from monthly samples of 40 hosts made during 12 or more months during 1985–1989.

Temporal relation between tick abundance and virus transmission

To determine the relationship between tick abundance and CCHF virus transmission, sheep were bled periodically and simultaneously examined for the presence of ticks. Approxi-

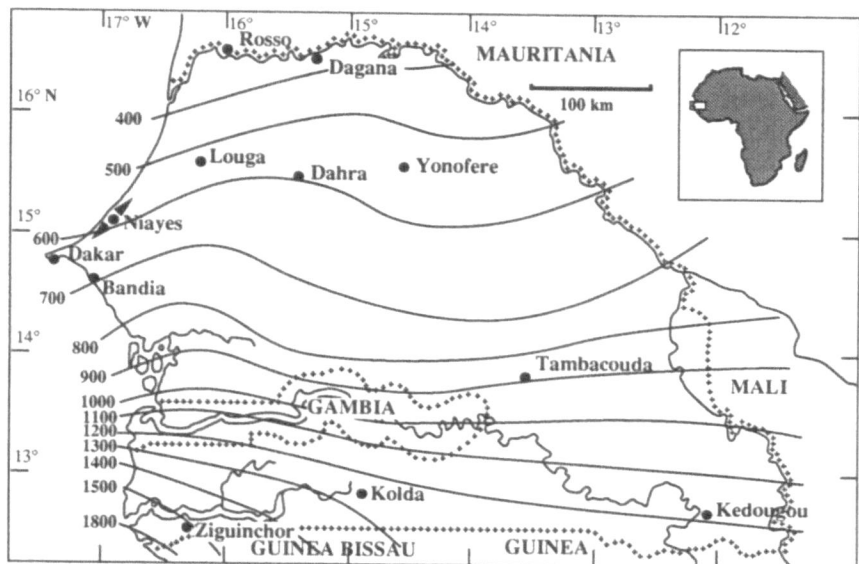

Fig. 1. Map of Senegal showing sites where observations were made, and isoyeths indicating the average annual rainfall [23]. Surrounding countries are noted

mately 300 individually-identified sheep were bled every 2 or 3 months during 1987–1989 at the Centre de Recherches Zootechniques at Dahra, as were another 200 sheep from freeranging herds in the village of Yonofere (Fig. 1). We estimated the prevalence of infection and attempted to correlate this with variation in tick abundance and other environmental variables.

Efforts to isolate virus were undertaken using ticks collected monthly from domestic ungulates in Yonofere and Bandia during 1987 through 1989. In addition to ticks from above mentioned collections, herdspeople from these and surrounding villages were given tubes for collecting ticks that they removed from their animals. Ticks were frozen at $-70\,°C$ until being pooled by species and herd, ground in Hanks' solution, centrifuged, and inoculated into suckling mice or cell culture.

Results

Spatial distribution of animal seroprevalence

A total of 942 sheep were bled between November 1987 and February 1988 from the 22 administrative regions that were studied [47]. The overall prevalence of IgG antibodies among sheep was 10.4%. Antibody prevalence was equal for male and female sheep, but increased with age from 2.1% of the youngest animals to 18.2% among the oldest age group [47].

The spatial pattern of CCHF virus transmission, as indicated by antibody prevalence, varied among the villages. Prevalence appeared highest in the north, decreasing to nil in the south. When seroprevalence was organized into 5 ecologic zones [27], three-quarters of the sheep tested from the northernmost, Sahelian zone showed evidence of previous infection, a rate

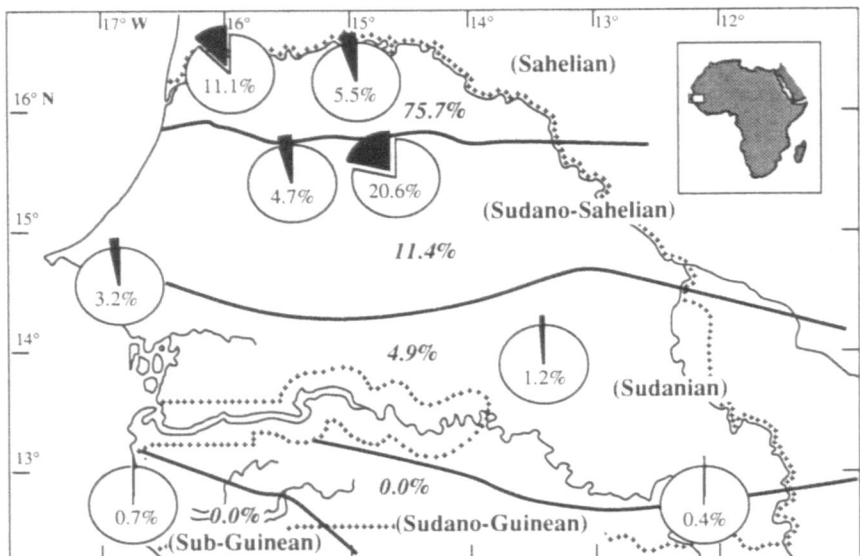

Fig. 2. Prevalence of IgG antibodies to CCHF virus in sheep and humans throughout Senegal. Human IgG prevalence is indicated by the circles that correspond to villages where samples were made. IgG prevalence in sheep is shown in italics as the average for all sheep sampled in each of 5 bioclimatic zones that are named in parentheses

which decreased to zero toward the southern sudano-guinean and sub-guinean zones (Fig. 2). Furthermore, antibody prevalence correlated negatively with rainfall, declining from a maximum of 31.3% at the 400 mm isohyet to 0% at the 1200 mm isohyet ($r^2 = 0.74$, n = 9, p < 0.003). Those regions experiencing the least rainfall had the highest rate of CCHF virus transmission. Attempts to explain this spatial pattern using geological formations, soil types, altitude, and sub-surface water revealed no such relationships. Thus, CCHF virus transmission was found to occur with greatest intensity in dryer, more sparsely vegetated Sahelian and northern Sudanian regions.

Comparison with results from previous serosurveys revealed that antibody prevalence for the sheep in our study varied within the range of average values reported from other sites in Africa (Table 1). The negative correlation between antibody prevalence and rainfall observed in our study was consistent with results from the 1969 Senegal survey by Chunikhin et al. [5] (Table 1). CCHF virus transmission in Senegal, particularly in the north, appears to have been especially intense during the 1980's.

Human seroprevalence

The prevalence of IgG for people aged 10 or more varied by as much as an order of magnitude among the villages sampled (Fig. 2). In northern sites along the Senegal River, 17 of 150 (11.3%) people from Rosso, and 5 of 91

Table 1. Prevalence of anti-CCHF virus antibodies among domestic ungulates from various sites in Africa

Country	Site	Domestic animal	Prevalence seropositive	No. tested	Ref.
Egypt	Wadi Natroun	C	14.3%	21	[7]
	Qena	C	13.3	13	[7]
	Cairo	S	18.2	66	[7]
	Cairo	D	8.8	34	[7]
Kenya	Eldoret	C	7.3	55	[1]
	Isiolo	C	8.3	96	[1]
	Kajiado	C	1.1	92	[1]
	W. Pokot	C	3.0	100	[1]
Mauritania	Selibaby	C	32.0	25	[35]
Nigeria	Bauchi	C	26.6	64	[44]
	Niger	C	29.5	535	[44]
	Kaduna	C	22.0	565	[44]
Senegal	(north)	C	11.5	235	[5]
	(central)	C	6.2	113	[5]
	(south)	C	15.4	26	[5]
	(s. west)	C	8.6	93	[5]
	(north)	S	5.8	512	[5]
	(central)	S	1.4	70	[5]
	(s. west)	S	0	70	[5]
	(central)	G	0	80	[5]
	(s. west)	G	1.4	70	[5]
Uganda	Mbarara	C	36.5	104	[19]
R.S.A.	Bloemhof	C	64.1	170	[43]
	(east)	C	26.5	6128	[40]
	(north)	C	28.4	8667	[41]
	Bloemhof	S	27.4	270	[43]
Zimbabwe	(north)	C	45.0	763	[41]

Domestic animals are cattle (*C*), sheep (*S*), dromedaries (*D*), and goats (*G*)

(5.5%) people from Dagana were seropositive. In the north-central Sahel plains, 5 of 106 (4.7%) and 27 of 128 (21.1%) residents from Dahra and Yonofere, respectively, had IgG antibody. In Bandia, further south along the coast, 3 of 92 (3.2%) adults tested showed evidence of previous infection. Low IgG prevalence was found in southern sites including 1 of 140 (0.7%) in Ziguinchor, 1 of 84 (1.2%) in Tambacounda, and 2 of 225 (0.9%) in Kedougou. The prevalence of human antibody against CCHF virus was highest in the north and lowest in the south, a pattern that corresponded to that of sheep.

Human disease

Sera from patients presenting with symptoms of hemorrhagic fever during May 1988 at the hospital in Rosso, Mauritania, on the northern Senegal border, were tested for antibody to CCHF virus and for antigen [11]. Among 8 such patients, 6 showed positive IgG titers; 3 of these 6 patients died. In one fatal case, IgM antibodies were detected and a strain of CCHF virus was isolated.

Subsequent investigations of people in 3 nomadic camps from which patients came showed that of 99 people tested, 36 (36%) had IgG antibodies, including 1 with IgM [11]. Among 17 sheep and goats that were tested in 2 villages, 5 (29%) had IgG and 1 (6%) also had IgM antibody. In addition, 46% of 120 such animals were infested with adult *Hyalomma impeltatum* or *H. dromedarii* ticks (data not shown). Thus, a period of intense transmission of CCHF virus, including human infection and disease occurred in the region around Rosso.

To determine whether such transmission occurred in other parts of the region, a survey of sheep from various sites in southern Mauritania was undertaken; IgG antibody rates ranged between 10% and nearly 50% [11]. Furthermore, about 7% of these sheep had IgM antibodies; intense transmission of CCHF virus apparently occurred throughout much of southern Mauritania during March through July 1988.

Vector tick distribution and abundance

The predominant ixodid ticks in our 3 northern Senegal study sites were *Hyalomma truncatum*, *H. impeltatum* and *Rhipicephalus guilhoni* (Table 2). These three species represent more than 95% of ticks infesting domestic animals in Dahra and Yonofere, and more than 80% of such ticks in Bandia. *H. marginatum rufipes*, *Amblyomma variegatum*, and other *Rhipicephalus* spp. comprised the remaining collections.

We reanalyzed studies by Gueye and collaborators [13–17], to compare their results with ours and again found that *Hyalomma* ticks were most abundant in sites of northern Senegal (Table 2). *Rhipicephalus evertsi evertsi*, and other *Rhipicephalus* spp. occurred predominantly in central and southern sites, such as Tambacounda and Kolda, while *Amblyomma variegatum* and *Boophilus* spp., absent in the north, were most abundant in the south. The distribution and relative abundance of ticks from our studies were similar to that described by Morel [25] in 1958 (Table 3).

Temporal relation between tick abundance and virus transmission

The prevalence of IgM in sheep at Dahra remained at less than 10% from September 1987 through February 1988, rising to nearly 40% in May 1988

Table 2. Infestations of adult ticks on domestic ungulates from various sites in Senegal

Zone region	Host[a]	Sample period	No. ticks total	mean	Percentage of[b] Ht	Hr	Hi	Hd	Hs	Re	Rs	Rg	Ru	Rl	Av	Bd	Bg	Ref[c]
Sahelian																		
Yonofere	S	5/87–5/89	4195	4.4	50.9	0.9	0.3	—	—	—	—	45.1	—	—	—	—	—	
Dahra	S	5/87–5/89	4218	4.1	44.1	0.1	51.2	<0.1	—	0.3	—	4.3	—	—	—	—	—	
Louga	S	1/84–12/84	1817	3.0	—	—	1.4	—	—	98.6	—	—	—	—	—	—	—	[13]
Louga	G	1/84–12/84	97	0.2	—	—	4.1	—	—	95.9	—	—	—	—	—	—	—	[13]
Sudano-sahelian																		
Bandia	C	4/87–3/89	1439	10.0	55.4	16.2	—	—	1.7	0.3	—	25.6	—	—	0.7	0.1	—	
Bandia	G	4/87–3/89	743	1.0	11.7	—	—	—	—	0.3	—	87.6	—	—	0.1	0.3	—	
Niayes	C	4/82–9/83	25315	35.2	30.0	8.2	—	—	—	<0.1	3.8	—	<0.1	—	27.0	29.1	—	[17]
Niayes	G	4/82–9/83	909	1.3	13.5	—	—	—	1.3	<0.1	<0.1	3.5	17.4	—	57.8	6.7	—	[17]
Sudanian																		
Tambacounda	C	10/83–12/84	2173	3.6	16.0	74.5	—	—	—	<0.1	<0.1	<0.1	<0.1	6.1	2.8	<0.1	—	[14]
Tambacounda	S	10/83–12/84	126	0.2	10.3	2.4	—	—	—	84.9	—	—	—	2.4	—	—	—	[14]
Tambacounda	G	10/83–12/84	32	0.1	15.6	—	—	—	—	65.6	—	—	—	18.8	—	—	—	[14]
Sudano-guinean																		
Kolda	C	1/86–3/87	9647	16.1	9.9	1.1	—	—	—	<0.1	<0.1	—	3.9	15.1	34.1	—	35.8	[15]
Kolda	S	1/86–3/87	207	0.3	4.4	0.1	—	—	—	—	—	—	10.1	10.6	25.6	—	9.2	[15]
Kolda	G	1/86–3/87	42	0.1	19.0	—	—	—	—	7.1	—	—	7.1	23.8	2.4	—	40.5	[15]

[a] Hosts are cattle (C), sheep (S), and goats (G)
[b] Tick species are *Hyalomma truncatum* (Ht), *H. marginatum rufipes* (Hr), *H. impeltatum* (Hi), *H. dromedarii* (Hd), *H. impressum* (Hs); *Rhipicephalus evertsi evertsi* (Re), *R. senegalensis* (Rs), *R. guilhoni* (Rg), *R. sulcatus* (Ru), *R. lunulatus* (Rl); *Amblyomma variegatum* (Av); *Boophilus decoloratus* (Bd), *B. geigyi* (Bg)
[c] Previously published work is cited in References. Our data from Yonofere, Dahra and Bandia are original
— Species not collected

Table 3. Relative abundance of ticks feeding on domestic ungulates in various bioclimatic zones in Senegal

Climatic zone[a]	Average annual precip.	Ticks[b]												
		Hd	Hi	Hr	Hs	Ht	Av	Bd	Bg	Re	Rg	Rs	Rm	Ru
Sahelian	250–500 mm	++	+++	++	+	+					o			
		×	××	×		××××	×			×	×××			
N. Sudan.	500–1000		+	+++	+	+++	++	++		+	++		+	
			×	××	×	××××	×××	×		×	×××		×	
S. Sudan.	1000–1250			+	o	++	++++	++++	++[c]	×		++o	++	++
				×		×	×××	×	××					
Guinean	>1250						+++	o	++			+++		+++
							o							

Relative abundances include: very abundant (+++), moderately abundant (++), rare (+), extremely rare (o), absent (blank)
+ Values from Morel [25]
× Values from our research [46; unpubl.]
[a] Zones and precipitation from Morel [25]
[b] Tick species include *Hyalomma dromedarii* (Hd), *H. impeltatum* (Hi), *H. marginatum rufipes* (Hr), *H. impressum* (Hs), *H. truncatum* (Ht); *Amblyomma variegatum* (Av); *Boophilus decoloratus* (Bd), *B. geigyi* (Bg); *Rhipicephalus evertsi evertsi* (Re), *R. guilhoni* (Rg), *R. senegalensis* (Rs), *R. muhsamae* (Rm), *R. sulcatus* (Ru)
[c] P. C. Morel, pers. comm.

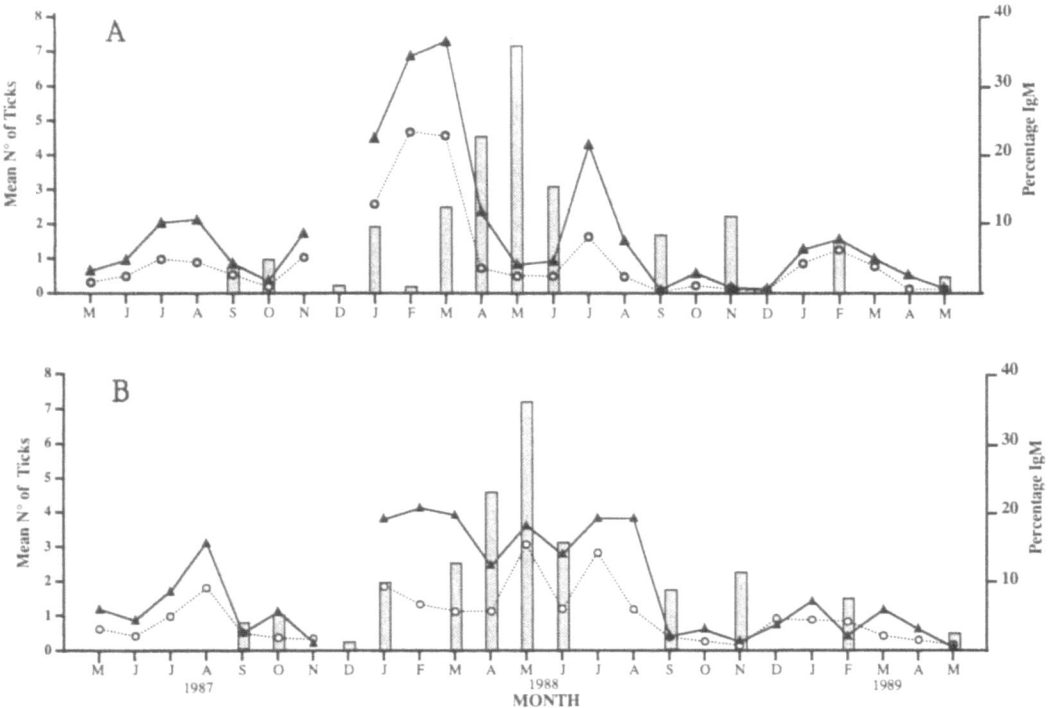

Fig. 3. Average number of adult male (▲ — ▲) and female (○ · · · · ○) *H. truncatum* (**A**) and *H. impeltatum* (**B**) on sheep sampled in Dahra, Senegal from May 1987 to May 1989, and the monthly prevalence of IgM antibodies to CCHF virus in sheep

(Fig. 3). IgG prevalence varied from 5% to 21%, rising in correspondence with that of IgM to a high of more than 60% in September 1988, then declining. In all, roughly two-thirds of more than 300 sheep were infected during this epizootic. Concurrently, numerous sheep became sick and a few died. Two of 4 sheep bled because they were ill had high titer IgM antibodies.

Adult tick abundance on sheep, simultaneously monitored throughout this same period, also changed (Fig. 3). A rapid increase in the abundance of adult *Hyalomma impeltatum* and *H. truncatum* feeding on sheep was observed prior to and during the period of increased transmission. Abundance of these ticks declined at about the same time that IgM prevalence diminished. The other tick that heavily infested sheep in this region, *Rhipicephalus guilhoni*, was absent at that time (data not shown). The prevalence of IgM among sheep in Yonofere remained essentially unchanged at less than 5% during the entire period.

Virus isolation from vector ticks

From January 1988 through June 1989, a total of 31,630 ticks of eight species were grouped into 1,838 pools and tested for arboviruses (Table 4). In addition to Wad Medani (57 strains), Dugbe (2), and Bandia (1) viruses, four

Table 4. CCHF virus isolations during 1988 and January through June, 1989 in ticks removed from cattle, sheep, and goats in Yonofere, Dahra, and Bandia, Senegal

Site	No. (lots) of ticks[a] tested							
	Ht	Hr	Hi	Hs	Re	Rg	Av	Bd
Yonofere								
1988	9755 (501)	342 (38)	131 (17)	—	1 (1)	4373 (231)[b]	—	—
1989	4095 (236)	128 (23)	5 (4)	—	3 (3)	981 (76)	—	—
Dahra								
1988	3260 (174)	27 (9)	3972 (212)[b]	—	31 (9)	104 (9)	—	—
1989	564 (37)	17 (3)	343 (30)	—	5 (5)	48 (9)	—	—
Bandia								
1988	327 (41)	130 (26)	—	10 (6)	5 (4)	653 (50)	80 (14)	2 (2)
1989	106 (19)[c]	53 (19)	—	1 (1)	—	97 (29)	—	—
Total	19766 (1008)	1019 (118)	4451 (263)	11 (7)	45 (22)	6256 (404)	80 (14)	2 (2)

[a] Tick species are *Hyalomma truncatum* (Ht), *H. marginatum rufipes* (Hr), *H. impeltatum* (Hi), *H. impressum* (Hs); *Rhipicephalus evertsi evertsi* (Re), *R. guilhoni* (Rg); *Amblyomma variegatum* (Av); *Boophilus decoloratus* (Bd)

[b] One strain of CCHF virus isolated

[c] Two strains of CCHF virus isolated

strains of CCHF virus were isolated. The CCHF virus strains include one from a pool of 17 male *H. impeltatum* taken from sheep at Dahra in March 1988, one from 20 male *Rhipicephalus guilhoni* removed from sheep in Yonofere during August 1988, and two strains from male–female pools of 9 and 10 *H. truncatum* removed from cattle at Bandia.

Discussion

The prevalence of antibodies in sheep varied among our study sites from intense to nil, suggesting that CCHF virus transmission was either spatially focal or temporally sporadic. In some sites, the prevalence of IgG indicated that nearly half of all sheep were infected at least once during their lifetime. Using an estimate of 32 months as the average age of sheep [47] 1 in 10 or 20 sheep become infected each year in sites where prevalence was high. This is likely an underestimate, as transmission probably was not temporally uniform. Intense epizootics of CCHF virus seem likely.

Transmission of CCHF virus was most intense in the northern, arid Sahel region of Senegal and decreased consistently toward the more moist, southern forest zones. A similar biogeographic relationship has been observed elsewhere in regions of enzootic transmission; sites in Africa, Eurasia, and the Middle East that experience the most intense transmission tend to be relatively more arid [45]. Transmission in Africa occurs primarily in arid savannah grasslands characterized by long dry seasons, and where *Hyalomma* ticks abound. In Eurasia, regions dominated by deserts, semi-deserts and steppes similarly support circulation of this virus. Perhaps the distribution and abundance of potential vectors, which are influenced by climatic conditions, in particular rainfall [2, 3], in turn determine the distribution of CCHF virus. Similarly, biogeographic zones differ in the presence and abundance of potential vertebrate reservoir(s) which may be capable of maintaining horizontal transmission. Alternatively, the periodic, long-distance migration of domestic animals that occurs in drier regions of Africa might increase exposure to questing ticks, thereby elevating the prevalence of infection. Whether greater transmission in semi-arid zones is due to differences in the species diversity of vector ticks, suitable reservoirs, amplifying vertebrate hosts, or other variables deserves further study.

Antibody prevalence increased with the age of sheep in those regions where transmission was relatively intense suggesting that exposure there was enzootic. However, our cross-sectional sample was not large enough to document local epizootics retrospectively. That antibodies were found in sheep of all ages in many regions suggests endemic transmission throughout much of the country. The absence of antibodies in our sample of sheep from southern Senegal probably reflects periodic or focalized transmission at a level too low for us to detect.

Human seroprevalence varied spatially in a manner similar to that of sheep. However, people are exposed, on average, during more years than are sheep, and people are more likely to be sampled other than where they had been exposed. Despite these differences, the pattern of human IgG prevalence also was greatest in the north and least in the south. Human risk seems to correspond with that of domestic animals [45], which may serve as useful sentinels of human infections.

The isolation of CCHF virus from a patient with a hemorrhagic fever and the detection of low titer IgM suggested that this person died of Crimean-Congo hemorrhagic fever, the first reported fatal case from West Africa [11]. Two other patients died with similar symptoms but showed no detectable IgM and low IgG titers. Thus, for the latter 2 patients, we did not determine that their fatal clinical syndromes were due to CCHF virus infection. Among contacts of the hospitalized cases, one family member had IgM antibody and more than one-third of contacts showed evidence of past CCHF virus infection. By comparison, a 1984 sample of healthy people from the same area indicated an IgG prevalence of 5.5% [35]. Furthermore, domestic animals throughout the region exhibited IgM consistent with recent transmission. The combination of recognized human cases, IgM and high prevalence IgG among case contacts, and IgM among domestic animals suggest that recent transmission of CCHF virus occurred in the region during May 1988, producing human infection and disease.

The paucity of human cases of CCHF in Senegal and Mauritania is enigmatic in light of seroprevalence rates indicating transmission at levels equal to or greater than that found in other regions of the world [45]. In southern Africa, at least 23 primary human cases (10 fatal) have been documented since 1981 [41], while seroprevalence among humans there was only 1.5% [40]. Many cases have been diagnosed in the Soviet Union, yet antibody prevalence is similarly low [45]. The hypothesis that West African strains of CCHF virus are less pathogenic to humans than strains in Eurasia or southern Africa has been suggested, but differences in the availability of health care and surveillance make such a hypothesis difficult to test. Access to clinical and diagnostic services is scarce in north-central Senegal thus limiting the correct identification of people with severe hemorrhagic fevers. Deaths in remote villages occur unreported or may be reported without etiology. These factors would lead to underestimating the amount of severe disease and number of deaths due to CCHF, though to what extent we cannot determine. Because CCHF virus appears weakly- or non-pathogenic for ungulates and disease in humans is rarely diagnosed, only large-scale epidemiological studies will accurately determine the true intensity of virus circulation and human disease.

The vector(s) of CCHF virus in Senegal remain poorly defined, although our results indicate that one or more *Hyalomma* spp. are important. Other studies have suggested a correlation between *Hyalomma* tick abundance and virus transmission [18], despite the fact that ticks from 6 other genera have been shown to be infected [3]. Our results demonstrated a positive correlation between the spatial patterns of *H. truncatum* and *H. impeltatum* abundance, and of the prevalence of infection in humans and sheep. Although *Amblyomma variegatum* and *Rhipicephalus* spp. were present where evidence of transmission was found, these ticks were more abundant in the central and southern sites where CCHF virus circulation was less. The predominance of *Hyalomma* spp. where transmission was most intense in humans and in sheep suggests that these ticks may serve both as enzootic and endemic vectors. That CCHF virus previously has been isolated from 2 other *Hyalomma* spp. as well as from ticks from 3 other genera [3] suggests that other ticks may also play a role in transmission.

The distribution pattern and relative abundance of ticks in Senegal, described originally in 1958 by Morel [25], was similar to that which we determined from our studies [46] (Table 3). *Hyalomma* spp. predominate in the dry Sahelian zone becoming progressively less abundant in the semi-arid Sudanian zone and the more humid habitats of the guinean zone. *Amblyomma variegatum* and *Boophilus* spp. predominate in the Sudanian and Guinean zones. Curiously, *Rhipicephalus guilhoni*, confused with *R. sanguineus* before 1962 [26], has apparently increased in relative abundance in the Sahelian zone since the 1950's. In general, the geographic distribution and relative abundances of these ticks exhibited a pattern that could be classified by bioclimatic zone.

A temporal relationship between the abundance of *H. truncatum* and *H. impeltatum* and increased transmission further supports our observation that these ticks play a role as important vectors. Tick abundance and anti-CCHF IgM prevalence in sheep were both relatively low, then rose precipitously, and later declined during a 2-year period of study. This focal epizootic in northern Senegal could not have resulted from transmission by other ixodid ticks, as only *Rhipicephalus guilhoni* occurs in abundance there and its distinct seasonal pattern peaked after most transmission had occurred [3]. In addition, virus transmission was apparently intense in other areas of southern Mauritania during this same period, as suggested by the high IgM prevalence in our sheep serosurvey [11]. Curiously, no epizootic was observed 150 km east in Yonofere despite the fact that adult *H. truncatum* but not *H. impeltatum* abundance, simultaneously rose to a high level there. Whether tick abundance, tick infection rates or other factors influenced this difference is not known.

H. truncatum appears to be effective as a vector of CCHF virus [24, 39] and certain *Hyalomma* ticks may also serve as reservoirs, in that transovarial transmission in *H. marginatum rufipes* and *H. marginatum marginatum* has been experimentally demonstrated [20, 22, 48]. Amplifying, horizontal transmission may be limited to a short period of viremia during which infected and uninfected ticks are feeding simultaneously. It seems unlikely that adults of these ticks horizontally transmit to immatures because they feed on different hosts. In the absence of transovarial transmission, domestic animals may play no role in the CCHF virus cycle, other than as a food resource for tick reproduction. Nevertheless, migration by these domestic animals, small mammal or bird population fluctuations, or other bioclimatic changes could lead to the observed temporal and spatial heterogeneity of CCHF virus transmission.

Acknowledgements

Funding for this research was provided by the U.S. Army Medical Research Institute of Infectious Diseases (USAMRIID), Ft. Detrick, Frederick, Maryland, U.S.A., through grant DAMD 17-87-G-7003. Additional support was provided by the Institut Pasteur through grants from the Ministere Francais de la Recherche Scientifique, and by Institut Francais de Recherche Scientifique pour le Developpement en Cooperation (ORSTOM). We thank Khalilou Ba, Abdoulaye Diouf, Elizabeth A. Dykstra, Magueye Ndiaye, Ibrahima Samb, and Rougy Sylla for technical assistance.

References

1. Butenko AM, Minja T (1979) Personal communication. In: Hoogstraal H (1979) The epidemiology of tick-borne Crimean Congo hemorrhagic fever in Asia, Europe and Africa. J Med Entomol 15: 307–417
2. Camicas JL, Robin Y, Le Gonidec G, Saluzzo JF, Jouan A, Cornet JP, Chauvancy G, Ba K (1986) Etude ecologique et nosologique des arbovirus transmis par les tiques au

Senegal. III. Les vecteurs potentiels du virus de la fievre hemorragique de Crimee Congo (virus CCHF) au Senegal et en Mauritanie. Cah ORSTOM Ser Entomol Med Parasitol 24: 255–264

3. Camicas JL, Wilson ML, Cornet JP, Digoutte JP, Calvo MA, Adam F, Gonzalez JP (1990) Ecology of ticks as potential vectors of Crimean-Congo hemorrhagic fever virus in Senegal: epidemiological implications. Arch Virol [Suppl 1]: 303–322

4. Causey OR, Kemp GE, Madbouly MH, David-West TS (1970) Congo virus from domestic livestock, african hedgehogs and arthropods in Nigeria. Am J Trop Med Hyg 19: 846–850

5. Chunikhin SP, Chumakov MP, Butenko AM, Smirnova SE, Taufflieb R, Camicas JL, Robin Y, Cornet M, Shabon Z (1969) Results from investigating human and domestic and wild animal blood sera in the Senegal Republic (West Africa) for antibodies to Crimean hemorrhagic fever virus. In: Proc XVI Scientific Conference of the USSR AMS Institute of Poliomyelitis and Virus Encephalitides, Moscow, October 1969, 2: 158–160 (in Russian) (in English: NAMRU-3, T 810)

6. Darwish MA, Hoogstraal H (1981) Arboviruses infecting humans and lower animals in Egypt: a review of thirty years of research. J Egypt Public Health Assoc 56: 1–112

7. Darwish MA, Imam IZE, Omar FM, Hoogstraal H (1977) A sero-epidemiological survey for Crimean-Congo hemorrhagic fever virus in humans and domestic animals in Egypt. J Egypt Public Health Assoc 52: 156–163

8. David-West TS, Cooke AR, David-West AS (1974) Seroepidemiology of Congo virus (related to the virus of Crimean haemorrhagic fever) in Nigeria. Bull WHO 51: 543–546

9. Digoutte JP (1985) Rapport sur le fonctionment technique de l'Institut Pasteur de Dakar, Instituts Pasteur Outre-Mer, Dakar, Senegal

10. Gonzalez JP, Baudon D, McCormick JB (1984) Premieres etudes serologiques dans les populations humaines de Haute-Volta et du Benin sur les fievres hemorragiques africaines d'origine virale. Organ Coop Coordin Grand Endem Inform 12: 113

11. Gonzalez JP, LeGuenno B, Guillaud M, Wilson ML (1990) A fatal case of Crimean-Congo haemorrhagic fever in Mauritania: virological and serological observations suggest epidemic transmission. Trans R Soc Trop Med Hyg (in press)

12. Gueye A, Camicas JL (1990) Distribution des tiques du betail. In: Centre Technique de Cooperation Agricole et Rural (eds) Elevage et potentialites pastorales saheliennes syntheses cartographiques. Senegal. IEMVT, Maisons-Alfort, France, p 20

13. Gueye A, Camicas JL, Diouf A, Mbengue M (1987) Tiques et hemoparasitoses du betail au Senegal. II. La zone Sahelian. Rev Elev Med Vet Pays Trop 40: 119–125

14. Gueye A, Mbengue M, Diouf A (1989) Tiques et homoparasitoses du betail au Senegal. III. La zone nord-Soudanienne. Rev Elev Med Vet Pays Trop 42: 411–420

15. Gueye A, Mbengue M, Diouf A (1989) Tiques et hemoparasitoses du betail au Senegal. IV. La zone sud-Soudanienne. Rev Elev Med Vet Pays Trop 42 (in press)

16. Gueye A, Mbengue M, Diouf A (1990) Tiques et hemoparasitoses du betail au Senegal. V. La zone nord-Guineenne. Rev Elev Med Vet Pays Trop 43 (in press)

17. Gueye A, Mbengue M, Diouf A, Seye M (1986) Tiques et hemoparasitoses du betail au Senegal. I. La region des Niayes. Rev Elev Med Vet Pays Trop 39: 381–393

18. Hoogstraal H (1979) The epidemiology of tick-borne Crimean Congo hemorrhagic fever in Asia, Europe and Africa. J Med Entomol 15: 307–417

19. Kirya BG, Semenov BF, Tretyakov AF, Gromashevsky VL, Madzhomba E (1972) Preliminary report on investigation of animal sera from East Africa for antibodies to Congo virus by the agar gel diffusion and precipitation method. Tezisy 17. Nauchn Sess Inst Posvyuashch Aktual Probl Virus Profilakt Virus Zabolev: 368–369 (in Russian) (in English, NAMRU3-T1073)

20. Kondratenko VF, Blagoveshchenskaya NM, Butenko AM, Vyshnivetskaya LK, Zaru-

bina LV, Milyutin VN, Kuchin VV, Novikova EM, Rabinovich VD, Shevchenko SF, Chumakov MP (1970) Results of virological investigation of ixodid ticks in Crimean hemorrhagic fever focus in Rostov Oblast. Mater 3 Oblast Nauchn Prakt Konf, Rostov-on-Don, May 1970: 29–35 (in Russian) (in English, NAMRU3-T524)

21. Lariviere J, Dupuy AR (1978) Senegal, ses parcs, ses animaux. Fernand Nathan, Evreux, France

22. Lee VH, Kemp GE (1970) Congo virus: experimental infection of *Hyalomma rufipes* and transmission to a calf. Bull Entomol Soc Nigeria 2: 133–135

23. Leroux M (1983) Climat. In: Pelissier P (ed) Atlas du Senegal. Jeune Afrique, Paris, pp 12–17

24. Logan TM, Linthicum KJ, Bailey CL, Watts DM, Moulton, JR (1989) Experimental transmission of Crimean-Congo hemorrhagic fever virus by *Hyalomma truncatum* Koch. Am J Trop Med Hyg 40: 207–212

25. Morel PC (1958) Les tiques des animaux domestiques de l'Afrique occidentale francaise. Rev Elev Med Vet Pays Trop 11: 153–189

26. Morel PC, Vassiliades G (1962) Les *Rhipicephalus* du groupe *sanguineus*: especes africaines (Acariens: Ixodoidea). Rev Elev Med Vet Pays Trop 15: 343–386

27. Ndiaye P (1983) Vegetation et faune. In: Pelissier P (ed) Atlas du Senegal. Jeune Afrique, Paris, pp 18–19

28. Niklasson B, Peters CJ, Grandien M, Wood O (1984) Detection of human immuno-globulin G and M antibodies to Rift Valley fever by enzyme-linked immunosorbent assay. J Clin Microbiol 19: 225–229

29. Robin Y (1972) Centre regional O.M.S. de reference pour les arbovirus en Afrique de l'Ouest. Rapp Inst Pasteur de Dakar

30. Robin Y (1973) Centre regional O.M.S. de reference pour les arbovirus en Afrique de l'Ouest. Rapp Inst Pasteur de Dakar

31. Robin Y (1974) Centre regional O.M.S. de reference pour les arbovirus en Afrique de l'Ouest. Rapp Inst Pasteur de Dakar

32. Robin Y (1975) Centre regional O.M.S. de reference pour les arbovirus en Afrique de l'Ouest. Rapp Inst Pasteur de Dakar

33. Robin Y. Camicas JL, Jan C, Heme G, Cornet M, Valade M (1978) Ecology of tick arboviruses in arid areas of Senegal. In: Cherepanov AI (ed) Transcontinental connec-tions of migrating birds and their role in distribution of arboviruses. Papers of the Symposium 1976, Novosibirsk (Akademgorodok). Nauka, Novosibirsk, pp 209–211

34. Saluzzo JF, Aubry P, McCormick JB, Digoutte JP (1985) Haemorrhagic fever caused by Crimean-Congo haemorrhagic fever virus in Mauritania. Trans R Soc Trop Med Hyg 79: 268

35. Saluzzo JF, Camicas JL, Chartier C, Martinez D, Digoutte JP (1986) Le virus de la fievre hemorragique de Crimee-Congo en Mauritanie. Cah ORSTOM Ser Entomol Med Parasitol 24: 129–137

36. Saluzzo JF, Digoutte JP, Camicas JL, Chauvancy G (1985) Crimean-Congo haemor-rhagic fever and Rift Valley fever in south-eastern Mauritania. Lancet 1: 116

37. Saluzzo, JF, Digoutte JP, Cornet M, Baudon D, Roux J, Robert V (1984) Isolation of Crimean-Congo haemorrhagic fever and Rift valley fever viruses in Upper Volta. Lancet 1: 1179

38. Saluzzo JF, LeGuenno B (1987) Rapid diagnosis of human Crimean-Congo haemor-rhagic fever and detection of the virus in naturally infected ticks. J Clin Microbiol 5: 922–924

39. Shepherd AJ, Swanepoel R, Cornel AJ, Mathee O (1989) Experimental studies on the replication and transmission of Crimean-Congo hemorrhagic fever virus in some African tick species. Am J Trop Med Hyg 40: 326–331

40. Swanepoel R, Shepherd AJ, Leman PA, Shepherd SP (1985a) Investigations following initial recognition of Crimean-Congo haemorrhagic fever in South Africa and the diagnosis of 2 further cases. S Afr Med J 68: 638–641

41. Swanepoel R, Shepherd AJ, Leman PA, Shepherd SP, McGillivray GM, Erasmus MJ, Searle LA, Gill DE (1987) Epidemiologic and clinical features of Crimean-Congo hemorrhagic fever in southern Africa. Am J Trop Med Hyg 36: 120–132

42. Swanepoel R, Sheperd AJ, Leman PA, Sheperd SP, Miller GB (1985b) A common-source outbreak of Crimean-Congo haemorrhagic fever on a dairy farm. S Afr Med J 68: 635–637

43. Swanepoel R, Struthers JK, Shepherd AJ, McGillivray GM, Nel MJ, Jupp PG (1983) Crimean-Congo hemorrhagic fever in South Africa. Am J Trop Med Hyg 32: 1407–1415

44. Umoh JU, Ezeokoli CD, Ogwu D (1983) Prevalence of antibodies to Crimean-haemorrhagic fever-Congo virus in cattle in northern Nigeria. Int J Zoon 10: 151–154

45. Watts DM, Ksiazek TG, Linthicum KJ, Hoogstraal H (1988) Crimean-Congo hemorrhagic fever. In: Monath TP (ed) The arboviruses: epidemiology and ecology. CRC Press, Boca Raton, FL, pp 177–222

46. Wilson ML, Digoutte JP (1989) Ecology and epidemiology of Crimean-Congo hemorrhagic fever virus transmission in the Republic of Senegal. Annual Report. U.S. Army Medical Research and Development Command, Fort Detrick, Frederick, Maryland

47. Wilson ML, LeGuenno B, Guillaud M, Desoutter D, Gonzalez JP, Camicas JL (1990) Distribution of Crimean-Congo hemorrhagic fever in Senegal: environmental and vectorial correlates. Am J Trop Med Hyg (in press)

48. Zgurskaya GN, Berezin SE, Smirnova SE, Chumakov MP (1971) Investigation of the question of Crimean hemorrhagic fever virus transmission and interepidemic survival in the tick *Hyalomma plumbeum plumbeum*, Panzer. Trudi Inst Poliomiel Virus Entsefal Akad Med Nauk SSSR 19: 217–220 (in Russian) (in English, NAMRU3-T911)

Authors' address: M. L. Wilson, Institut Pasteur, B. P. 220, Dakar, Senegal.

Arch Virol (1990) [Suppl 1]: 341
© by Springer-Verlag 1990

Detection and differentiation of tick-borne encephalitis virus strains by nucleic acid hybridization

V. A. Shamanin, A. G. Pletnev, and **V. I. Zlobin**

Novosibirsk Institute of Bioorganic Chemistry, Siberian Branch
U.S.S.R. Academy of Sciences, Novosibirsk, U.S.S.R.

Accepted January 16, 1990

Nucleic acid hybridization with cloned tick-borne encephalitis (TBE) virus, strain Sofyin, ^{32}P-labeled cDNA as the probe, was used to detect TBE viral RNA; the detection limit of the test was 10 pg Sofyin RNA. The assay was used to detect newly isolated and prototype virus strains, the detection limit being $4.3 \log_{10}$ suckling mouse intracranial LD_{50} of infectious virus. The probe cross-reacted with viruses of the TBE antigenic complex but did not react with unrelated flaviviruses of the West Nile and dengue antigenic complexes. Other viruses of the TBE antigenic complex and other TBE virus strains were differentiated from TBE Sofyin by the melting temperature of RNA-DNA hybrids.

Field-collected ticks were tested for the presence of TBE virus RNA. Results of hybridization assays were in agreement with virus isolation for 33 positive and 136 negative samples; for 12 others hybridization assays were positive and virus isolation negative, and in 3 samples hybridization assays were negative but virus isolation positive. Results of hybridization assays also were in agreement (87% correlation) with antigen detection ELISA.

Synthetic deoxyoligonucleotides complementary to Sofyin genomic RNA, used to differentiate TBE virus strains from different geographic areas, revealed genetic heterogeneity of TBE virus isolates. The number of probes hybridizing with a particular strain was assumed to be a quantitative measure of similarity to strain Sofyin. Probes complementary to different regions of the virus genome had different specificity from antigenic complex- to strain-specific. A pattern of hybridization of TBE virus strains with a panel of 11 oligonucleotide probes correlated significantly with the source of virus strain but correlated only slightly with the geographical distribution.

Authors' address: V. A. Shamanin, Novosibirsk Institute of Bioorganic Chemistry, Siberian Branch U.S.S.R. Academy of Sciences, Lavrentev Prospect 8, Novosibirsk 630090, U.S.S.R.

Arch Virol (1990) [Suppl 1]: 342

Structure and organization of tick-borne encephalitis virus genome

A. G. Pletnev

Novosibirsk Institute of Bioorganic Chemistry, Siberian Branch
U.S.S.R. Academy of Sciences, Novosibirsk, U.S.S.R.

Accepted January 16, 1990

We have cloned and sequenced the genomic RNA of tick-borne encephalitis (TBE) virus, strain Sofyin. The complete genome is 10,480 bases in length with a single open reading frame extending from nucleotides 127 to 10,365 encoding 3,413 amino acids. The 5'- and 3'-noncoding region is in the form of a stem-and-loop structure. A polyprotein precursor apparently is cleaved proteolytically by a mechanism resembling that proposed for expression of polyproteins of other flaviviruses. The deduced TBE virus gene order is 5'-C-preM(M)-E-NS1-NS2A-NS2B-NS3-NS4A-NS4B-NS5-3'. The genomes and polyproteins of TBE virus and other flaviviruses appear to be similar, although they are transmitted between vertebrate hosts by different vectors. Comparison of sequence homology of polyproteins of flaviviruses suggests that TBE virus is most closely related to yellow fever virus. The hydrophobic profiles of polyproteins of the flaviviruses are highly conserved. The nonstructural proteins NS2A, NS2B, NS4A, and NS4B are extremely hydrophobic; these proteins likely are associated with cellular membranes. Proteins E, NS1, NS3, and NS5 are the most conserved and may be involved in activities related to viral replication. When TBE virus replication complex was labeled by treatment with an ATP derivative and then with ^{32}P-GTP, the most efficient labeling took place with protein NS5, suggesting that it functions in the synthesis of nascent RNA.

Author's address: A. G. Pletnev, Novosibirsk Institute of Bioorganic Chemistry, Siberian Branch U.S.S.R. Academy of Sciences, Larentiev Prospect 8, Novosibirsk 630090, U.S.S.R.

Arch Virol (1990) [Suppl 1]: 343
© by Springer-Verlag 1990

Detection by time-resolved fluoroimmunoassay of antibodies in patients with tick-borne encephalitis

N. A. Lavrova[1], **V. G. Pomelova**[1], and **P. Halonen**[2]

[1]D.I. Ivanovsky Institute of Virology, U.S.S.R. Academy of Medical Sciences, Moscow, U.S.S.R.
[2]Department of Virology, University of Turku, Turku, Finland

Accepted January 16, 1990

A total of 148 serum and 58 cerebrospinal fluid (CSF) samples was collected from patients with illnesses clinically compatible with tick-borne encephalitis (TBE) and fed on by ticks in the summer of 1988 in the Perm region (eastern European portion of the U.S.S.R.). These specimens were tested by enzyme immunoassay (EIA) and time-resolved fluoroimmunoassay (TR FIA) for IgM and IgG class antibodies to TBE virus. Individual wells in plastic 96-well panels were coated with commercially-available (capture) antibodies to human IgM or IgG; specimens were screened at 1:1000 (serum) or 1:100 (CSF). The adsorbed samples were then reacted with normal or TBE virus antigens and detector (secondary) anti-viral antibody conjugated to either peroxidase (for EIA) or europium (for TR FIA).

IgM antibodies were detected in serum samples of 37 (25%) patients by EIA and in 41 (27.7%) by TR FIA; IgG antibodies were detected in 16 (10.8%) patients by EIA and in 25 (16.9%) by TR FIA. Early specific antibodies were detected within the first week of illness in 14 (9.7%) patients by EIA and in 23 (15.3%) by TR FIA.

Antibody titers were much lower in CSF than in serum samples. TR FIA appears to be a somewhat more sensitive technique than EIA. Serum antibody detected by EIA did not exceed 125,000, whereas antibody in the same specimens tested by TR FIA were detected at titers to 625,000.

Authors' address: N. A. Lavrova, D.I. Ivanovsky Institute of Virology, Academy of Medical Sciences of the U.S.S.R., Gamaleya Street 16, Moscow 123098, U.S.S.R.

Arch Virol (1990) [Suppl 1]: 344

Monoclonal antibodies used to study antigenic characteristics of Powassan virus strains isolated in the Primorye Territory, eastern U.S.S.R.

G. N. Leonova

Research Institute for Epidemiology and Microbiology, Siberian Branch
U.S.S.R. Academy of Sciences, Vladivostok, U.S.S.R.

Accepted January 16, 1990

Twelve isolates of Powassan virus from the Primorye Territory of the eastern U.S.S.R. were studied by hemagglutination-inhibition (HI) and precipitation in agar. Reactions indicated antigenic relationships of some of the field isolates of Powassan virus with not only prototype Powassan virus but with tick-borne encephalitis (TBE) virus as well.

All strains were tested by kinetic HI with 4 monoclonal antibodies (MCAs; 1E11, 2D1, 5G10, 2H3) to glycoprotein E (envelope) of TBE virus, strain Sofyin, prepared at the Research Institute of Biochemistry, Siberian Branch U.S.S.R. Academy of Medical Sciences. Poor or no antigenic relationships with these MCAs were detected with most strains; most frequently, reactions were detected only with MCA 2H3. A high specific activity for MCA 2H3 (1:80 titers) was shown with a strain isolated from *Dermacentor silvarum* ticks; reactivity was somewhat less (1:20–1:40) with strains isolated from homeothermic animals.

The occurrence of such strains may be a result of mixed Powassan and TBE virus infections of arthropod or vertebrate hosts in common natural foci of these ecologically similar (vectors, hosts) viruses.

Author's address: G. N. Leonova, Novosibirsk Institute of Bioorganic Chemistry, Siberian Branch U.S.S.R. Academy of Sciences, Selskaya Street 1, Vladivostok 690600, U.S.S.R.

Arch Virol (1990) [Suppl 1]: 345

Syrdaria Valley fever, a new virus disease in Kazakhstan

S. K. Karimov[1], **D. V. Lvov**[2], and **T. V. Kiryushchenkol**[1]

[1]Institute of Epidemiology, Microbiology, and Infectious Diseases, Ministry of Health,
Alma-Ata, Kazakh S.S.R.
[2]D.I. Ivanovsky Institute of Virology, U.S.S.R. Academy of Medical Sciences,
Moscow, U.S.S.R.

Accepted January 16, 1990

In 1973 a newly recognized virus was isolated from whole blood of a febrile patient in Kyzl Orda region, Kazakhstan, and its role in the etiology of fever in 5 other patients was determined. In all 6 patients the disease course was similar: acute febrile episode with headache, elevated body temperature (39–40.3 °C) lasting 5–7 days, and appearance of abundant polymorphous rosette-petechial rash on days 3–4 of the illness were typical features. Retrospective epidemiological studies showed that the patients were fed on by ticks while working in the fields 4–7 days before the appearance of the first signs of illness. In the summer of 1974, ticks were collected in the area of the Syrdaria river and the Ili river flood plains (Kyzl Orda and Alma-Ata regions). Seven strains identical to the prototype (strain Kaz-3) were isolated from ixodes ticks *Dermacentor daghestanicus* and one from *Hyalomma asiaticum* ticks.

We studied certain biological and antigenic characteristics of strain Kaz-3. By electron microscopy of ultrathin sections of brain tissue from infected suckling mice virions 25–27 nm in diameter and otherwise typical of picornaviruses were observed. Although tested against antibody to 116 arboviruses and others viruses, including encephalomyocarditis virus, antigen of Kaz-3 reacted in complement fixation tests only with mouse immune ascitic fluid against Sikhote-Alyn virus, a recognized picornavirus. Antigenic relationships between Kaz-3 and Sikhote-Alyn viruses were not detected by neutralization tests, thus suggesting that Kaz-3 is distinct from other recognized viruses. We suggest the name Syrdaria Valley fever virus for this newly recognized agent.

Authors' address: S. K. Karimov, Institute of Epidemiology, Microbiology, and Infectious Diseases, Ministry of Health, Pasteur Street 34, Alma-Ata 480002, Kazakh S.S.R.

Arch Virol (1990) [Suppl 1]: 346
© by Springer-Verlag 1990

Isolation of a Getah-like virus (family *Togaviridae*, genus *Alphavirus*, Semliki Forest antigenic complex) from mosquitoes collected in Mongolia

V. P. Andreev[1], D. Abmed[2], S. D. Lvov[3], G. A. Dmitriev[1], L. Kupul[2],
V. L. Gromashevsky[1], A. D. Avershin[1], T. M. Skvortsova[1], N. G. Kondrashina[1],
O. V. Voltsit[1], T. N. Morozova[1], and E. A. Gushchina[1]

[1]D.I. Ivanovsky Institute of Virology, U.S.S.R. Academy of Medical Sciences,
Moscow, U.S.S.R.
[2]Institute of Hygiene, Epidemiology, and Microbiology, Ministry of Health,
Mongolian People's Republic, Ulan Bator, Mongolia
[3]N.F. Gamaleya Institute of Epidemiology and Microbiology,
U.S.S.R. Academy of Medical Sciences, Moscow, U.S.S.R.

Accepted January 16, 1990

Six strains of a single virus type were isolated from 21,400 female hematophagous mosquitoes collected August–September 1988 in eastern Mongolia in an area of high-grass steppe and flood plain. *Aedes dorsalis* mosquitoes predominated in the steppe landscape and *A. vexans nipponii* and *A. dorsalis* in the flood plain. All strains were isolated from mosquitoes collected in flood plains near the Kerulan and Khalkhingol rivers.

By electron microscopy of ultrathin sections of infected cell cultures the isolates appeared to be members of the family *Togaviridae*. Serologic studies provided data allowing us to identify them as strains of a Getah-like virus (family *Togaviridae*, genus *Alphavirus*, Semliki Forest antigenic complex).

Hemagglutination-inhibition tests of serum samples from residents of the area revealed positive reactions with antigens of the prototype of the Getah-like (10.9%), tick-borne encephalitis (5.2%), West Nile (1.6%), and Japanese encephalitis (1%) viruses.

Authors' address: V. P. Andreev, D.I. Ivanovsky Institute of Virology, Academy of Medical Sciences of the U.S.S.R., Gamaleya Street 16, Moscow 123098, U.S.S.R.

Arch Virol (1990) [Suppl 1]: 347

Immune blotting studies of California encephalitis antigenic complex viruses recently isolated in the U.S.S.R.

E. A. Vladimirtseva[1], **E. A. Bakulina**[1], **E. A. Bystrova**[1], **S. D. Lvov**[2], and **A. M. Butenko**[1]

[1]D.I. Ivanovsky Institute of Virology and [2]N.F. Gamaleya Institute of Epidemiology and Microbiology, U.S.S.R. Academy of Medical Sciences,
Moscow, U.S.S.R.

Accepted January 16, 1990

We recently applied an immune blotting (Western blotting) method, first described by Towbin et al. [1] in 1979, to the identification of California encephalitis antigenic complex (family *Bunyaviridae*, genus *Bunyavirus*, California serogroup) viruses isolated in the U.S.S.R. We studied 3 strains, isolated from mosquitoes collected in eastern U.S.S.R. on Sakhalin island (strain LEIV 11483), in northern U.S.S.R. on the Taimyr peninsula (strain LEIV 11552), and near Moscow in the western U.S.S.R. in the Kalinin region (strain LEIV 12812). The Sakhalin and Taimyr strains were shown to be closely related to snowshoe hare (SSH) virus. The Kalinin strain appears to be a reassortant of Tahyna (TAH) and SSH viruses, close to SSH virus in its antigenic properties and in the molecular characteristics of its L protein and to TAH virus in the characteristics of its G1, G2, and N proteins. By immune blotting, antibodies to all TAH virus proteins were detected in serum samples from a patient convalescing from a clinical infection with TAH virus and in serum from an African green monkey (*Cercopithecus aethiops*) experimentally infected with TAH virus.

Reference

1. Towbin H, Staehelin T, Gordon J (1979) Electrophoretic transfer of proteins from polyacrylamide gels to nitrocellulose sheets: procedure and some applications. Proc Natl Acad Sci USA 76: 4350–4354

Authors' address: E. A. Vladimirtseva, D.I. Ivanovsky Institute of Virology, Academy of Medical Sciences of the U.S.S.R., Gamaleya Street 16, Moscow 123098, U.S.S.R.